[H.N.S.C. No. 105–50]

UNITED STATES POLICY REGARDING THE EXPORT OF SATELLITES TO CHINA

JOINT HEARINGS

BEFORE THE

COMMITTEE ON NATIONAL SECURITY

MEETING JOINTLY WITH

COMMITTEE ON INTERNATIONAL RELATIONS

HOUSE OF REPRESENTATIVES

ONE HUNDRED FIFTH CONGRESS

SECOND SESSION

HEARINGS HELD

JUNE 17, 18 AND 23, 1998

U.S. GOVERNMENT PRINTING OFFICE

55–784 WASHINGTON : 1999

For sale by the U.S. Government Printing Office

Superintendent of Documents, Congressional Sales Office, Washington, DC 20402

ISBN 0-16-058445-0

HOUSE COMMITTEE ON NATIONAL SECURITY

ONE HUNDRED FIFTH CONGRESS

FLOYD D. SPENCE, South Carolina, *Chairman*

BOB STUMP, Arizona
DUNCAN HUNTER, California
JOHN R. KASICH, Ohio
HERBERT H. BATEMAN, Virginia
JAMES V. HANSEN, Utah
CURT WELDON, Pennsylvania
JOEL HEFLEY, Colorado
JIM SAXTON, New Jersey
STEVE BUYER, Indiana
TILLIE K. FOWLER, Florida
JOHN M. McHUGH, New York
JAMES TALENT, Missouri
TERRY EVERETT, Alabama
ROSCOE G. BARTLETT, Maryland
HOWARD "BUCK" McKEON, California
RON LEWIS, Kentucky
J.C. WATTS, JR., Oklahoma
MAC THORNBERRY, Texas
JOHN N. HOSTETTLER, Indiana
SAXBY CHAMBLISS, Georgia
VAN HILLEARY, Tennessee
JOE SCARBOROUGH, Florida
WALTER B. JONES, JR., North Carolina
LINDSEY GRAHAM, South Carolina
JIM RYUN, Kansas
MICHAEL PAPPAS, New Jersey
BOB RILEY, Alabama
JIM GIBBONS, Nevada
BILL REDMOND, New Mexico
KAY GRANGER, Texas
MARY BONO, California

IKE SKELTON, Missouri
NORMAN SISISKY, Virginia
JOHN M. SPRATT, JR., South Carolina
SOLOMON P. ORTIZ, Texas
OWEN PICKETT, Virginia
LANE EVANS, Illinois
GENE TAYLOR, Mississippi
NEIL ABERCROMBIE, Hawaii
MARTIN T. MEEHAN, Massachusetts
ROBERT A. UNDERWOOD, Guam
JANE HARMAN, California
PAUL McHALE, Pennsylvania
PATRICK J. KENNEDY, Rhode Island
ROD R. BLAGOJEVICH, Illinois
SILVESTRE REYES, Texas
TOM ALLEN, Maine
VIC SNYDER, Arkansas
JIM TURNER, Texas
F. ALLEN BOYD, JR., Florida
ADAM SMITH, Washington
LORETTA SANCHEZ, California
JAMES H. MALONEY, Connecticut
MIKE McINTYRE, North Carolina
CIRO D. RODRIGUEZ, Texas
CYNTHIA A. McKINNEY, Georgia

ANDREW K. ELLIS, *Staff Director*
DAVID TRACHTENBERG, *Professional Staff Member*
PETER BERRY, *Professional Staff Member*
MICHELLE SPENCER, *Research Assistant*

HOUSE COMMITTEE ON INTERNATIONAL RELATIONS

ONE HUNDRED FIFTH CONGRESS

BENJAMIN A. GILMAN, New York, *Chairman*

WILLIAM GOODLING, Pennsylvania
JAMES A. LEACH, Iowa
HENRY J. HYDE, Illinois
DOUG BEREUTER, Nebraska
CHRISTOPHER SMITH, New Jersey
DAN BURTON, Indiana
ELTON GALLEGLY, California
ILEANA ROS-LEHTINEN, Florida
CASS BALLENGER, North Carolina
DANA ROHRABACHER, California
DONALD A. MANZULLO, Illinois
EDWARD R. ROYCE, California
PETER T. KING, New York
JAY KIM, California
STEVEN J. CHABOT, Ohio
MARSHALL "MARK" SANFORD, South Carolina
MATT SALMON, Arizona
AMO HOUGHTON, New York
TOM CAMPBELL, California
JON FOX, Pennsylvania
JOHN McHUGH, New York
LINDSEY GRAHAM, South Carolina
ROY BLUNT, Missouri
KEVIN BRADY, Texas
RICHARD BURR, North Carolina

LEE HAMILTON, Indiana
SAM GEJDENSON, Connecticut
TOM LANTOS, California
HOWARD BERMAN, California
GARY ACKERMAN, New York
ENI F.H. FALEOMAVAEGA, American Samoa
MATTHEW G. MARTINEZ, California
DONALD M. PAYNE, New Jersey
ROBERT ANDREWS, New Jersey
ROBERT MENENDEZ, New Jersey
SHERROD BROWN, Ohio
CYNTHIA A. McKINNEY, Georgia
ALCEE L. HASTINGS, Florida
PAT DANNER, Missouri
EARL HILLIARD, Alabama
BRAD SHERMAN, California
ROBERT WEXLER, Florida
STEVE ROTHMAN, New Jersey
BOB CLEMENT, Tennessee
BILL LUTHER, Minnesota
JIM DAVIS, Florida
LOIS CAPPS, California

RICHARD J. GARON, *Chief of Staff*
MICHAEL H. VAN DUSEN, *Democratic Chief of Staff*

CONTENTS

CHRONOLOGICAL LIST OF HEARINGS

1998

Page

DOCUMENTS SUBMITTED FOR THE RECORD—CONTINUED

UNITED STATES POLICY REGARDING THE EXPORTS OF SATELLITES TO CHINA

HOUSE OF REPRESENTATIVES, COMMITTEE ON NATIONAL SECURITY, MEETING JOINTLY WITH COMMITTEE ON INTERNATIONAL RELATIONS, *Washington, DC, Wednesday, June 17, 1998.*

The committees met, pursuant to notice, at 9:37 a.m. in room 2118, Rayburn House Office Building, Hon. Floyd D. Spence (Chairman of the Committee on National Security) presiding.

OPENING STATEMENT OF HON. FLOYD D. SPENCE, A REPRESENTATIVE FROM SOUTH CAROLINA, CHAIRMAN, COMMITTEE ON NATIONAL SECURITY

Chairman SPENCE [presiding]. The meeting will please be in order.

I'm pleased to open the first of 2 days of joint hearings between the Committees on National Security and International Relations on U.S. satellite export policy toward China.

Let me first welcome my fellow chairman, Ben Gilman, and all of our colleagues from the Committee on International Relations to our hearing room for what I expect will be a productive review of this important national security and foreign policy issue.

The policy question that we will examine for the next two days appears to be simple: Is it in the United States' national interest to allow the use of Chinese space launchers to place American commercial satellites in orbit? The United States' aerospace industry, this administration, and, to some extent, the Bush administration before it, believe that the answer is yes. This position is based on a belief that United States commercial competitiveness and the need to protect market share demands continued access to cheap foreign space launch services.

However, when you look at the complexities of this issue, and weigh the competing argument, I firmly believe that the compelling answer to the policy question is no. And why? Quite simply because China possesses nuclear weapons and the missile on top of which those nuclear weapons are sitting are the very same type as those being marketed to the U.S. satellite industry as providing cheaper access to space. Therefore, every time we export a satellite to China, and pay the Chinese government to launch it, we are directly or indirectly enhancing the ability of China to improve its space and military launch capability.

The arguments on both sides set up the classic export control policy conflict: How to advance U.S. commercial interests without jeopardizing national security. This policy conflict is neither new nor unique to satellite exports. We currently face similar debates in the case of other high-technology exports such as supercomputers or encryption software and hardware. Our two committees have been at the forefront of the congressional debate in this area for years. While we have not always agreed on narrow questions or process or policy, I think it is fair to say that we both agree that there must be a well-established and effective national security component to U.S. high-technology export control policy.

The question of how to strike the proper balance between commercial and national security interests has prompted lengthy and sometimes heated debate among many of us in this room. However, while that remains an active and ongoing debate on numerous fronts, the reason we are meeting here today under these unusual circumstances is that we appear to have a compelling case—satellite exports to China—that demonstrate how the administration's export policy has tilted too far in one direction at the direct expense of United States' national security.

Ironically, while the U.S. Government export policy appears to have for once attained a remarkable level of consistency, unfortunately, I believe it to be in the wrong direction. Since 1992, U.S. industry has applied for nine so-called waivers of existing sanctions prohibiting the export of satellite technology to China. Since 1992, the Clinton administration has approved exactly nine waivers. Perhaps even more disconcerting than the administration's consistency is the fact that a number of these exports prompted the expression of legitimate national security concerns as part of internal administration deliberations. Yet in each instance, the decision was made in favor of commercial or other interests and the export was allowed. The administration's track record in this regard displays a single-minded determination to evade sanctions and to ignore or downplay national security concerns in the interest of exporting sensitive American technology in the name of engaging with China.

To help us address some of these complex questions, we have scheduled two days of hearings. Today we will hear from a panel of outside experts to help us better understand the process, policy, and technical issues associated with this topic. Tomorrow, we will hear from a panel of administration witnesses to examine past and present U.S. policy in this area.

Today's witnesses are Ms. Shirley Kan, Analyst in Foreign Affairs for the Congressional Research Service; Dr. Henry Sokolski, Executive Director of the Nonproliferation Policy Education Center; and Dr. Gary Milhollin, Director of the Wisconsin Project on Nuclear Arms Control; and Dr. Joan Johnson-Freese, Professor of International Security Studies at the Air War College.

Let me welcome all of you, but before turning to our witnesses, let me first briefly touch on several administrative matters and then recognize my colleagues Chairman Gilman, Mr. Skelton, and Mr. Hamilton, for their opening remarks. Members should be advised that for the next 2 days' worth of hearings, we will be operating under the rules of the Committee on International Relations. For the purpose of recognizing members for questions, we will follow the same process everyone is used to, as our respective committee rules are similar.

Further, as the combined membership of our committees totals up to 101 members, it will be a challenge to conduct these hearings in a manner that ensures that each and every member is afforded the opportunity to be recognized. Therefore, Mr. Gilman and I will diligently observe the 5-minute rule and I urge the cooperation of all members in closely adhering to this limit in order to accommodate as many members' questions as possible.

And finally, I also ask the cooperation of our witnesses in observing the 5-minute limit as closely as possible given the potential number of members who will want to ask questions.

Let me now recognize Mr. Gilman, the chairman of the Committee on International Relations, for any opening remarks he might like to make.

STATEMENT OF HON. BENJAMIN A. GILMAN, A REPRESENTATIVE FROM NEW YORK, CHAIRMAN, COMMITTEE ON INTERNATIONAL RELATIONS

Chairman GILMAN. Thank you, Chairman Spence. I am pleased to join with you and your Committee on National Security in opening these joint hearings today by our two committees looking into President Clinton's policies regarding the export of U.S. commercial satellites for launching in China. These hearings stem from news reports that have raised serious questions with regard to whether the administration's policies foster a climate in which critical U.S. missile technology may have been provided to the Chinese. The administration's overall policy with regard to China has been to emphasize the value of trade and commerce. The President uses the phrase "constructive engagement" to describe that approach. And in pursuit of that policy, the President transferred responsibility for licensing the export of U.S. commercial satellites for launching in China from the State Department to the Commerce Department. One result of that change was an interagency recommendations on whether a satellite export license should be issued were framed primarily in terms of their value to Sino-American commercial ties rather than upon their impact on U.S. national security.

In the 1992 Presidential campaign, Governor Clinton attacked President Bush for coddling dictators with an "undisguised contempt for democracy, for human rights, and for the need to control the spread of dangerous weapon technology." It was unimaginable then that President Clinton's administration would face questions about compromising our national security at the hands of those very same Chinese leaders. Yet, in May 1997, a highly classified Pentagon report reportedly concluded that scientists from two leading American satellite companies, Loral Space and Communication and Hughes Engineering, provided expertise that significantly improved the guidance and reliability of China's ballistic missiles. Moreover, documents released by the White House last month disclose that the Justice Department had concerns about issuing a waiver in February 1998, permitting Loral to export a satellite for launch in China and the Clinton administration knew that.

According to a memo prepared for the President by the national security adviser, the Justice Department has "cautioned that a national interest waiver in this case could have a significant adverse impact on any prosecution that might take place." However, despite that advice by the Justice Department, the President did grant Loral a waiver. Many of us are concerned that in its desire to promote the commercial interests of key United States companies, the administration may have undercut its own efforts to limit the spread of missile technology to China which today is the world's leading exporter of weapons of mass destruction.

The administration has insisted that nothing untoward has occurred, that no inappropriate action or actions have been taken that resulted in any harm to U.S. national security. Others of us are not so certain.

Today's witnesses, whom the chairman has already introduced, will give us the benefit of expertise that they have developed over many years of following these issues and we look forward to their testimony.

Thank you, Mr. Chairman.

Chairman SPENCE. Thank you, Mr. Chairman. Let me now recognize Mr. Skelton, the ranking minority member of the Committee on National Security for any opening remarks he would like to make.

STATEMENT OF HON. IKE SKELTON, A REPRESENTATIVE FROM MISSOURI, RANKING MEMBER, COMMITTEE ON NATIONAL SECURITY

Mr. SKELTON. Thank you, Mr. Chairman. Henry Sokolski and Gary Milhollin are well known to the Washington security policy community and need no introduction. Shirley Kan and Dr. Joanne Johnson-Freese are, on the other hand, are less well-known. Ms. Kan is an analyst for the Congressional Research Service at the Library of Congress and will put the issues we are addressing today in context. Dr. Johnson-Freese is a Professor of International Security Studies at the Air War College at Maxwell Air Force Base. Her expertise is space policy and she's written widely on this subject. Her latest book is "The Chinese Space Program: A Mystery Within a Maze." We especially look forward to her testimony today as someone who can speak on these matters with a bit of scholarly detachment. She's accompanied today by her son, J.B.

This extraordinary hearing, with our colleagues from the International Relations Committee, is designed to examine two issues: Whether our 10-year-old satellite export policy has resulted in the transfer of sensitive space and missile technology to China and the reasons behind the decision to transfer licensing jurisdiction over commercial satellites from the State Department to the Department of Commerce.

These presentations will help us set the stage for an additional hearing tomorrow that will allow the administration representatives from the Departments of Defense, State, and Commerce to provide the two committees further information to help us in our efforts to understand the current policy.

There are actually four, Mr. Chairman, actually four separate issues which have become entangled over the course of the last few months. The first concerns the present satellite export policy, one initiated under the Reagan administration and continued under the Bush and Clinton administrations. The second concerns the decision to transfer licensing jurisdiction over commercial satellites from the State Department to the Commerce Department. The third concerns the specific case of Loral and the possible release of export-control information in the review it conducted subsequent to the Feb. 15, 1996, crash of the Chinese space launch vehicle. And the fourth issue concerns the February waiver of the launch of another Loral satellite from China.

Today we have the opportunity, Mr. Chairman, to begin the kind of thoughtful deliberation on these serious and complicated issues. Thank you.

Chairman SPENCE. Finally, let me recognize Mr. Hamilton, the ranking minority member on the Committee on International Relations for any opening statement he'd like to make.

STATEMENT OF HON. LEE H. HAMILTON, A REPRESENTATIVE FROM INDIANA, RANKING MINORITY MEMBER, COMMITTEE ON INTERNATIONAL RELATIONS

Mr. HAMILTON. First, let me commend Chairman Spence and Chairman Gilman for holding this joint hearing. This is an important and a complex issue. It's very appropriate that we look at it and I want to extend my thanks to each of the witnesses for their willingness to help us out and to participate this morning.

As I see it, we have three issues before us. The first is the wisdom and the effectiveness of the current U.S. export control regime. The second is the integrity of how that regime was implemented. And the third are the merits of the underlying policy of licensing U.S. satellites for export and launch on Chinese rockets.

First, with regard to the export control regime now in place for regulating U.S. commercial satellite exports, it is a comprehensive one. So far as I am aware, it is a good one, although I want to hear from the witnesses on that point. Every satellite license request gets reviewed by Commerce, State, Defense, and the Arms Control and Disarmament Agency. Defense often asks other agencies for further review. Every recommendation to the President for a national interest waiver to allow the launch of a satellite on a Chinese missile has been a unanimous recommendation by these agencies. Every license for a satellite launch on a Chinese missile has required a Defense Department-approved technology control plan and a transportation security plan.

Under the current export control regime, in place since 1996, every launch in China has had a Department of Defense monitor present. Every launch of a U.S. satellite in China has also required a separate State Department license for technical assistance reviewed by the Defense Department, in addition to a Commerce license reviewed by State, Defense, and ACDA.

The second issue is there are serious allegations that this system of export controls was circumvented, especially in the case of a launch failure review. If the licensing requirements and controls on technology that make up the backbone of our export control regime are not properly followed, or not properly enforced, then the integrity of the regime is undermined and U.S. national security interests are at risk. We should pursue these allegations as we are doing today. That is our oversight responsibility. The Department of Justice should look into them as I understand it is doing. Other U.S. agencies have already reviewed the matter, but we should proceed carefully. Export controls are immensely complicated and arcane. They involve judgments about advanced technologies, complex legal questions on a subject that most of us have not dealt with on a regular basis, at least until recently. I understand that in this environment there are very strong reasons to rush to judgment before all the facts are known or their implications fully appreciated. But there is a great deal at stake here, both for the United States and for one of its most important international relationships. I urge my colleagues to withhold judgment until these matters have been fully investigated.

Finally, I believe there are powerful arguments why the United States should, and should not, export satellites for launch in China. Obviously, the most important argument against such a policy is the possibility that it might assist China in advancing its ballistic missile program. The question of whether and how U.S. satellite exports might have contributed to China's missile program is of interest to every member of this institution. There are also significant arguments on behalf of a prudent satellite export policy. First, there is not enough domestic launch capacity for the requirements of the U.S. satellite industry. China has an important share of the international market. China launches are significantly less expensive than other alternatives. Second, the satellite industry provides enormous benefits to American citizens and businesses who use phones, fax, Internet, and video every day, and to the forces of democracy around the world, especially in China, that are bolstered by their contact with the outside world.

Third, we have focused on what the Chinese may have learned from us. But we have also learned a great deal about them. We have an unprecedented opportunity to learn not only about Chinese rockets, but many other aspects of China to which we would otherwise not have access.

Fourth, this policy has created important incentives for China to respect international norms on the export of missiles and related technology. While China has moved some in the right direction, I think all of us would agree that it has still got a long distance to go. We will have less leverage, the argument would go, with China if we cut ourselves out of the launch market. China will turn to foreign satellite suppliers and foreign markets for its hardware.

And, finally, of course, U.S. satellite exports are consistent with the policy of engagement. Mr. Chairman, I appreciate the opportunity to make these remarks. I commend you and Mr. Gilman again for letting us proceed with them and I look forward to the testimony.

Chairman SPENCE. Thank you, Mr. Hamilton. Now, for our panel of witnesses. I urge each of you to summarize your presentations and, without objections, your full statements will be submitted for the record. Let's begin with Ms. Kan.

STATEMENT OF SHIRLEY A. KAN, ANALYST IN FOREIGN AFFAIRS, CONGRESSIONAL RESEARCH SERVICE

Ms. KAN. Thank you. Good morning. I'm Shirley Kan of the Foreign and National Defense Division of CRS. I'm here to present a statement to provide background on the foreign policy and national security issues concerning Chinese launches of U.S. origin satellites and Chinese missiles. I request that my statement and three CRS reports, which were provided earlier, be entered into the record.

Important for our understanding is the Chinese company at the heart of various concerns. China Great Wall Industry Corporation is the only Chinese State-owned company that offers launch services to foreign customers. It is also a subsidiary of China Aerospace Corporation, which heads the Chinese aerospace complex and oversees both space and missile research and development. China Great Wall uses the Long March series of rockets to launch satellites. The Long March rockets are also produced as China's DF–4 and DF–5A ICBM's. U.S. sanctions have been imposed twice on China Great Wall for missile proliferation in Pakistan.

China gains several benefits from launching satellites: foreign capital, technical expertise, and international prestige. Also, some satellites or parts exported to China for launch have been for Chinese use. However, China has also been embarrassed by a number of costly failed satellite launches. In 1992, the engine for a Chinese rocket failed to ignite while attempting to launch a satellite and another rocket exploded and destroyed a second satellite. Both satellites were built by Hughes Electronics Corporation. In 1995, a Long March rocket exploded and destroyed another satellite built by Hughes. In 1996, yet another Chinese rocket exploded and destroyed the Intelsat satellite built by Loral. Aside from the dramatic explosions, other problems have prevented Chinese rockets from successfully launching satellites into their correct orbits. More recently, since 1997, China had success in all of its satellite launches. Initially China's response to the failures was to blame foreign companies. China later shifted to a willingness to work with U.S. companies to determine the causes of the explosions.

Let me now turn to the several issues surrounding the recent controversy on the use of Chinese rockets to launch satellites. The immediate legal issue which prompted the controversy concerns whether Loral Space and Communications provided expertise to China that it could use to improve its ballistic missiles. The New York Times, in April, cited a classified Pentagon report that reportedly concluded in May 1997 that American expertise was transferred to China that significantly enhanced the reliability of its ballistic missiles and that U.S. national security was harmed. Since last September, the Justice Department has been conducting a criminal investigation into whether Loral violated export control laws. Loral says that after the explosion of a Chinese rocket that destroyed the Loral-built satellite in February 1996, it formed an independent committee to review the Chinese findings of the cause of the explosion. The company acknowledges that the committee provided a report to the Chinese without first getting approval from the State Department. Loran says that it promptly informed the Departments of Defense and State of having provided the report and is cooperating with the Justice Department and with Congress. Loral says that the report did not transfer significant or sensitive technology or information.

A second issue concerns how the Clinton administration has shaped and implemented U.S. policies on China, on satellite exports to China, and on missile nonproliferation. Specifically, the administration reviews satellite exports to China in making decisions on two sets of sanctions, one for the Tiananmen crackdown and one for missile proliferation, in conjunction with export control process.

The administration had transferred control over the licensing of satellites from the State Department to the Commerce Department in November 1996.

Some, including the President, have argued that engagement with China is a better policy than isolation of China. Other says that the issues are how the policy of engagement is carried out and whether the exports of satellites need to be included in that policy. In part to promote ties with China, the Reagan administration, in September 1988, first notified Congress of its decision to approve licenses for the export of satellites to China. That administration acknowledged the need to protect legitimate U.S. national security interests, and the U.S. commercial launch industry, while promoting relations and trade with China.

Over the years, some in Congress, and satellite manufacturers in the United States, have supported the use of Chinese-launched vehicles to launch U.S. origin satellites. President Bush first issued the post-Tiananmen—first issued waivers of the post-Tiananmen sanctions. President Clinton continues the long-standing policy of engagement with China, and includes satellite launches as part of that policy. The White House argues that satellite launches expand multifaceted relations, they have benefits for commercial competitiveness, nonproliferation goals, and the spread of democracy.

Let me clarify that, for satellite exports to China, considerations of Presidential waivers for post-Tiananmen sanctions and missile proliferation, have been added to the export control process. U.S. sanctions imposed in 1990 for the Tiananmen crackdown have meant that Presidential waivers allowed by the law were needed before licenses can be approved. As shown in the first chart that is also in your statement, Presidents Bush and Clinton have issued 13 waivers for 20 satellite projects, by reporting to Congress that it is in the national interest to do so.

[See chart A in the appendix on page 363.]

Those waivers have been issued fairly routinely except for the consideration of the Justice Department's concerns about the latest waiver and about China's missile proliferation activities. After China transferred technology for the M–11 short-range ballistic missile to Pakistan, the Bush administration, from June 1991 to March 1992, denied exports to China Great Wall and another company of high-speed computers, satellites for launch, and missile equipment and technology.

The Clinton administration determined that China again transferred M–11 technology to Pakistan and imposed sanctions from August 1993 to November 1994, denying exports of missile equipment and technology to Chinese aerospace companies, including China Great Wall. During those two period of sanctions, one presidential waiver was issued, in July 1994, for the export of satellites for launch by China Great Wall. Some debate has arisen as to whether the administration is again required to impose sanctions, since officials are concerned about reports that China has continued to transfer missile equipment and technology to Pakistan and Iran.

A third and broader issue concerns the underlying debate over the national interest, weighing the benefits of trade and commercial competitiveness versus the risks to national security. Some are calling for a stop to the export of satellites to China. Industry says that such exports benefit American companies, consumers, and the country.

The White House says the licenses or waivers for satellite exports to China do not authorize the transfer of sensitive missile technology. Administration officials stated that licensing procedures and strict security measures preclude any assistance to Chinese missile programs. The policy is intended to allow American exporters to compete while protecting U.S. national security interest.

Some also consider the benefits of the Pentagon having access to the secretive Chinese aerospace and missile complex. According to the March 23 Washington Times, a secret NSC memo discussed the use of satellite launches as a carrot for China to curb missile proliferation. Some supporters also argue that satellite broadcasts help to promote the spread of democracy. Finally, the Chinese argue that commercial space launches are part of what they call defense conversion, defense industries doing civilian business.

Critics say that allowing China to launch U.S.-origin satellites indirectly subsidizes and assists China's missile research and development. The technology and know-how used in launching satellites are similar to those used in ballistic missiles. And there is a chart on that side that was provided by the CIA to demonstrate those technologies.

[See chart B in the appendix on page 364.]

Second some say that although an individual piece of technology or technical assistance may not be significant, assistance given over time has a cumulative effect of conveying valuable expertise. Third, U.S.-China cooperation has been characterized by a tension between a certain unreliability of Chinese rockets and the interest of American companies to have successful launches. Working with Americans and Europeans, Chinese engineers may have learned valuable quality controls and other lessons. Some also say that allowing the Chinese to launch satellites rewards or ignores their missile proliferation and any U.S. expertise gleaned from launching satellites may then be transferred by China to other countries. Others say that while satellites may promote a more open Chinese society, they may also be used by the Chinese military. Last, some experts point out that China's defense industries have not converted, but diversified into both military and civilian programs.

As for potential benefits to China's ballistic missile program, some observers point out that launching satellites has not provided breakthroughs and China has deployed ICBM's since 1980, before launching U.S. satellites. As seen in my second chart that is also provided in the statement, the Chinese have a diversified missile force in what they call the *Second Artillery*, consisting of intercontinental, medium-range, submarine-launched, and short-range ballistic missiles.

Other experts, however, point out that China has experienced problems with its missiles and is improving the reliability and accuracy of its missiles. The *Second Artillery* is expected to deploy after the turn of the century, new ICBM's possibly with MIRV technology, and a new submarine-launched ballistic missile. MIRV technology is similar to that used by China to launch multiple satellites.

This concludes my presentation of background. Thank you.

[The prepared statement of Ms. Kan can be found in the appendix on page 194.]

Chairman SPENCE. Thank you, ma'am. Dr. Sokolski.

STATEMENT OF HENRY SOKOLSKI, EXECUTIVE DIRECTOR, NONPROLIFERATION POLICY EDUCATION CENTER

Dr. SOKOLSKI. Good morning. My name is Henry Sokolski and I am executive director of a nonprofit educational organization known as the Nonproliferation Policy Education Center. From 1989 to early 1993, I was Deputy for Nonproliferation Policy in the Bush Pentagon. Those sanctions that Ms. Kan talked about, were ones my office promoted and the first memos trying to block the transfer of jurisdiction of the satellites from State munitions controls to Commerce, a copy of which you have in your testimony packet, were signed out by me.

I'd like to ask that a copy of that memo, a one page question-and-answer sheet which will basically track my testimony today, two charts, and a recent article be placed in the record——

Chairman SPENCE. Without objection.

Dr. SOKOLSKI. Along with another thing which I just discovered and pulled and that is a very detailed web page presentation called *Dragon in Space* which is quite popular among experts and I think it will be useful for your staff to have a copy of that. So I ask that those be placed in the record.

[The information referred to can be found in the appendix beginning on page 351.]

Chairman SPENCE. Without objection.

Dr. SOKOLSKI. I think we're here actually for much less exalted reasons than have been presented. I think it really boils down to two questions—I'm saying this on the basis of all of the phone calls that I get since April. The two questions that seems to propel most of the press and, frankly, a lot of politicians, are this: Did Loral-Hughes, in their investigation and the handing over of the report, end up giving militarily significant technology to the Chinese? And the second question, which is perhaps even more popular and in demand to get the answers is, has the Clinton administration in promoting satellite technology transfers, done anything any much more different than what Bush or Reagan did?

Now, tomorrow you will hear testimony from the administration and, undoubtedly, they will spend a great deal of time trying to explain why the answers to those questions is "no." I suppose that, since these are the questions that brought us here, I am duty-bound to give you my own personal answer which, I guess, succinctly would be "yes" instead of "no" to those questions. But, as somebody who has worked on these issues with knowledgeable people since 1988, I would be just as quick to point out that I don't think those questions warrant your attention, at least not initially and maybe, arguably, not at all. The reason why is pretty clear, and I think some of you have been doing a great deal of good work in pointing out the deficiencies of those popular questions.

The first problem is if we focus on the blameworthy question, who's more blameworthy or more responsible, is there a symmetry of responsibility, it begs the question of whether or not something was done or is being done that's objectionable. With regard to Loral-Hughes, I have to tell you focusing on the events of early 1996 will not get you the answer to what may or may not have been transferred in Spring of 1996. And the reason why is the only way you can know what of significance was transferred is to view what was transferred in 1996 in the context of what has been transferred, or not, since our first commercial launch with China in 1990.

The proper question then, I think, is this: What technology of military systems have U.S. satellite technology transfers since 1990 afforded the Chinese and, if they have afforded any militarily significant assistance, what, if anything, should be done to prevent such assistance now and in the future?

Now I want to be clear, in a general way, about what we're talking about when we talk about satellite technology transfers. We're not talking about hardware—this is not a black box. Nor are we speaking about secrets that can be written down in a book and locked up or handed over. We're not even talking about know-how in the sense of a kind of instruction book on how to perform a particular task. Instead, using kind of biblical language, we are not trying to convey to China a fish or even an instruction book on how to use a rod and reel but, rather, what we, I believe, may be conveying is one-on-one, hands-on transfer of a lifetime of knowledge about how to fish. The very best angling information. What we are handing over then is not, I guess, know how, but as the Chinese once told a contractor, "We have all know how we need. What we need is know why." And the know why is to understand what are the problems that have to be mastered to put a fragile satellite in a particular spot in space. What are the kinds of interactions between all the systems that have to be considered? What is the kind of engineering that you have to do to get around those problems, not just in the instance case, but in aerospace problems of similar character, as they say, on other launches of other satellites and other payloads, including, and what we're worried about, military payloads.

Now, when you talk to knowledgeable people, I think you find that the answer to the question, "What technology of military significance have our U.S. satellite transfers afforded the Chinese?," the general answer, I think you'll find when you talk to folks both Democrat and Republican, is a concern that probably a great deal has been transferred. And the reason why isn't because there is specific knowledge exactly what was handed on what date by whom. But rather because of the knowledge of four things, some of which has been mentioned in the previous testimony by Ms. Kan.

First, we know that the contractors who make the satellites have a keen interest in making sure that very sensitive technology know-how gets into the hands of the Chinese to make sure that their satellites are properly launched. Now, we'll get back to this in a moment.

Second, we know that the Chinese, both for commercial and military reasons have a very keen interest in trying to get this information.

Third, we know by any standard the controls over these kinds of intangible technology transfers have become more lax, not tighter, over time, and, third and finally, and what these charts that are in your packet will help demonstrate, we know what the Chinese have learned since 1990, what they have mastered, and we know that what they have mastered is of military significance.

I'd like to go over each one of those. First, contractors in their interest in sharing know-how that may be of military significance. Contractors certainly have a keen interest in seeing to it that their satellites, one, don't get blown up on the launch pad; two, don't get shattered by excessive vibration due to rigid launch designs, such as the kind of rigid launch designs as you found in early versions of the Long March. They certainly have an interest in not seeing their satellites torn apart by jolts of acceleration due to faulty stage separation and poorly designed kick motors of Chinese design. Nor do they want to see their satellites spun improperly once they're released from the platform by dispensers or kick motors. And, if there is damage done to the satellite, they have just as keen an interest to make sure that the insurance companies find out and know that it wasn't their fault that there was a failure.

Second, as I mentioned, the Chinese have a keen interest in trying to get this information. First because they want to have a successful business. You'll notice on the chart, there is a line drawn at 1996 and you'll notice that there are some explosions on the chart prior to 1996 and that the reliability rate through 1996 is only 75 percent. That is a prescription for going out of business. Why? Insurers do not want to bet the farm on someone who only succeeds 75 percent of the time in launching these satellites. These satellites cost a great deal of money. More important, they make a great deal of money, much, much more than the launch. It should be noted that while there were price advantages to launching Chinese in 1990, they have largely evaporated. We're talking now by all accounts a difference of about 10 to 15 percent. The launch, however, is the least of the revenue streams that the satellite maker and the customer realizes. Since the satellite costs in access of $100 million, sometimes even up to $200 million and makes that or even a multiple of that over its lifetime.

[See chart C in the appendix on page 365.]

Now, with that kind of reliability rating, 75 percent, a lot of insurers would not insure. You notice that it's gotten better and I'll come back to that at the end. It's now at 100 percent for the last 10 launches. In addition, of course, and it was mentioned, they like getting money. If the business fails, they don't get the money. They've gotten about a half billion in hard currency from the United States. That's money they can use.

In addition—and this is the important point—the technology necessary to put a satellite precisely in space helps deal with a number of military problems that the Chinese have been quite vocal about. First of all, since the late 1980's, the Chinese have been writing both in classified and unclassified realms about their desire to hit targets precisely and use non-nuclear weapons. And I'm not talking about chemical or biological weapons, but plain old, garden variety explosives.

Why would they want to do this? If you have a renegade province, and you want to influence their behavior, it would be nice to be able to do so without threatening to destroy them literally. One country comes to mind. It's call Taiwan. They're very concerned about developing this technology for their M family missiles, which they fired in 1995 and 1996, off the Taiwanese coast.

The second thing they're very interested in is U.S. development of missile defenses. Now, this may seem somewhat fantastic to us, as we move so slowly to develop missile defenses, but the Chinese are frightened and have been quite vocal for the last 3 years that the United States do nothing that would negate the power of their missile forces. Missile defenses with the current DF5, which is an old monster, would do the job.

They are, therefore, developing new types, as Ms. Connors pointed out, of missiles that will have multiple independently targetable reentry vehicles of the kind R and RMX, post-boost vehicles that can change the predictable direction or course of a missile warhead to evade defenses. Together, they hope to overwhelm and evade any defenses.

Finally, they're interested in developing command and control systems and intelligence systems that will require sophisticated satellites of their own. They need a soft and sophisticated launcher to play out that game. Developing kick motors for satellites, developing satellite dispensers that are precise, giving them higher reliability for their launches, all play in the hands of these military objectives.

Third, a comment about controls over transfers, and why I believe they have gotten laxer, and others who have tracked this issue believe they've gotten laxer. We only have experience in the Bush Administration with four launches. We have much more, 14, in the case of the Clinton Administration.

Luckily, in the beginning, we played a fairly tight game. There were things that leaked, and we knew what they were. Now, however, under commodities jurisdiction, it is up to the contractor to request monitoring of conversations prior to launch. If they don't ask, the monitoring does not occur.

More important, we know not just from Loral Hughes, but two other launch failures that were mentioned, one in 1992 and APSTAR II that occurred in 1995, that the insurers did determine who was at fault, but we have yet to find where that report resides in the government, whether there ever was a written report, what was conveyed by the Chinese and the contractors with one another, to determine who was at fault and what of military significance, if anything, may have been shared. That suggests to me there has been a lax—much more lax—control effort in place.

Finally, as I've come back to these charts, we do know what the Chinese have mastered. If you take a look at the first chart—and I've tried to be ecumenical here—you'll notice I have things in the Bush and the Clinton Administration. You'll notice with the first chart, Chinese satellite transfer is an increasingly dangerous business, that they have learned and been able to validate attitude control with regard to their M family missiles. They have been able to make sure that reliability questions with regard to certain engines in their launcher are taken care of.

They have in fact launched successfully a kick motor of their own design, and they have mastered their own multiple satellite dispense, a very precise character. And those two kinds of technologies having been demonstrated to the satisfaction of U.S. contractors propels them into the military realm, discussed in the military benefit category.

[See Chart D in the appendix on page 366.]

Finally, you'll notice the second chart, which took a great deal of effort to do. My staff deserves kudos. You'll notice that there are quite a lot of launch failures. If you go back all the way to 1970 moreover, the launch reliability is only 78 percent—it's not much better. But you'll notice that something happened in 1996. Those launches were at their most sophisticated launch systems, ones that have been only recently developed, and of the most sophisticated kinds of satellites, multiple satellites, and the most advanced communication satellites. You'll notice there are no failures.

Is that coincidence, or is that the result of technology transfers, cumulative? That is the question that you will have to determine to your own satisfaction. It will require deposing all of the monitors that worked in the government, looking at these launches, and the conversations. It will require deposing their superiors—whether or not they enforced infraction reports. It will require probably talking to contractors and their counterparts, and maybe even granting them immunity to get the truth.

One thing I would urge you all to do, and hope would be the case, and that is to heed the words mentioned in the opening statements, which is not to jump to easy solutions. Many people are now focused on changing jurisdiction as if that's the fix. Word out on the street is, summits going to occur, the fix will be in just to change the jurisdiction to State, and this whole thing will blow over in about three or four weeks. I don't believe that's the case, or should be the case. And I really believe that you folks have taken the key measure and making a marker by saying that perhaps we should suspend all satellite transfers until we get to the bottom of this.

I would hope in conference that you would stick to that position, until you get to the bottom of all of this, and get the answers that you believe are correct.

That concludes my statement. Thank you very much.

[The prepared statement of Dr. Sokolski can be found in the appendix on page 200.]

Chairman SPENCE. Thank you.

Dr. Milhollin.

STATEMENT OF GARY MILHOLLIN, DIRECTOR, WISCONSIN PROJECT ON NUCLEAR ARMS CONTROL

Dr. MILHOLLIN. Thank you, Mr. Chairman. I'd like to begin by apologizing for my late arrival. I was not informed that the time for the hearing had been changed until this morning when I was putting my shoes on. The Red Line then proceeded to behave like a Chinese space launcher, and broke down twice.

I'd like to concentrate on—this morning—on three decisions that the Administration has made, which I believe have had the effect of giving Chinese companies a green light to sell missile technology to countries like Iran and Pakistan.

I think in a sense there are two questions that the committee should look at. First, the question that has been announced, that is, was ballistic missile useful technology, transferred to the Chinese? But a second question, and possibly a more important question is, does our satellite export policy—does the result of that policy create a condition in which we've given—we've insulated—Chinese aerospace companies from possible missile sanctions for their own proliferation behavior?

So, whether or not the United States transferred ballistic missile technology to China, it's pretty clear to me that the result of our policy is that China has been able to transfer ballistic missile technology to Iran and Pakistan, without any serious worry about losing its most valuable source of revenue, which is U.S. satellite launch contracts. I think that is possibly a more important question than the question of technology transfer to the Chinese.

As I said, there have been three decisions that are important. First, the most complicated one—and I certainly agree with Representative Hamilton when he says that the export control laws are arcane—I'll try to explain this in a non-arcane way. The first decision was to transfer licensing jurisdiction from the State Department to the Commerce Department over satellite export.

The second decision was not to enforce existing law on sanctions.

Third—the third decision—was to invite China to join the Missile Technology Control Regime. Now although that sounds like a worthy goal, that is to get the Chinese in the Regime, the effect of letting them join when they're still proliferating, would mean that they would be insulated in the future from any possible application of U.S. missile sanctions. That's an effect of joining the Regime that most people don't appreciate.

Our sanctions laws are missile sanctions laws, not be confused with Tiananmen sanctions. I'm talking about missile sanctions, the sanctions that apply to foreign companies that proliferate missiles. They're based on a simple idea. That is, that you cannot import U.S. missile-related technology or items with one hand, and proliferate missile technology with the other to countries like Iran and Pakistan. That's always been the principle behind our sanctions laws. Unfortunately, in my view—well, unfortunately—I won't even qualify that—the Administration has abandoned this principle. I'll explain why this is.

First, China has changed it's export strategy. It no longer exports full, complete missiles. It is, as the CIA would say, reducing the signature of its activities. It's exporting missiles component by component—piecemeal. Well, what happens to a company that does that? Under U.S. law, a company is sanctioned, but only if—I'm sorry—but the sanctions only apply to the export from the United States of missile-related technology.

Then you get into the definition of what is missile-related technology. Well it just so happens that the Commerce Department defines it one way, and the State Department defines it another way.

The State Department defines missile-related technology as technology even if it's embedded in a satellite. So even if a satellite has missile items in it, under the rules in the State Department, it's still sanctioned if the recipient is guilty of missile proliferation behavior.

In the Commerce Department, as you might expect, things are defined differently. A missile-related item loses its character, under the Commerce Department view, once it is embedded in a satellite. So even if the recipient of the satellite is selling—is proliferating—missile technology to Iran and Pakistan, the Commerce Department wouldn't block the export of the satellite under its rules.

So moving the satellites from State to Commerce has the effect of insulating Chinese aerospace companies from future applications of U.S. sanctions for their own missile proliferation behavior.

Since missile launch contracts are the most lucrative thing that a Chinese aerospace company can get from the United States now, if you take satellites out of the sanctions picture, you've removed any real effect the sanctions could have. And as I said before, you're giving a green light to the Chinese to proliferate, because you're taking any risk of punishment away.

Now it's important to realize that the same companies who launch our satellites are the companies that sell missile technology to countries like Iran and Pakistan. Who are they?

China Great Wall Industries launches satellites. It is a part of China Aerospace Corporation—I'm sorry, the China Aerospace Industry Corporation. These companies launch satellites on China's Long March rockets. The United States in the past, in 1991 and 1993, sanctioned both of these companies for missile proliferation activities. These companies are continuing their missile proliferation activities today.

So, the recipients of our satellite exports are the very same companies that are guilty of spreading missile technology throughout the world. The effect of putting money in their pockets, just makes it easier for them to do that job.

The second point I'd like to make is that we are not applying existing sanctions laws—we, the United States. I too have some tables. They're attached to my testimony. There are three. The first is a catalog of China's dangerous exports to Iran. The second is a catalog of China's—a list of China's—dangerous exports to Pakistan. And the third is a list of China's dangerous exports to the rest of the world. These include nuclear, missile, and chemical biological related exports.

China is still the most serious proliferation threat in the world. If China continues its present course of proliferation exports, it's going to make it impossible for the United States and its allies to halt the spread the weapons of mass destruction.

We have already heard testimony about the series of sanctions that the United States applied to China in the early 1990's. They had essentially no effect. China is still proliferating today.

By the autumn of 1996, the intelligence community—our intelligence agencies—had completed what are known as findings of fact or statements of fact about China's missile transfers to Pakistan. These were prepared with a lot of effort by our best experts and they are airtight. They've been sitting in the State Department since the autumn of 1996.

Roughly one year before that, our government lawyers completed a legal analysis, which showed that you don't need—that is, to apply missile sanctions, you don't need a photograph of a missile with made in China written on the side. Under our laws, conspiracy or an attempt to proliferate is enough for the application of sanctions.

If you put these findings of fact and the legal analysis together, the case is overwhelming, and it has been since 1996.

Last week, I testified with Mr. Gordon Oehler, who just left the CIA's non-proliferation shop. He was director of the non-proliferation center at the CIA. He testified before the Senate Committee on Foreign Relations that in his mind and in the mind of the intelligence community, that there was unanimity that these findings were correct, and that under the law and the facts the case had been made, and that sanctions should be applied. He was very clear in his testimony.

So what we have then is the State Department simply not completing the administrative process. What's supposed to happen after these findings are reviewed, there's supposed to be a memo that goes to the Under Secretary of State, who is the official that applies the sanctions. Those steps have never been taken. The process has simply been short-circuited.

So, it seems to me it's a question of democracy. Congress passes laws. The executive branch shouldn't be able to just enforce whichever law it likes; it should follow the law. And I think these committees should insist that the Administration follow the law in this case.

I think it's interesting the Administration's no longer saying there's not evidence to apply sanctions. It's now saying that it has "not yet made a determination to apply sanctions." Which is not the same thing. The State Department is really saying look, we just haven't completed the administrative process on this.

The last point I'd like to make about present sanctions laws is the Iran-Iraq Non-proliferation Act of 1992. It refers to conventional weapons. If a foreign company provides destabilizing numbers and types of advanced conventional weapons to either Iran or Iraq, it gets sanctions, under existing law.

In 1995, the China Precision Machinery Import Export Corporation sold Iran advanced anti-ship cruise missiles. Everybody agrees that these are advanced conventional weapons, but the State Department said that they were not of "a destabilizing number and type." Which is odd, considering that our admiral in the Gulf complained about them, and that they do in fact threaten both our sailors and our ships.

It's interesting to notice that if sanctions had been applied in that case, they would have been applied to the parent of the guilty party. And the parent of the guilty party was China Aerospace Corporation, the same company that launches our satellites. So here we have another decision really to insulate a company that just happens to be the company that's launching our satellites—insulate that company from sanctions.

I wanted to also mention China in the MTCR. That's covered in my testimony. What you need to know is that if China comes into the MTCR, the Missile Technology Control Regime, without proving that it has really stopped selling missiles, what happens is that sanctions can't be applied to companies in China. That is a tremendous boon for China, which I think it's imprudent to offer China at this time.

I'd like to conclude by drawing your attention to sort of time line. In 1991, the Bush Administration applied sanctions for obvious missile proliferation by China. In 1993, the Clinton Administration applied sanctions to Chinese companies for obvious missile proliferation behavior.

That behavior has continued, but suddenly in 1995 and 1996 our policy changes. China was supplying the cruise missiles in 1995. It was also supplying other missile components in 1995 to Iran and to Pakistan. In 1996, the Administration ignored these findings of fact that I've talked about. Also in 1996, the Administration moved satellites from the State Department to the Commerce Department.

Why did this happen? Why in 1996 and 1995, did we suddenly get a big change in behavior by the Administration? I don't know the answer, but I think the committee ought to find out—or, the committees, I should say—the committees should find out. That's a very important question. What caused this seed change in our attitude?

Our government suddenly decided that sanctions were no longer part of our foreign policy, particularly with respect to satellite launches. We know the result of this change. I've just described it. It really enables Chinese satellite companies to proliferate to Iran and Pakistan with no threat of punishment from the United States.

Now India watched all this happen. India watched China help Pakistan, watched our government try to figure out how not to do anything about it. I think the Indians concluded that Uncle Sam was against the spread of the bomb and against the spread of missiles unless it costs us something. I think it shouldn't surprise us, in light of that, that our non-proliferation policy lacks a certain amount of credibility.

Thank you.

[The prepared statement of Dr. Milhollin can be found in the appendix on page 203.]

Chairman SPENCE. Thank you. Dr. Freese.

STATEMENT OF JOAN JOHNSON-FREESE, PROFESSOR OF INTERNATIONAL SECURITY STUDIES, THE AIR WAR COLLEGE

Dr. JOHNSON-FREESE. Thank you. Good morning. My name is Joan Johnson-Freese. I'm a professor of the International Security Studies at the Air War College at Maxwell Air Force Base. But I would like to make it clear that I'm here this morning as an academic, and as an individual, and in no way represent views of DOD or the Air Force.

I'm here because I wrote a book on the Chinese space program, so what I would like to do is take a few moments and tell you a little bit about that experience and the findings to try and offer you something new. The testimony that's already been given gives you a significant amount of information.

And I guess to keep you so that you don't have to read the book, I'll cut right to the recommendation, and tell you that one of the three recommendations that I made—and I completed the book last November, before this was an issue of great controversy and debate—one of the three recommendations that I offered in my book——

Chairman SPENCE. Before you do, could you speak more directly into the microphone, please.

Dr. JOHNSON-FREESE. Is this better? As an academic, I'm used to projecting, so I guess I thought I could.

One of the three recommendations that I made in the book was to selectively increase the export of licensed satellite technology to China, because it had been such an effective policy in gaining U.S. leverage in the area of national security.

I'm a space policy person. I've been writing on international cooperation and competition in space for about 15 years. This was my fifth book in that field. When I decided to take on the project, I did it because there is a significant literature gap in the field. There's materials put out by the Chinese that give information on what they are doing in a most cursory manner, and that's it.

One of the things I do in the summers is teach at an organization called the International Space University, which couldn't be more open and more international. In fact, two of my students from there are in the room today, I'm proud to say. And even in that international forum there is no material available on the Chinese space program. There was nothing.

So I wanted to write this book to talk about not only what the Chinese are doing, because I think you pick up Janes and find that out, by why and how they are doing it. And it's there that I came across the undisputable fact that the Chinese space program has, and to a significant extent remains, totally opaque to the outside world.

Progressing from there, when I finally got the manuscript to my editor, she called me and said, Joan I hope you realize that a third of your manuscript deals with Chinese culture, politics, and history, before you ever talk about space. And I would suggest that I would have been remiss not to include that detailed summary of where the Chinese are coming from, how they operate, before getting into what they are doing, or you get a very distorted picture—very distorted.

We cannot assume that rules that apply in the Washington Beltway, Paris, Bonn, or even Tokyo, apply in Beijing. And I guess the best way I can explain it to you is through a description that was given to me in China with regard to not only their bureaucracy, but their culture as a whole. If we think that the United States—if we compare the United States to—strategically and how we operate to chess, where each team has 16 players—or each player has 16 pieces—the Chinese play wa che, where each player has 256 players.

They make us look like amateurs at bureaucracy. The complications, the ties, that go through their space program are phenomenal. Hence, the title, "A Mystery Within A Maze." Breaking through that has been a task, which as much as we've done, can be attributed to the commercial business that's being done in China in space. We are learning more about what they are doing, how they are doing it, and why they are doing it through commercial space activities than we have ever learned before. There is mutually beneficial sharing of information going on.

When I reached my conclusions—when I was trying to draw conclusions, I was not looking at it as a non-proliferation issue. Most of the discussion this morning has looked at the China satellite export policy from the perspective that it is a nuclear export—or, nuclear non-proliferation—issue. I would suggest that is a major component.

I would also suggest there is an economic component. We would be remiss not to consider the 1.3 billion people in China and the market that that involves. We would be remiss not to include the human rights issue. We would be remiss not to think of Korea, Taiwan, and the entire North Asia issue.

But, all told, when I came down to doing my analysis, I looked at the question, has the United States gained more leverage through our commercial satellite export policy than we had prior. And unequivocally, the answer was yes.

Therefore, drawing on my background as a space policy person, I looked at what has the United States done in the past controlling sensitive technology, which by the way, is available on the commercial market. The satellite technology that we are talking about has been on the international commercial market since Intelsat.

This is not a ball game that the U.S. totally controls. This is a market issue where there are many countries who are very hopeful that we will pass policies more restrictive toward China, because they would love to move in and take up the commerce.

But it's not solely a commercial either. All this has to be looked at in a very balanced way. China is at a very precarious point of development. If you look at China post-cold war, and again, I guess I'm drawing on my experience as an international security professor here too, China post-cold war.

During the cold war, when the United States and the Soviet Union were kind of at each other, China was feeling pretty good. They were feeling pretty stable and secure. The cold war to them has meant externally a more precarious situation. Internally, those 1.3 or 1.4 billion people are a lucrative market to us. They are a massive employment problem to the Chinese, not to mention food and housing.

The way the government sees internal stability as being maintained and fortified is through economic development. Space has the unfortunate and sometimes fortunate—this is both a blessing and albatross—a nature of having absolutely symbiotic material. One of the first things I do in an undergraduate class is ask my students what's the difference between a rocket and a missile. We love rockets; we don't like missiles. Well, what's the difference—very little technologically. That's a situation we have to deal with.

Also, like the United States and the Soviet Union, now Russia, the Chinese have a program grounded in the cold war paradigm, where the military—and there were no civilian aspects at that time—the technology is tied as are the industries. I think we would be kidding ourselves that the industries in the United States and many other developed countries don't deal with both commercial space and military sales. To trace things to organizations in China and point at them is unique in that field, I think, is a misnomer.

What we have to do is show China that it is in their benefit to make the transition—the post-cold war paradigm transition—that the United States has made and that the United States is working very closely with Russia to get them to move from filling their economic coffers through missile sales to filling their economic coffers through commercial satellite launches and through domestic use of the same space technology, which we are so fearful of.

Moving to the issue of the Commerce Department, and whether or not the appropriate agency to be dealing with the technology, I would suggest this is part of the inter-agency process. What we've done is move it from a lead agency with State to a lead agency at Commerce. The other players have not lost their voice. I would suggest it is very unusual to get unanimous decisions as we have, and it would be unheard of to not have dissenting views.

Finally, I would return to the question of the larger—is this policy in the National security interest of the United States. And I can only tell you that from the research that I did and the experience I had, I found the answer to be yes. And I would urge you to continue it.

Thank you.

[The prepared statement of Dr. Johnson-Freese can be found in the appendix on page 216.]

Chairman SPENCE. Thank you, ma'am. I think what we've heard this morning proves the old adage and our maker has endowed us with minds that can look at the same sort of facts, and arrive at conclusions 180 degrees apart from one another.

I won't ask any questions at this time. I'll yield to Mr. Gilman.

Chairman GILMAN. Thank you, Mr. Chairman. As we know, the President is departing next week for his summit meeting in Beijing, and like all summit meetings, each leader is going to want to conclude a number of new agreements and try to declare the meeting thereafter a success.

In the non-proliferation area, I have one point. Our Nation had put a number of proposals on the table, including supporting China's membership in the Missile Technology Control Regime. But I understand those efforts have been abandoned. Also on the table were proposals to increase a number of Chinese launches of U.S. satellites in the new Scientific and Space Cooperation Agreement.

And I'm going to address this to the whole panel. In your view, what should the President be saying to Jiang Zemin regarding missile proliferation? What kind of agreement should we be making with the Chinese, if any agreement? And, Dr. Milhollin, let's start with your testimony—you've touched on the subject.

Dr. MILHOLLIN. Well, I guess from my previous remarks, I would say that we should not make any kind of an agreement with the Chinese that gives them a green light to continue their present proliferation activities.

The decision China has to make is whether to be part of the solution or part of the problem. They've been part of the problem up to now. I think that other countries, Argentina, Brazil, some of the countries in central Europe, have made the decision they want to be part of the solution.

So, I think that the President should say that China can expect no special treatment from us on sanctions or on satellite launches until it proves that it's willing to become part of the solution to the proliferation problem, rather than part of the problem.

Chairman GILMAN. Thank you, Dr. Milhollin. Dr. Sokolski.

Dr. SOKOLSKI. I think it's important to get people to live up to what they've already agreed to. There's plenty there. And in this case, what they have agreed to do, they are not doing. They are signatories in the NPT.

I can remember appearing in front of your committee 24 months ago, and we had the very same discussion, the very same worry about the MTCR being opened up to China. We had the very same concerns that China wasn't living up to NPT pledges or it's MTCR pledges that were bilaterally made to the Bush and Clinton Administrations.

So, let's not make more work. Let's focus on what we've done and get them to agree to actually follow through on what they said. We've seen also, by the way, that there pledges about Iran are in dire need of attention, because they are not following through there at all.

Chairman GILMAN. Thank you. And Ms. Kan, did you want to respond?

Ms. KAN. Thank you. One of the descriptions of the U.S.-China relationship is that we are moving toward a strategic partnership. I think a lot of people may agree that strategic partners would agree on the same international guidelines for curbing missile proliferation. And that the Missile Technology Control Regime, to which China has committed, is a set of international guidelines, and not just the U.S. set of goals, as well as the fact that curbing proliferation is not only in the U.S. interest, but also in China's interest.

Chairman GILMAN. And Dr. Freese.

Dr. JOHNSON-FREESE. I would suggest that if we do not encourage commercial space activities with China, that we can expect that their arms sales will increase. Therefore, we ought to do everything we can to encourage the commercial space activity field in China, and the MTCR.

Chairman GILMAN. And thank you. To Drs. Sokolski and Milhollin, do you agree with Dr. Freese if we have additional leverage over China because of our space cooperation—if we have such leverage, if we used it properly?

Dr. SOKOLSKI. I would agree that we have tremendous leverage. And in fact, we did use it in the Bush Administration to get, I guess accurately described by Dr. Milhollin, as pledge that was not lived up to. But it was in writing. That leverages all of the things I describe that China wants, a successful commercial space launch industry and one that can help their military efforts.

Now I want to take exception, however, to what's been portrayed with regard to how this leverage only works one way. It turns out that 75 percent—I repeat, 75 percent—of all the commercial payloads are U.S. made or have U.S. made parts in them, and therefore cannot be transferred unless the United States gives permission. A vast majority of the remainder are European space agency payloads, and these countries in Europe are inclined to launch off of easel launchers. And it's very interesting to see, when you take a look at all 60 !aunches in China, that there are not launches of satellites by other nations, but rather, purchases of satellites by China from other nations.

So I would say that if we used the leverage of our satellite commerce, we would definitely get the attention of the Chinese. We have not used it, and I think Dr. Milhollin is correct. This was something which, again, I was arguing 24 months ago that we needed to use the leverage. But we could use it.

Chairman GILMAN. Thank you, Dr. Sokolski. Dr. Milhollin.

Dr. MILHOLLIN. It's hard for me to see how you gain leverage with respect to someone if you give that person or that country everything it wants with no conditions.

I think we do have leverage, but only if we have credibility in the sense that the Chinese actually believe that we would withhold something they want if they don't behave according to the rules. So far, we've proved over and over again that the Chinese, can in effect, thumb their noses at us and we cave.

I think that we have to establish our credibility in order to have leverage, and in order to do that, we need to be prepared to withhold satellite launch contracts unless China behaves with respect to proliferation.

Chairman GILMAN. Thank you. And Dr. Milhollin, your testimony describing how the Clinton Administration has immunized exports of U.S. satellites from our missile sanctions laws is excellent.

And in my view, it's a combination of the change in licensing jurisdiction from State to Commerce, and this Administration's interpretation of how missile sanctions laws are applied. That seems to be the key to understanding how our national security may be at risk by the export of U.S. satellites. It's a critical point, and we intend to pursue that with the administration tomorrow.

And I want to thank our panelists for their very important testimony today. Thank you, Mr. Chairman.

Chairman SPENCE. Thank you, Mr. Gilman. Mr. Hamilton.

Mr. HAMILTON. Mr. Chairman, thank you. I'll just ask one question really, because I know there are many my colleagues have. What access has China gained to military technology because of these satellite launches that is meaningful, that is harmful to the U.S. national security? I've heard things about improving their guidance systems, their missile accuracy, and improving their launch capabilities and so forth.

So, in your judgment, can you tell me what—because of this pattern we've had of supplying these satellites, what harm specifically has been done to the United States national interest?

Dr. SOKOLSKI. I think, as I've tried to demonstrate with the charts, having mastered how to put fragile payloads into space precisely and reliably, and figuring out how to do multiple payloads precisely, and to use kick motors, is directly transferrable to the development of MIRVS, which are these multiple independently targetable reentry vehicles, post-boost vehicles, and to have accuracy so that you don't need to use weapons of mass destruction to achieve military missions——

Mr. HAMILTON. And it's your view, Dr. Sokolski——

Dr. SOKOLSKI. Yes.

Mr. HAMILTON. That they would not have been able to do that had it not been for the U.S. satellite launch program?

Dr. SOKOLSKI. My contention is that by doubling their experience, which is what we've done, and giving them all their experience practically with sophisticated multiple satellites launches, when the case of Iridium, we have at least indirectly propelled them to master these very important military capabilities.

The question before your committees will be how much and whether or not it was preventable? Now let me be clear. It could be that after you do all your investigations, sir, you will come to the conclusion that we want to make money, we want to continue, or that there's no way to prevent the leakage of this technology. And that will impact directly whether you decide to bar certain kinds of commerce or not.

But, you have to run that to ground, and that will require the deposing of specific individuals, names of people that I've shared with this committee staff.

Mr. HAMILTON. Mr. Chairman, maybe the other witnesses, if they want to, could comment on that.

Dr. MILHOLLIN. I would comment. I think that, just in the broad sense, you have to remember that an ICMB is just a space launcher with a warhead instead of a satellite on top. If the space launcher doesn't work, the ICBM doesn't work. Technologies for staging, for propellance, for reliability, for guidance—they're all interchangeable. So if you're helping somebody make a reliable, effective space launcher, you can't help but be aiding that person in making a good ICBM. It's the same thing.

And second, I think you have to remember the financial aspect of it. Big military programs cost money. China has a limited amount of money. Even we have a limited amount of money. The big countries that made ICBM systems with nuclear warheads on top used space launches to defray the expense of the total package. And that's what China's doing. China—what we're doing by putting money into the pockets of the space launch companies in China is to make it cheaper for the Chinese to build good ICBM's just from the financial point of view. That's inescapable. Money is fungible.

So, I guess my comments would be those two.

Dr. JOHNSON-FREESE. I would just raise the point that launching payloads on rockets does not inherently mean that you are conveying military technology. The fact that we are launching international space technology on Russian launchers does not mean that we are helping the Russian military. I think it's also important to point out that the Chinese have been working on an ICBM program for a very long time. Now, the technology that they are alleged to use to improve their guidance systems, their targeting ability—first of all, if we're talking about American cities and a countervailing strategy, you don't need to be precise with a nuclear weapon. I think we all acknowledged that. If we're looking at nonnuclear weapon targets, I don't think the Chinese—I don't see it in the Chinese benefit to use them against renegade provinces or anything else that they see as a "cash cow." That would be like saying we're going to take out Hong Kong which is one of their prime funding sources. They aren't going to do anything in that regard.

There are ways to get at Taiwan—economic ways that they are very, very adept at—that they don't have to move to the military.

Dr. SOKOLSKI. May I rejoin? I think this is where the dialog comes out and information comes across. First of all, let's not just talk about the DF–5 because I think that if the administration, in fact—that's that missile there—the one they have, it's the ICBM. The range is the same as it was in 1980. The accuracy is the same. So if you just talk about that—hey, no problem. Certainly, if all you're trying to do is destroy a portion of Los Angeles—not an issue.

If you're trying to make sure that your missile investments in a new system called the DF–31 or a DF–41 which is listed—well on one of the charts in your package—you're going to want to make sure it can work 10 years from now when there may be some kind of missile defense for a number of places, including the United States.

Second of all, you're going to want to make sure that you can at least credibly threaten to do missile strikes, even though you would never want to do them against places like Taiwan. I think it's important to understand that when you develop a multiple satellite dispenser, you have developed a MIRV. Let me explain why.

Do you know where our multiple independent re-entry vehicle systems came from? They were not developed by the military. They come from TRW and they were developed to—you guessed it, to dispense satellites. When you dispense a satellite, you have to dispense it accurately and you have to spin it a certain way. Turns out when you dispense a warhead accurately, you have to spin it accurately. It's exactly interchangeable. The post-boost vehicle and the kick motor phenomena are also interchangeable. If you want to evade someone's missile defenses or if you want to have high accuracy, you have to have a post-boost vehicle. By the way, the M-family missile—the one that they're trying to make non-nuclear has such a system and they're trying to develop it further.

So what I'm trying to convey here is it's easy to block out the complexity of the proposition that we're helping them if you pick select pieces of the picture. If you look broadly, you have reason to worry.

Mr. HAMILTON. Thank you, Mr. Chairman. Mr. Chairman, may I ask unanimous consent that the statement by Congresswoman Lois Capps be entered in the record, please?

Chairman SPENCE. Yes, without objection.

[The prepared statement of Mrs. Capps can be found in the appendix on page 193.]

Chairman SPENCE. Mr. Stump? Mr. Hunter?

Mr. HUNTER. Thank you, Mr. Chairman. Thank you very much to our witnesses for your excellent testimony.

Let me ask you, Dr. Sokolski, and all the panelists—let me simply ask you this question. You had a diagram that showed that the Chinese missiles have become more reliable over the last several years.

Dr. SOKOLSKI. Yes, I believe that's still on the floor there—maybe Dr. Shuey can put it up. There we go.

Mr. HUNTER. So they've gone from a sub-75 percent reliability to close to 100 percent reliability?

55-784 99 - 2

Dr. SOKOLSKI. Well, since the last accident in August 1996, they haven't gone to about 100 percent, they have 100 percent. Now you can look at this and say well but that's not a statistically important number—10 launches; or you can look at those launches and say yes, but they're the most complex satellites and the most advanced and experimental of the launchers that China has done. If they were going to be failures, you would think they would occur there. So half full, half empty. This is where the rubber meets the road. You have to figure this out.

Mr. HUNTER. So it's your conclusion that their missiles have become more reliable.

Dr. SOKOLSKI. There's no question.

Mr. HUNTER. Dr. Freese, do you argue with that statement?

Dr. JOHNSON-FREESE. Their launchers have become more reliable.

Mr. HUNTER. OK, secondly, is it your opinion, Dr. Sokolski, that they become more reliable partly as a result of America's satellite program with the Chinese and the interaction that has taken place as part of that?

Dr. SOKOLSKI. Well at the hazard of making a bad joke—you're a great proponent of SDI. You know that not all the experiments of SDI or ballistic missile defense always work, but we always learn something every time there's a failure. By that token, they've been learning quite a lot because they've had quite a lot of failures, and they have been claiming that they have fixed those failures with regard to their latest—I think it's the Long March 3—is it E or B? Well, the latest one. They had problems. They claimed they fixed it. So if you also take a look at the black dots on that chart, those are U.S. waived satellites. You'll notice that if you subtract that, you have half as many launches, twice as much experience, and they get paid for it. Yes, I think we helped.

Mr. HUNTER. OK, we helped.

Chairman GILMAN. Will the gentleman yield a moment?

Mr. HUNTER. Sure.

Chairman GILMAN. How much do they get paid for each launch?

Dr. SOKOLSKI. Well, initially, there was an enormous price advantage. I mean, they were giving this stuff out for slave labor prices. But our Trade Department and Rep did her work and badgered them, if you will, into raising their prices so that they wouldn't be—it's not that different now. I think now some of the launches is as expensive as $85 million.

Chairman GILMAN. $85 million per launch?

Dr. SOKOLSKI. Some of them.

Mr. HUNTER. Dr. Milhollin, do you agree with the two statements that we've basically established here? One that their missiles have become more reliable; and two, that that's partly the result of American interaction with the Chinese with respect to our satellite program with them.

Dr. MILHOLLIN. Yes, I agree with that.

Mr. HUNTER. Dr. Freese, do you agree with that—that we've helped them to some degree? Help make their missiles more reliable with our program?

Dr. JOHNSON-FREESE. I would agree that because they want to launch our satellites, they knew they had everything riding on making that launcher reliable.

Mr. HUNTER. Well, do you think in the least the moneys that we've given them have been helpful to their ballistic missile program when they have been able to buy more things.

Dr. JOHNSON-FREESE. Clearly, one-third of the commercial launch moneys that they get go back to the kind of parent organization which deals with both rockets and missiles.

Mr. HUNTER. So to some degree, we have helped to make their ballistic missiles more reliable with our satellite program?

Dr. JOHNSON-FREESE. I couldn't say where the money went to directly.

Mr. HUNTER. OK. Dr. Freese, you stated that you thought we should convince China to move to "fill their coffers" to move to domestic sales of domestic items. You know they enjoy about a $40 billion trade surplus with us for such domestic items as "Tickle Me Elmo" dolls and lawn sprinklers, and just about everything else that you can find at a Home Depot. How much more cash do we have to get them before we can move them off of missile sales?

Dr. JOHNSON-FREESE. Well, I would suggest that that's part of their domestic issue that they're dealing with. They have problems, too.

Mr. HUNTER. Well, but my point is—you said that we need to move them into domestic sales to get money. They get $40 billion more each year from us than we get from them. They have a lot of American dollars as a result of a lot of domestic sales. That apparently has not deterred them from selling missile technology. How much more do you think we have to—they have to make off of us in these domestic sales before they will move away from missile sales?

Dr. JOHNSON-FREESE. I'm saying that for all the domestic sales that we cut back, they will make them up elsewhere. We can count on that. They will—not in "Tickle Me Elmo" dolls, but in launch technology. It would either be a rocket or a missile, and we get to choose which we want them to launch.

Mr. HUNTER. OK. Last question for all of you. Very simply, we passed a ban on exports. Basically, the House said we're going to put the brakes on until we sort this mess out. Do you agree that we should maintain that ban until we re-scrub American policy? Start with Ms. Kan and go right on down the line.

Ms. KAN. CRS will not make a judgment on your policy.

Mr. HUNTER. OK, thanks.

Dr. SOKOLSKI. AMPAC will. Folks, how many times has the Congress voted correctly—super computers, nuclear cooperation, and now, satellites? How many times have you gone into conference? How many times have you left with all of your clothing? I hope you are fully clothed and you stick to this position, and that you convince your counterparts, in the Senate, that this is a serious matter. That both Republicans and Democrats have a great deal politically at stake in taking care of. No one is going to believe the credibility of Congress, the importance of National Security, or even the respect of law, in my opinion, if by the end of this year, China SAT–8 flies without as much as a flutter of concern. Or if all you do is change the jurisdiction in the next 4 weeks or six, or that's the word out on the street and nobody really has to worry about whether or not this commerce may be upset.

Dr. MILHOLLIN. Mr. Hunter, I think this is a credibility issue for us. I think if we just go forward as if nothing had happened and do business, as usual, everyone will conclude that we really don't care about nonproliferation. So I think the House and Congress should persevere. I think that our satellite—commerce with China should be suspended until we come up with a policy that we can feel comfortable having.

Dr. JOHNSON-FREESE. I think the policy is sound. I think it's very good that you're looking into whether or not the control system is working appropriately. I would suggest, too, it's a credibility issue but I think we need to look at what is it that we want? Is this satellite policy that's been construed as a nonproliferation policy that turned economic policy? I would suggest that, like most other space fields, it's become part of a broader, national security policy. In that regard, the policy has been successful. If you just look at it focused on nonproliferation, just look at it as economics, just look at it as human rights, then we really do have a credibility issue to everyone else whose trying to decide what is our priority. I think we'd be remiss to look at it too narrowly.

Chairman GILMAN. Mr. Sisisky.

Mr. SISISKY. Thank you, Mr. Chairman. Dr. Freese, in your testimony—I think it's yours that I read—it was very interesting the competition among all nations of the world almost to sell China whatever they want. Dr. Sokolski, or one of the gentlemen, talked about the wheels. As I have it here—a German firm exports momentum wheels to China and its DFH–3 communication satellites; a Canadian company has formed a joint venture in China to build onboard satellite electronic ground facilities; and Russia and France sell high resolution imagery. That in itself is a problem. I'm not opposed, I want you to know, about freezing until we find out—get to the bottom of it—but we're going to open up a whole periphery of countries selling them—the technology.

Dr. SOKOLSKI. Well, it's not as bad as you think. First of all, take a look at what's happened with that satellite project with the Germans. They've had a heck of a time getting that to work. It hasn't—they've had two launches.

Second, the Russian assistance and I encourage you folks to take a look at that. One of the things—when I worked for Ambassador Jim Lilley in the Defense Department—we tried to get the Bush Administration to look more seriously at the beginning of massive sales of hardware with technicians with know-how to bolt things together for the Chinese. I rather suspect, however, that what was missing and why the Chinese are so eager to work with us is no why. It's a little subtle, but not that subtle. I think there is a reason why they like working with the American-made satellites. It's not an accident. As far as others rushing in, there are only so many pairs of jeans the Russians can take from China for their goods. That is not going to produce a lot of sales. They do not have commercial satellites. The Germans have their hands full.

Dr. MILHOLLIN. Sir, I'd like to add just one comment. I think when we talk about other countries rushing in, China is not a unique case. German companies are selling high technology now to Iran that we won't sell. Russia is building reactors in Iran that nobody else would build or certainly that we won't build. I think it's dangerous to assume that U.S. behavior should be the same as the lowest common denominator in the world, less we lose a market. This really is a moral issue. We see it that way with respect to Iran, Libya, Syria and lots of other countries. I think we should simply suspend this judgment that we've made because it's China.

Mr. SISISKY. The only argument that I could give to that is that with American technology it's also Americans there watching what's happening that will not be taking place. Did you want to comment, Dr. Freese?

Dr. JOHNSON-FREESE. I agree with what you're saying. I would simply urge you to look at this estimate of the issue. If the consequences—if the United States were to pull back, would that actually mean that China would lose technology—access to technology? Or would it just mean that they would get it from elsewhere with less controls than we have now?

Mr. SISISKY. The basic question is what changes would you make in current satellite export policy and/or procedures to reduce the risk to compromise the U.S. national security—if you can do that pretty fast.

Dr. MILHOLLIN. I would subject satellite exports to the same rules that all other U.S. exports are subjected to which is that our laws on missile sanctions should apply to the recipient of a satellite. If somebody wants to import a satellite from the United States and that company is proliferating missiles to Pakistan and Iran, that company should be excluded. That's our law and it should be applied. If we just did that, that would solve a lot of problems.

Mr. SISISKY. Any other comments?

[No response.]

Chairman SPENCE. Thank you. Mr. Bateman.

Mr. BATEMAN. Thank you, Mr. Chairman. Let me first, very briefly, get on the record a strong sense of resentment almost to I think one of the dumbest national policy decisions any nation could ever pursue. That is to assist in the development of space-launched technology that puts $500 billion in somebody else's pocket, instead of having encouraged and promoted a robust commercial space-launched capability within the United States of America. Quite frankly, while I'm very sympathetic to 1.4 billion Chinese people, I have little or no sympathy for the regime which governed them or giving them an advantage commercial, or military, or strategic.

Having said that, Dr. Freese, I read in the first paragraph of your written statement that in the research on your book from the Chinese space program a mystery within a maze that you felt you had encountered "encountered deliberate subterfuge." In view of that—in view of a summit conference between Boris Yeltsin and a Chinese regime in which their communique said that they were going to participate in a special strategic relationship because it was contrary to their interest and their policy to have the United States be the world's only superpower, I am amazed and appalled that you would come to the conclusion that it's to our mutual benefit to advantage the Chinese missiles targeted on the United States. I find it something that calls for more than the statement of the conclusion that there's mutual benefit. I'd like to know what benefit have we received for that which we have given. I frankly am not aware of any. If there is some, please inform me of what it is and where it is that we have more leverage by continuing that which we have foolishly and improvidently done that gives us more leverage.

Dr. JOHNSON-FREESE. I would suggest that U.S. administrations since Nixon have realized and worked toward the goal that U.S. foreign policy cannot be successful without engagement with China.

Mr. BATEMAN. Dr. Freese, I am not quarreling with engagement. I even support MFN and always have with China. That is not the question. But you can't take engagement and take it on everybody else's terms without any constraints that affect vital matters of our national security. What I ask of you is that you give me some detail as to where we have been advantaged by what we have done, vis-a-vis transfer of satellite technology to the Chinese. I want some specifics. Not we need to be engaged, that's not the issue.

Dr. JOHNSON-FREESE. I beg to differ, sir. I think engagement—it's only through engagement that we can get away from the subterfuge. I encountered it immediately. I learned most of the information that I got for my material—for my book—from other country's companies; from representatives; from those who have been there; who have worked through the Quan Chi system; who have penetrated the system. It's only through being there that we're going to learn more.

I would suggest—I know you're going to hear from the intelligence community that we're gaining more intelligence than we've had in the past through our contacts through companies. We are learning how to work the system. We are taking democracy there. Democracy is a system that's seen, not read about. I think that's one of the major things that we're getting. If we expect to talk human rights to a culture with 5,000 years of continuous history of which none of it involves democracy, we have to show them—we can't tell them.

The experiences that I had, in the research there with individuals, cannot be given in specifics in terms of leverage. It can only be given in terms of engagement—a larger engagement policy which I would suggest if we focus on nonproliferation as the sole element of national security policy in the United States would be dysfunctional. It is a large part. It is a critical part. It's not the only part.

Mr. BATEMAN. Mr. Chairman, if I have a moment left. In your original testimony, you make the statement that "Chinese culturally understand the premise of reciprocity and are willing to negotiate it to preserve their own precarious, domestic balance. It is up to the United States to make sure that the Chinese are then clear as to what we consider our priorities in particular instances of negotiation—nonproliferation, human rights, market access, demarketization, Taiwan, political influence in Asia, military modernization and other issues." Which of those do you think we ought to give away in our engagement, in our negotiations with the Chinese to understand reciprocity?

Dr. JOHNSON-FREESE. None of them. I think—I simply don't think that we can say if you don't ship missiles to Pakistan, then we expect concession in all the following areas. What I'm saying is I think that we ought to make our priorities clear to them in specific instances.

Mr. BATEMAN. Well you language suggests that we've got to select a priority from that list. I don't know of anything that you've added to your—you have on your list that I would abandon vis-a-vis the Chinese.

Chairman SPENCE. Mr. Gejdenson.

Mr. GEJDENSON. Thank you, Mr. Chairman. First on Mr. Sisisky's point, I'd like to ask unanimous consent to place in the record a UPI story. One starts with a statement that Americans have benefited greatly from this relationship in an intelligence area. "U.S. Government and industry officials, speaking on condition of anonymity, describe a steady flow of intelligence about Chinese rocketry over the past 3 years." In the next to last paragraph, it says The CIA recently stated that there was no harm to American national security.

Now, we have several things going on here. We have an honest debate. I've fought with Mr. Hunter for years on these kinds of issues. Mr. Hunter thinks that we should not export anything more modern than a wheelbarrow, essentially or maybe a little more than that.

Mr. HUNTER. Maybe Connecticut. [Laughter.]

Mr. GEJDENSON. Maybe Connecticut, he'd do it. [Laughter.]

That has been a consistent an honest policy of his and I understand it. Now, part of what we understand here also is the Bush Administration licensed nine satellites. The Clinton Administration licensed 11 satellites. I think two of those didn't launch. So where basically Bush and Clinton have engaged in almost an identical policy—frankly boggles my mind. Not so much on the technology side, but for some reason this great country—starting in the Nixon era—doesn't seem to understand that human rights in China are a problem. My parents survived Hitler and Stalin. I think human rights are a problem whether they're in Europe or they're in China. I frankly voted against MFN and will continue to do so based on the fact that this country shouldn't give its good housekeeping seal of approval for what happens in China as far as human rights.

On the technology side, there are important battles. I think the gentleman, Mr. Milhollin, is wrong. The MTCR doesn't take away the ability to punish China. It says frankly if the Government's involved in it, not a company, but a Government company then you can punish an MTCR member. That's clearly, if the facts are correct, is what the law says. So that's not exactly a killer. The other countries involved here are Japan, Canada, France, England, Germany, and Australia. So there's a question here of if we control critical technologies that nobody has, I don't think we should sell them to any of these countries. There's a question of what our actions would do. If we decided tomorrow to never launch another satellite in China, what impact would that have on the Chinese military. My argument would be that it would be very little.

Let me just say something about this administration. This administration has frankly been too slow in many of these areas. The first fight I had with the Clinton Administration was over telecommunications to China. I imagine some of the gentlemen on both sides of the aisle disagreed with me at the time. The Chinese wanted to import 565 switches. The Clinton Administration said no. What happened? The Israelis sold them 625s and the Chinese developed their indigenous 565 switches. We didn't stop the technology from getting there. We lost market share and we lost control. I am not for selling them everything, but I'm certainly not letting the State Department be the arbitrator of it.

The reason that the definitions are important is when the State Department was doing licensing—a toilet on a satellite was considered a munition. So if you made a toilet that had some other commercial application, you couldn't sell it without getting a military export license because it had been part of some larger system. You know, when I sat on the Foreign Affairs Committee at the end of the Bush Administration, Mr. Mosbacher, who was Secretary of Commerce—who wasn't known exactly as a radical—got into this incredible fight with Dick Cheney because Mr. Mosbacher wanted to decontrol 286 computers. Mr. Cheney said letting people have 286 computers would undermine American national security.

I would venture to say there are people in this room that have never seen a 286 computer. While that fight was going on, you could buy 286 chips in every Radio Shack in the world, including the ones in Beijing. So what should we focus on? It seems to me that there's a legitimate focus to see if we've lost control of information when the Loral company, without apparently its legal counsel's approval, transferred some information. I mean, some people may want to make that political. It seems to me a breakdown in morale in our oversight, but you can't make it a political battle. If you really want to come up with a good policy here and not damage the American economy, we have to focus on what's really at issue here. The Bush Administration and the Clinton Administration, much to my regret, have embraced the Chinese government to too far a degree. I can't find a hair of difference between the two when it comes to policies on human rights and trade with the Chinese.

I think people want to get to the Chinese on these issues ought to be voting against MFN because that's the only message that they really understand. Let me, before asking one or two quick questions, just add one last thing here. [Laughter.]

If you want to get the Chinese attention, Mr. Hunter had the numbers pretty close to right, you stop Chinese imports to the United States. That's when you get their attention. They—except for critical technologies that aren't available anyplace else, what we don't sell them—I think when President Bush imposed embargo on nuclear power plant equipment—Chancellor Kohl couldn't get out of Germany fast enough. He was in China the next—you know, week—trying to sell them the very same technology. Understand what we have is replicated globally. Understand that we succeed economically at the front end of technology. It seems to me if we can focus on what happened in this instance, we may have a better system for dealing with what information we need transferred and which information we don't. I see I've run out of time. I guess I'll save my questions for the next round.

Chairman SPENCE. That was a good question. [Laughter.]

Mr. Weldon.

Mr. WELDON. Thank you, Mr. Chairman. Let me thank our witnesses for being here. Let me say that we appreciate your comments. None of us want to try to characterize the testimony of any of you or think of the intentions or the motives of Members of Congress, which leads me to raise a question, Dr. Freese. I have to get clarified in my mind, so on the record I'd like you to answer it. I have here a featured Op-ed piece that you wrote in the June 8–14 issue of Space, which I'm sure you're familiar with.

In the second paragraph, I'm going to quote you. "I view the legislation passed May 20th by the U.S. House of Representatives only as evidence that partisan politics is alive and well in Washington. It has very little to do with anything else." The votes that you are referring to, Dr. Freese, were 417 to 4, 414 to 7, 412 to 6, and 364 to 54—with only 44 Democrats opposing it. Is that partisan politics?

Dr. JOHNSON-FREESE. I view this as a partisan politics issue to a certain degree.

Mr. WELDON. No, I ask is that partisan politics? Those votes? I mean is that your definition of partisan as a professor?

Dr. JOHNSON-FREESE. Yes.

Mr. WELDON. 417 to 4 is partisan politics?

Dr. JOHNSON-FREESE. Is the vote or is the issue? The vote wasn't, the issue yes.

Mr. WELDON. Well, I assume what we do is repeat the vote and to cast the opinion for our 600,000 constituents on key issues. I think that trivialized this whole debate. This Congress is serious about proliferation. When this Congress, back in November, voted overwhelmingly to pass the bipartisan Iran Missile Sanctions bill—which Jane Harmon, on this side, was the lead sponsor of and Mr. Gilman, the primary sponsor of, that bill passed with 398 votes. Was that partisan? When we passed that bill on the House floor last week with a vote of only 22 members opposing it—was that partisan? I think to characterize this debate as partisan trivializes the whole issue. I'm offended by it, to be honest with you.

Let me get in some of substantive questions. First of all, I agree with Dr. Sokolski that this is not just an issue of the Loral case alone. This is a pattern of lack of enforcement, in my opinion, of arms control agreements. I support arms control agreements, as you know Henry, and I've been working with you. In fact, 3 years ago I tried to get Bob Bell to accept some recommendations that you made to give the administration more flexibility in enforcing arms control agreements so their hands weren't tied. Bob Bell refused that—said they didn't need it.

I want to put into the record, Mr. Chairman, hopefully without objection, a CRS report—not a Republican document, not a partisan document—done by CRS which I've put in the House record before.

[The information referred to can be found in the appendix on page 268.]

Mr. WELDON. It documents, in the case of China since November 1992, 21 actions—21 actions that CRS has deemed to be possible violations of a number of arms control agreements and documents which this country adheres to. In the case of Russia, since the early

1990's, 17 specific cases. In only 3 of those 38 cases were sanctions imposed and in each of those cases, the sanctions were eventually waived.

My point is we have a problem in this country because when violations occur, we do not take the action necessary to stop the proliferation. Our House National Security Committee will remember when I've held up an accelerometer and gyroscope on at least six occasions in this hearing room—with Russian markings—that we intercepted three times—120 sets going from Russia to Iraq. These are devices to improve and enhance their guidance systems. We knew they had these devices. In fact the CIA currently has them in their possession. We never imposed sanctions. We were told by the Russians that they would impose criminal actions against the proliferators. Last November, our best estimates is that the entire case was dropped and there were no actions brought against anyone proliferating in Russia.

The problem is similar in both countries. As someone who will today at 1 o'clock, testify in favor of MFN for China—who believes in engagement—but enforced engagement. Engagements where we call the Chinese when there are violations that occur. We have not done a good job.

Now let's talk about national security implications because you refer—I want to get very specific. Is it not true, Dr. Sokolski and Dr. Milhollin, that we knew the Chinese were trying to acquire MIRV technology from the Russians and the Ukrainians? Is it not true that, in fact, we demarched the Ukraine because Russia was in fact in the process of buying access to the SS–18 to give them partially MIRV capability?

Dr. SOKOLSKI. That is correct.

Mr. WELDON. When we officially—do you agree with that Dr. Milhollin?

Dr. MILHOLLIN. I'm not aware of that specific case.

Mr. WELDON. When we stopped that transfer of the SS–18 which would have given the Chinese this MIRV capability, in effect we took action to block the transfer of very specific sophisticated technology. Is it now my understanding, Dr. Sokolski, that when—I believe it was for Motorola—the Loral satellites, there were two or three, the Chinese were having problems with having a multiple capability of putting those satellites into orbit. Therefore, it was with our help that they—using what was in fact a commercial-use—developed the capability to dispense those satellites into space. Is that true, Dr. Sokolski?

Dr. SOKOLSKI. It is true and I want to make something clear. They, the Chinese, did attempt a launch of multiple satellite dispensing earlier—I believe it was in 1981. But by their own commentators and Dragon in Space said that it was very, very crude. What they did with Iridium most recently with a series of launches is anything but. It is extremely precise.

Mr. WELDON. Now, is it not true that when you give a country like China MIRV capability what we know to be a current inventory of 18 to 24 ICBMs, instantly it is multiplied by—is it a factor of 10 perhaps?

Dr. SOKOLSKI. I would focus probably much more on the new series than the DF–5, although it's something that has to be monitored. The reason why is they're interested in having road mobile solid rockets that can be very hard for us to find that are MIRV to do the business of those old rockets—the DF–5. They may—that's something you should ask——

Mr. WELDON. But to simplify this issue for all of us. Isn't the issue here, that China has now obtained a technology that will give them a significant factor increase of being able to deploy nuclear warheads.

Dr. SOKOLSKI. Yes and probably just as important, they will have an easier time coping with the possibility of missile defenses in the next decade.

Mr. WELDON. One final question. Isn't it true—and we're going to ask this question tomorrow—there's an encrypted device—encrypted capability that was in the Loral satellite failure? Isn't it—let me ask this question. If it comes out tomorrow that the black box that was contained in the Loral satellite crash was—when it was uncovered. There have been press reports that the encrypted portion of that device which is very sensitive for the U.S. in terms of our satellite control capability, that that device—that that encrypted portion was missing. If that case is true tomorrow, would this in fact cause additional damage to our national security?

Dr. MILHOLLIN. I think that is a question you have to handle in closed session, but my inclination is to be just as concerned, at least as the New York Times, when they reported that that is what they believe their sources are arguing.

Mr. WELDON. So if that chip is in fact—was missing—and my understanding is that the Chinese held back the U.S. Response Team to go out and recover the damage from that failure for 5 hours. If that piece is missing, would you agree, Dr. Milhollin, that would cause increased threats to our national security—or could hurt our national security?

Dr. MILHOLLIN. I think if that piece is missing, you have to re-evaluate your confidence in the Chinese as partners.

Mr. WELDON. Thank you.

Chairman SPENCE. Mr. Skelton.

Mr. SKELTON. Let's ask a very, very basic question, Dr. Freese. What's out there? What's on the open market? Mr. Sokolski has talked about the Chinese learning clean-room technology for preparing only satellites for warhead packages. He's talked about improving nose cone, altitude and engine controls. He's talking about making advances in cushioning the vibrations of rockets and improving the dispensing systems for satellites and warheads. With that, let me ask you, Dr. Freese, two questions based upon your experience and your knowledge. Have the Chinese made these kinds of advances and have they done so with American technology? Two, how much of this information is available in open public sources or from foreign countries?

Dr. JOHNSON-FREESE. To answer them in reverse order, I would suggest that again going back to the Intelsat example since Intelsat is an international consortium formed at U.S. urging which by its organizational mandate requires both international components and international launches. Clean rooms and the other material that you named is all part of commercial launch technology. Therefore, any of the Intelsat satellites which have been—would involve the utilization of this technology—it is inherently international. Much of it as has been pointed out as a tribute to the U.S. satellite industry is U.S. made—has been U.S. made. That is changing. Japanese technology is becoming more and more involved, as is European technology.

To then go to the question of whether American know-how has been involved or been transferred to the Chinese—I would have to very honestly say, it's absolutely true. I would think not unexpected that satellite manufacturers whose satellites are in vast—vastly more expensive than the launcher they're being launched on—want the launchers to work and are going to make sure that those who are launching their materials are doing everything they can according to satellite manufacturers. Probably more important insures requirements that all those areas of technology that you mentioned are up to international standards to make the launch successful. So are the U.S. satellite manufacturers going to insist that they have the best chance possible for a successful launch? Yes.

Chairman SPENCE. Thank you. Mr. Buyer? Mrs. Fowler? Oh, Mr. Buyer. Mr. Rohrabacher will be on the list. You're down the line a few more. We'll get to you.

Mr. BUYER. Thank you. One area that is puzzling to me, ma'am, as I listened to some of your responses is that we can't let—whether it's U.S. or global insurers or manufacturers of satellite technology control our own domestic security policies. The goals are incongruent. They have obligations to stockholders. We have obligations to the general welfare of our citizens and at times there are often conflict. So I understand what you're saying, but we have to judge this in a different prism and profit is not the bottom-line. So if it's profit, I understand where you're coming from Doctor, but that's not the prism that these two committees are judging this through and that's how we find ourselves in a pickle.

Dr. JOHNSON-FREESE. May I respond?

Mr. BUYER. No, that's not a question, it's a statement. My question is I've also heard some—at least Mr. Gejdenson had mentioned about—well, we're basically just following the policies of old whether it was the Reagan or Bush Administration, we're just kind of following those policies. I've got several questions here for you. I'd like to know how does this administration's export control policy regarding U.S. technology—whether it's with guidance systems or the satellites—differ from the policies of the Reagan/Bush Administrations. I'd also like to hear your comments about the changes in the waiver process. If in fact we had these transfers of technologies—we used to have safeguards and what happened to the safeguards? Or if—are they still there, but how come there such a more freer flows of these waivers? So how does the administration's export control policy regarding satellite or rocket technology track with the other export control policies in this administration? Would you agree that this trend during the administration has been toward a greater overall de-control in that policy? I'd like to hear your opinion. Do you believe that this is a positive or a negative development? Please, Dr. Sokolski.

Dr. SOKOLSKI. I will give part of the answer. I'm sure the other panelists can add to it.

When I came into the Pentagon in 1989, my office alone with only a handful of people—less than 10—were processing and looking at 5,000 items relating to missile, nuclear and related items. I believe now, the figure for all of the government looking at licenses is less than 10,000 and may in fact be even substantially less than that. So clearly there's been a move away from licensing of a kind where the government is making key decisions.

With regard to the satellites, I think it's worth noting that what we know about the Bush Administration is that four launches were made. They were made under what are called State emissions control and probably just as important, those controls were strictly enforced to this extent. Every inch of the way, the contractor had a shadow and it was a DOD monitor who was a rocket engineer. He followed and took notes. When the fellow misspoke, in the opinion of the monitor, he was reported on. More important than that, the superior who got the reports actually read them and picked up the phone and made complaints. In a diplomatic way, leveraged the behavior of the contractor. What we now see happening by the time you get to 1995, however, are satellites like APSTAR II being entirely under Commerce's control where accidents occur probably had to be discussions going on between the insurer and the Chinese, but perhaps no record of what was said at all; and Commerce's controls being instituted on a number of satellites, starting in 1996, where the monitoring of those discussions between the contractor and the Chinese is at the option of the contractor because the contractor gets to choose when he thinks he is going out of the bounds of the guidelines and needs the monitor.

So I think those are significant differences.

Dr. MILHOLLIN. Mr. Buyer, if I might respond, I see one big difference between the Clinton and Bush policies on satellites, but to understand that you have to keep in mind the difference between Tiananmen sanctions and missile sanctions. I'm sure you understand that. Tiananmen sanctions prevent the export of satellites. They have to be waived before a satellite application can go forward, but missile sanctions are different because there the policy is whether you want to countenance dangerous export behavior by the recipient of the satellite. And there I see a big difference between the two administrations. I'm not aware that the Bush administration ever sought to insulate the recipient of a U.S. satellite contract from missile sanctions for the proliferating behavior of the recipient.

In terms of decontrol generally, I think that now we're controlling only about a tenth as much high technology as was controlled at the end of the Cold War. I think the applications are now down from about 50,000 to about 10,000 per year of high technology being controlled for the whole world. So what we've seen is a drastic reduction in the amount of export control going on in the United States for high technology that has been on lists that were drawn up by experts with the idea that these items could contribute to weapon of mass destruction programs. We're just not doing anything like the control we did before.

There's also I think a big difference in the Bush and Clinton administrations with respect to the political levels at the Pentagon. During the Bush administration, the political levels at the Pentagon believed in export control. I don't think that Secretary Perry or any of his subordinates or his successor really believe in export controls on dual-use items. I mean, Secretary Perry said as much and the policy of the Pentagon has been not to oppose exports that the Bush administration fought. So those are—I see—for better or for worse, I've been watching this for a while and those are the differences that I see.

Mr. BUYER. Thank you.

Chairman SPENCE. Mr. Berman.

Mr. BERMAN. Thank you very much, Mr. Chairman. I'd like to cover a bunch of things as quickly as possible and so. Mr. Skelton talked about four issues, that this hearing was about four issues: licensing export of commercial satellites—I think Dr. Sokolski has made some fascinating points about why we shouldn't think of the licensing of satellites as detached from the transfer of missile technology and I want to read more carefully what he said. But then the issue of the shift from State to Commerce, the Loral transmission, and the giving of the waiver.

This all started as far as I'm aware in terms of a public consciousness, when the New York Times on April 4 of this year wrote an article where Jeff Gerth said, talking about the launch that went wrong, the investigation, the report by Loral to the Chinese on what went wrong. And then he goes on to say, "These exchanges, officials believe, may have gone beyond the sharing of information the companies have been permitted, giving the Chinese crucial assistance improving the guidance systems of their rockets very specifically." And, of course, that relates to guidance systems for missiles.

He then goes to say, "In February, with the investigation of this incident well underway"—that's February of this year, that's what people have been alarmed about, this decision—"Clinton gave Loral permission to launch another satellite on a Chinese rocket," business as usual, but, then he says, "and provide the Chinese with the same expertise that is at issue in the criminal case." Which is, supposedly, the transfer of missile guidance—with guidance technology.

I think he's wrong about that. I think he made a mistake. I have not read his correction of that mistake and I just want to know if any of you know, the license—the waiver that was granted and the license that was issued was simply for the transmission of the Loral satellite and the things related to it. It wasn't for the subject matter of this investigation. Isn't that true?

Dr. SOKOLSKI. Congressman Berman.

Mr. BERMAN. Yes.

Dr. SOKOLSKI. You are correct. Once the expertise was given, kind of like chastity lost, it was already gone. But the administration did not, in my understanding, explicitly then reissue it.

Mr. BERMAN. Right. So the point of this article, which was, by licensing that which the Justice Department was investigating, we have somehow—and I imagine we would have—undermined the investigation, is an unfair—you can say a lot of things about a lot of these issues, but it is unfair to say the administration licensed the transfer of missile guidance technology.

Dr. SOKOLSKI. Well, my understanding of what the Justice Department—and clearly you're hearings or investigations will bring them into scope and you'll be able to get their views—but my understanding is the Justice Department's concerned that it would be hard to argue that there was a national security concern with the previous case if indeed yet another satellite of identical character is licensed. And I think that's where they——

Mr. BERMAN. Well, we let people who are accused of shoplifting still buy food. You know, I mean, I'm not sure. I mean, I can see the argument, but I can also see a lot of answers to the argument.

But now I want to go to the issue of, well, Dr. Milhollin, well, both of you, actually. The reason the Defense Department was so good in the Bush administration was because Sokolski was there then. That didn't mean he won, because Congress always won those fights anyway. So whether Defense yelled more when they lost in the Bush administration, than when they yell—and whether they yell now is—may be an academic kind of a debate.

But on the—we have two kinds of sanctions and we keep going back and forth. One is Tiananmen Square sanctions. I assume those were—those sanctions said, no munitions list items to China. Is that correct, Ms. Kan? OK. Satellites are included on the munitions list?

Ms. KAN. They were until November, 1996.

Mr. BERMAN. All right, but—so each time before that time there was a waiver of that Tiananmen Square sanction, right?

Ms. KAN. Right.

Mr. BERMAN. Because we know we gave them the satellites between 1989 and 1996.

Ms. KAN. The Tiananmen Square sanctions still apply though.

Mr. BERMAN. That's right.

Ms. KAN. Although the satellites have been moved out of the munitions list, because the sanctions also apply specifically to the export of satellites for launch from China as well as munitions——

Mr. BERMAN. I see. There's a specific satellite-related sanction.

Ms. KAN. Yes.

Mr. BERMAN. Now, missile technology control regime sanctions. Those aren't—that isn't a munitions list sanction, as I understand it; that's a list of items on the annexes that involve missile technology, missile components, missiles, category one, category two, a whole list of items. Is the sanction for violating or for proliferating that you cannot get any of those missile-related items? Dr. Milhollin?

Dr. MILHOLLIN. If it's a category two violation, you cannot get a missile-related item. If it's a category one violation, that is, you export a whole missile, then—if my memory serves—you cannot get munitions items, which are a broader category than simply missile-related items.

Mr. BERMAN. All right. But when you say—well, when you say sanctions aren't in effect, the sanctions are in effect. We sell—as far as I know—do we sell anything other than satellites which are on these lists? That's the only thing we sell to the Chinese, right?

Dr. MILHOLLIN. What I said is that the administrative process was not completed for a sanctions finding under existing law.

Mr. BERMAN. But we found it—but they found sanction.

Dr. MILHOLLIN. But, since these sanctions were imposed in 1993 and lifted in 1994, the export behavior which is sanctionable continued, but there has been no sanction applied—no missile sanction applied since 1993. That's my point.

Mr. BERMAN. That's right. They have not renewed it because the Chinese pledged once again to adhere to the MTCR.

Dr. MILHOLLIN. They pledged in 1994, but they haven't fulfilled the pledge.

Mr. BERMAN. They pledged in 1991 and broke that pledge.

Dr. MILHOLLIN. Yes.

Mr. BERMAN. Yes. Then my final question: I don't understand why you would not want a China that agreed now to share information on every export it made of any item on either the munitions list or the missile technology control list to become part of the missile technology control regime. What better way for enforcing nonproliferation behavior than to get them to have to share with us all of the decisions of companies under their jurisdiction exporting these items?

Dr. MILHOLLIN. There are a number of reasons. First, they wouldn't share all of that information with us.

Mr. BERMAN. Well then they're not being good members of the MTCR.

Dr. MILHOLLIN. No. Members of the MTCR share denials. When they deny something to somebody, they send a memo around to the other members indicating that they denied it. I've had foreign diplomats——

Mr. BERMAN. I think they share more than that.

Dr. MILHOLLIN. No.

Mr. BERMAN. I think they pool——

Dr. MILHOLLIN. No.

Mr. BERMAN. When there are questions, serious questions, about a particular export, they're supposed to bring to the other members of MTCR that information so they can discuss it.

Dr. MILHOLLIN. In fact, the primary communications that occur under the MTCR are denial notices. I've been told by foreign diplomats that they were worried that if China ever became a member and got the denial notices—if it, for example, got a notice from France that France would refrain from selling something, that the Chinese would immediately move in and make that sale.

Mr. BERMAN. I see. I see.

Dr. MILHOLLIN. That's a big risk in bringing—that's one of the risks in bringing China in. The other risk is that I think China would fall into what's called a false adherent category. It would be like the Iranians who belong to the—and the Iraqis who both belong to the NPT, the Non-Proliferation Treaty. I mean, the fact that they're members of the treaty allows, especially Iran, to import things from, for example, Russia, import reactors. And everybody knows that they're planning to divert and that they're not real members.

And so I think that if you let the Chinese in, given their current behavior, that they will become false adherents and it will make it easier for them to achieve their proliferation aims. That's why I oppose it.

Chairman SPENCE. We're going to have to break for this vote and we'll come right back in a few minutes. We do three short votes, two short votes after this one, and be right back.

[Recess.]

Chairman SPENCE. The meeting will please be in order and Mrs. Fowler.

Mrs. FOWLER. Thank you, Mr. Chairman. I want to thank you and Chairman Gilman for holding this very important hearing today and the one we're going to have tomorrow too. As one who's been tracking the transfer of technologies with military applications to the People's Republic of China for several years now, I remain deeply troubled by the administration's policies in this regard. I think that the transfer of missile technology related to this Loral-Hughes situation really is only the tip of the iceberg.

Earlier this decade, for instance, the administration licensed the sale of an entire production facilities' worth of sophisticated machine tools to China over the opposition of senior analysts at the Defense Department's Defense Technology Security Administration, or DTSA. This production facility, which was owned by McDonald-Douglas had previously been used to build parts of peacekeeping missiles, B–1 bombers, and C–17 air lifters. DTSA analysts accurately predicted that some of this equipment would be diverted to military production and, indeed, it was only months after the transfer that we learned that some of these machine tools had been diverted to a cruise missile and fighter aircraft factory in Nanching.

The administration also allowed the transfer of gas turbine hot section technologies to China, which is critical for enhancing the performance and longevity of combat aircraft and cruise missile engines. It has transferred supercomputers that can be used for important military applications, including nuclear test modeling, to China, including 46 supercomputers over a span of some 15 months in 1996 and 1997. And, furthermore, it supported the sale of telecommunications technologies that have had the impact of reducing U.S. access to intelligence information concerning China.

Though in the case of these most recent developments concerning this transfer of missile technologies by Loral and Hughes, transfers to a Chinese-owned state company that had twice in this decade been sanctioned by the U.S. government for transferring M–11 missile technologies to Pakistan. The DTSA has, according to the press reports I've seen, characterized this as the United States national security has been harmed. A report also, dated April 9, 1996, by the Air Force Air Intelligence Center—and I only can refer to it in a general way now because it is classified—but I gather has reached this same conclusion. Nevertheless, the White House proceeded to authorize and allow additional launches despite the ongoing Justice Department investigation of these events.

Though I have a couple of questions here for both Dr. Sokolski and Dr. Milhollin. Last Saturday in The New York Times, Jeff Gerth wrote a piece that indicated that the PLA possesses dedicated transponders on the U.S.-built satellites that have been sold to China. Thus, it appears that the United States has significantly augmented the Chinese military's communications capabilities and, in fact, I understand that the communications that, between these satellites and the Chinese military, are encrypted from the ground. So it doesn't matter when the administration says, oh, these satellites don't have encryption technology within them, the encryption is coming from the ground and that is causing our intelligence people some problems.

Do either of you know whether there was any Defense Department analysis of this possibility before the transfer of these satellites occurred and I'd be interested in your opinion as to the impact on our national security as to these transponders being in use?

Dr. SOKOLSKI. Let me say what I know and let me say what I can say in open session to encourage you folks to get the answers. First, there is no question but that if a country like China's encryption is relatively strong and you allow them to put massive amounts or volumes of message traffic through a low-cost means like a satellite, you could actually lose your ability to keep track of message traffic or fall back—not lose—but you could get overburdened trying to break down the traffic.

Second issue that's worth pursuing is how narrow—this is a little technical—how narrow and how many are the number of the antennas on the satellite. Let me restate that, because I think I stumbled there. Commercial satellites have an interest in having antennae that are very narrow beam width, they're called. And the beam width is simply a fancy way of saying it will throw energy back down onto earth in a small area so that, instead of having a broadbeam transponder, where what you have is covering all of the United States with ABC News on TV, you have an uplink that shoots back down and only hits Arlington and that's all it hits. Now, the narrower the beam, the harder it is to intercept the beam because you have to get right up the barrel, in a sense.

One of the questions that needs to be pursued, undoubtedly with NSA officials who I'm sure were involved in some way in looking at these, is what their relative ability is to deal with the mass of the message traffic, the level of encryption, and the antenna. Now I was told by someone who ought to know—and I can't reveal his name publicly—that you will never get a straight answer, but you should have straight questions. And I think there's a lot to be had in that.

As for what the Defense Department did and didn't do, I would ask directly of the NSA and the DOD authorities.

Mrs. FOWLER. Thank you.

Dr. MILHOLLIN. I don't have an answer to your direct question, but I do have a comment. The article you referred to also covered the question whether the United States, when it makes a civilian export such as a satellite, receives adequate assurances about its use. I think it's important to recognize that if you just take an atomic bomb apart, virtually every component of an atomic bomb is made with dual-use equipment. And the same is true of a nuclear-capable missile. So for the United States to avoid helping other countries make these things, we must have assurances that when we sell a civilian product that is particularly useful for that, that we have assurances it's not going to be diverted to military ends.

As I read Mr. Gerth's article, the administration is saying, well you just can't do that for satellites. I'm a little skeptical. Why should we put, again, we seem to have a pattern of putting satellites in a special category of items. When we export other dual-use items, if we can't get the assurances, we don't make the export. Because it would be reckless for our national security to export something that's particularly useful for military purposes without getting assurances it's not going to be diverted.

So either, I think, we should get assurances that we can be comfortable with or we ought not to make the exports. We ought to treat satellites just like we treat all other dual-use commodities.

Mrs. FOWLER. Thank you. Thank you, Mr. Chairman.

Chairman SPENCE. Thank you, ma'am. Mr. Taylor.

Mr. TAYLOR. Thank you, Mr. Chairman. And I want to thank our panel, even though we have a little difference of opinion as to just how bad this is. I think that all of you have done a good job of pointing out that the mentality in Washington that prevails way too often, and that it that greed is good, has its limits and national security is obviously the greatest case where greed is not good. This is a case of greed. Somebody saved some money by launching those satellites rather than launching them here and the national security implications are very obvious from what you've told us.

I noticed in yesterday's New York Times, there was about a three-page article on a U.S.-Russian—and if I'm not mistaken—Finnish effort to launch satellites from a maritime platform near the equator using Russian missiles, U.S. technology, and—if I'm not mistaken—a Finnish offshore platform. My question is, given the fact that we are cooperating with one former nemesis and certainly a potential nemesis and that much of this information is available on the international market and that I do condemn what has happened, I guess a fair question to ask is how much of the critical information that we now know has been given to the Chinese could have only been acquired from the United States? Or, another way of putting this is, could this same information that is now going to be used on this joint venture with the Russians and the Finns make its way to China by just taking one more additional step and those controls that you are advocating that we put in place with China, should they really also be put in place universally?

Dr. SOKOLSKI. Good questions. The right questions. Many of the monitors who work the China watches go to Russia too and I think you're exactly right. If we have Russian missile proliferation—and we do—if we have Russian technology that maybe is helped by U.S. cooperation, if you really want to do this investigation right, you will probably have to spend some time thinking about the Russian dimension.

Let me explain one aspect of it that may complicate things. During the break a number of reporters came to me and said, so, basically, you guys have to decide whether you want to ban this trade or let it go more or less along the lines it's been going. I said, well, no. Really your choices are much wider and they involve the Russians.

In about 48 months, some of these other launch options, including American ones, I might add—Congressman Bateman is not here—and Ukrainian ones will come online. The price difference, as I mentioned before, is not that great for the Chinese. They're within 10 or 15 percent now. There will be an oversupply. One of the questions you have to ask as our representatives making, if you will, law and policy, is do we ban trade? Do we limit trade with regard to China and the satellites to only new satellites that might convey new kinds of integration information? Do we simply go ahead with what we're doing under some tighter form of controls?

The key objective is timelines. And there are two timelines that I think you have to pay attention to. The first is, can we buy enough time to have launch alternatives? You would hope American, but at least American-Russian-whatever to Chinese so that we could leverage the behavior. And the second timeline is one that is I guess beyond my pay grade, but one that certainly isn't beyond yours. I worked for a very bright man who was assistant secretary of defense, Harry S. Rowan, and he wrote an article about a year ago called, "2015: China Goes Democratic." He's an optimist. I like being optimistic. And the question you've got to ask is can we get to 2015, whenever that is, when China goes democratic, without a dispute that turns into military competition or military conflict?

Delay of their improvement of these missile capabilities and military satellite payloads has to meet those two timelines and you have a variety of options to think through with regard to how to address these problems, if you're convinced that there is a leakage.

Mr. TAYLOR. May I get back to my first question. How much, in your opinion, of what they acquired—what the Chinese acquired do you think the Chinese could have only acquired from the U.S. through this Loral-Hughes transaction?

Dr. SOKOLSKI. I don't know about Loral-Hughes, but if you take the entire history from 1990, I think a case could be made—and you need to check this by listening to DIA, which disagrees with CIA, and have fun with it. Get all the information and make your own judgments—that the kind of technology that they have generally gotten from Russia is hardware and know-how and that, if there is a concern here, it's more that they may have gotten know-why.

Now that can change. And I think you're on to a very significant point. As we bring up the level of quality in launchers in Russia, they're very willing to sell for, as I said, jeans, for crying out loud. So it's something to be really seriously looked at and factored into your considerations.

Mr. TAYLOR. Please.

Dr. MILHOLLIN. With respect to your question, I'm afraid I can't give you a laundry list of things that the Chinese might have received from us that they could not have received from others. One thing that comes to mind is high-speed electronics. This is not specifically connected to Loral, but because of the administration's policy of slashing export controls, the Chinese have been able to import a great number of supercomputers. The last time I checked it was 77, but that's old data. I think by now—I'm just guessing—I would say the number now is probably pushing 100.

Since China refuses to let the United States verify the use of such exports—it's the only country that refuses to allow us to go in and see where things are—I think we have to assume that its missile program and nuclear program have benefited tremendously from American electronics. And those electronics are not available elsewhere except from Japan and the Japanese and ourselves have an agreement upon export control concerning supercomputers. And, unfortunately, I'm sad to say this, the Japanese have been more strict than we have. So there's a case, there's a clear case, of an indispensable and very powerful technology that's gone to China from here that could not have been obtained elsewhere.

You also mentioned whether our laws should be universal and catch the Russians as well as the Chinese. I think that's an excellent question, because I think it shows a principle which is that to have credibility, it does have to be universal. If we're seen as simply applying our non-proliferation policy only when it doesn't hurt or when it's consistent with greed, then the policy has no credibility and other countries aren't stupid. They see what's happening and so when we go to jawbone them about one of their exports, they just laugh at us.

And I think in the case of Russia, there was an understanding with the Russians that we were prepared to put money into their space program, their rocket program, if they were prepared to do the right thing with respect to their own exports. That was the quid pro quo. Unfortunately, we see a pattern here which is similar to the one in China. We're delivering our part, but the Russians aren't delivering their part. They're flagrantly helping Iran and we don't seem to have any—we, our government, doesn't seem to have any successful formula for stopping it. I think that's one of the most serious foreign policy failures we now are experiencing.

Dr. SOKOLSKI. By the way, I would add one other commodity which I think Dr. Milhollin would probably agree with me that we give that they cannot get from Russia: hard currency and lots of it. They get paid perhaps, if I'm right, to learn. It's not that way with the Russians. They have to pay to learn.

Mr. TAYLOR. Dr. Freese.

Dr. SOKOLSKI. And that's a pretty important difference because money is a strategic commodity when it comes to making weapons.

Mr. TAYLOR. I'm going to save that quote for the MFN debate. [Laughter.]

Dr. SOKOLSKI. Right.

Chairman SPENCE. Mr. Rohrabacher.

Mr. ROHRABACHER. Thank you very much. I would like to note that you can't expect the Russians to stand back and not sell technology that is threatening to the West and threatening to themselves if the United States is engaged in transferring technology. I mean, the Russians are just as afraid of Communist rockets, probably more afraid of Communist rockets than we are and when we see a collapse of standards on our part, it's bound to affect the Russian decisionmaking.

I was afraid that this hearing was going to end with a candy-coated impression about the damage to our national security that has been caused by the actions of American corporations. So I'm happy; I've got a chance now to jump in here. But this is not, not about the decision to launch American satellites from Chinese rockets. This idea that we're just talking about some expertise they developed by letting them have a few more chances to try launching rockets and the experience they got off it—that's absurd. That's like saying, oh, well, we're upset that we pitched them some balls and they got more batting practice at it and, thus, that's what we're concerned.

We're not concerned about what they learned from practice. We're concerned about an actual transfer of technology that the American taxpayers paid for. You know, I supported the idea that American satellites should be launched on Chinese rockets. We were given all kinds of guarantees and then we end up finding that the administration, whether it was intentional or not, but there's been a total collapse of the standards and safeguards that we were told would be in place to prevent a transfer of technology to the Communist Chinese if we let them launch our satellites. So what we're talking about is an upgrade of Communist Chinese—of a Communist Chinese nuclear weapons delivery system by American technology. That's why we're concerned. That's why there's a hearing.

It's not whether they have more practice. They could have launched a number of missiles on their own and practiced without having a U.S. satellite on top. And it's not about satellites; it's about the upgrading of Communist Chinese missiles with a result that the American people are now in greater danger of nuclear warheads landing in our country. That's why we're upset.

Mr. BERMAN. Would the gentleman yield?

Mr. ROHRABACHER. No, not yet. No, not yet. I've got to finish my point. Thank you, Howard.

First and foremost, know, Dr. Freese, those of us who supported launching the satellites on top of the Communist Chinese rockets did not believe that American companies would then be permitted—or should have expected that American companies would do everything they could to ensure to meet to the insurers' or the manufacturers' standards their desires to have that launch.

No, we didn't think that would happen, because that would be a betrayal of our country's national security interest. We were, in fact, guaranteed just the opposite in briefing after briefing that there would be no technology transfer and, in fact, the Communist Chinese rockets would not be upgraded. And we're not talking about, the Chinese now have greater satellite technology, we're worried that they have better rocket technology. And we don't expect American companies to go overseas and provide a belligerent dictatorship, the greatest human rights abuser on this planet, with technology that would put us in greater danger.

The upgrading of those Chinese rockets was not a result of them getting more practice; it was the result of them getting technology. And it came, again, from a betrayal of our country's national security interests when Americans, Americans, transferred specific technology and specific know-how to the Communist Chinese and when this administration, Howard, when this administration issued a relaunch permit despite the knowledge that technology had been transferred that put our security in jeopardy. That's where this administration's decision come into place. It undercut the prosecutions of those who are putting our country in jeopardy.

Now this is not, as I say, about just one faxed report, which of course our people trying to throw sand in the air are saying, oh, it was just one faxed report that was sent before it was approved. That's what we're talking about. No, that's not what we're talking about either and at least Dr. Sokolski does say it's over a long period of time that we have to examine.

But let's take a look. In the last 6 years, Chinese missiles have gone from having a success rate of zero, not 75 percent, of zero success rate to today they have 100 percent success rate. What we have to get to is, what happened? Now let me tell you. And, plus, they have 100 percent success rate and now we know they've got MIRVing on top of it, which permits them to send multiple warheads towards the United States. I think we can rest assured that the blueprints for all of these changes in these Chinese rockets that permit them to have 100 percent success rate, that these blueprints were not left under their pillow by the tooth fairy. So let's ask some serious questions as to what technologies were transferred.

Ms. Freese, or the rest of you who claim to have studied this and claim to be experts, please let me know if—are there now exploding bolts, which are necessary for stage separation and for MIRVing—do the current Chinese rockets or the Chinese rockets that are now being built, are they privy to exploding bolts? Do they have those exploding bolts technology? Does anyone know? What about laser magnetic ring gyros? Ms. Freese? No? You don't know? I thought you were a specialist in aerospace. I thought you knew about the space program. What about—OK, we already talked MIRV technology—they didn't have it 10 years ago; now they do.

What about these discussions, these technological discussions? Aren't we talking more about then just sending of one report? Aren't we talking about a process that included a wide range of consultation with our greatest experts and the expertise had been provided by the American taxpayer? Isn't that what we're talking about here?

Chairman SPENCE. Time's up. Let them answer the question.

Mr. ROHRABACHER. Yes, sir.

Chairman SPENCE. We've got to get on.

Dr. SOKOLSKI. Well, I think it's important to understand something. As I said before, it doesn't take much to convey a lot when you're talking about a launcher, a rocket system of this type. There are about a quarter of a million parts in even a solid-rocket program that have to work perfectly with one another. So when there is a failure, it isn't 250,000 possible things that might—you have to look at, but rather, arguably—this is a bit of an exaggeration—250,000 to the 250,000th power. Because everything has to work with everything else just so.

So when someone leans over and says, what about those O-rings? You have not conveyed very much in words, but you have conveyed an enormous amount by encouraging the person not to focus on all those many millions of other possible explanations for why something went wrong. And so mere conversation becomes highly levered in tech transfer. And that's the reason why the first four launches were very carefully monitored prior to launch with regard to all conversations.

Dr. MILHOLLIN. Sir, I would make a couple of points. First, one of the reasons why the United States has opposed Russia's decision to build reactors in Iran is for the reasons that you have expressed.

The assurances are that the plutonium made by the reactor is going to be inspected; that the reactor is civilian; that it's a known technology, everybody has one; why should we be worried? The answer to that is that because if you create this relationship, this umbrella under which these things happen, in which scientists from both countries are meeting continuously and talking to each other, it's very difficult to be confident that militarily usable nuclear information isn't also being transferred.

I believe, myself, that the fact that the reactors are being sold to Iran is moving Iran forward on the path to the bomb. That's what our intelligence services believe. It's very hard to have peaceful nuclear cooperation with a country like Iran and, for the same reason, I think it's difficult to have sort of peaceful, non-missile cooperation with a country like China in the space area. What we're looking at here is almost a concept of missiles for peace. Once you create the umbrella, once you create the relationship, it's very difficult to know what's being exchanged.

Mr. ROHRABACHER. But we did have the standards under the Bush administration before and I guess the testimony you have today is those standards collapsed under the Clinton administration.

Dr. JOHNSON-FREESE. If I might, I'd like to make three points in response to the question. The first is, with regard—since the focus was on the MIRVs, I think it is highly likely that in the future China will have MIRVs. I think in looking at that, we ought to look back too and look at the Chinese program, the indigenous program that developed without any help——

Mr. ROHRABACHER. They already have MIRVs. They got them—we already know they've got MIRVs. The last launch they had on May 2nd was a MIRVed launch. So they've got MIRVs and we want to know where it came from.

Dr. JOHNSON-FREESE. It was a payload launch. Again, it was a dual payload launch which, again, when the technology is transferred from commercial to military, I think that was the issue. Has it been; will it be? What I would like to point out is if you look at the question in terms of could the Chinese have gotten MIRV technology without any help from the United States, Russia, or anyone else, the question is—or the answer is likely yes. There may well have been an acceleration of the timeline because of more practice, but I don't think that anyone who's looked at the Chinese space program would disagree that they built phenomenal capabilities on their own. During the cultural revolution, they launched their first satellites.

Furthermore, if the technology transfer issue is what we want to focus on, that's controls and we would raise the issue that the DOD can still have people there. If they don't we need to ask why. That's a very important issue and one I would urge you to focus on. The controls, are they being applied?

And, finally, I would point out that in the Intelsat 708 accident review where the technology from the United States was alleged to have been given, we learned an awful lot from that too. That was the first time—and this is where I go back to my comment earlier, moving from opaqueness to transparency—that was the first time we had access to a Chinese accident report. We got information there too and I think from that we were able to gather how far advanced they were in certain fields and I'm sure with the intelligence community you will go into, they were further along in some areas than we had thought. That was something we gained out of it.

And that is something that the policy, whether or not it's being implemented correctly, whether or not the controls and the assurance are being carried out, have been done appropriately is a critical question. But the policy has gained us access to information too.

Chairman SPENCE. Mr. Menendez.

Mr. MENENDEZ. Thank you, Mr. Chairman.

Dr. Sokolski, let me ask you, isn't—having listened to your testimony and your answers—isn't ultimately your conclusion, although I know you don't offer it as such, but isn't it ultimately your conclusion that, in essence, not permitting satellite transfers or at least the transfers of the most advanced satellite types is ultimately the solution?

Because you talk about that we've doubled their experiences by the number of launches, we've provided them with hard currency, we've given them—or if we haven't given them, they've had access—to a whole host of information and technology that they wouldn't have otherwise. So if Defense is going to be our single priority, as it may very well should be, isn't your ultimate conclusion that we shouldn't be participating in this at all? Isn't that what you're, in essence, saying in the answer to one of the questions in the second page there, that there's a technology transfer problem, isn't put all the commercial satellites back under State controls the preferred fix? And then the last sentence there says, finally, if the customer, for example, the People Liberation Army is intent on diverting technology, even the best controls cannot prevent it.

Dr. SOKOLSKI. Right. Well, you'll notice in the next paragraph, just to aggravate everybody, I said, well, if nothing was transferred, then all's well. I probably could say the opposite too. And I'm trying to be serious here. If everything of value has already been received by the Chinese, you could, I think, reasonably reach the conclusion, let's go ahead with the trend, because now all the damage has been done. Maybe you don't even want to do controls. You might get behind the administration. That is your call.

What I'm trying to do here by pointing out all these possibilities is to broaden the spectrum of options that an intelligent person could come to based on your investigations. What I'm trying to propose to you is that that investigation has to be very thorough and serious because these choices are going to get a lot of concerns.

Mr. MENENDEZ. Well, let me just simply ask you, then, if providing greater experience beyond that which has been provided, providing more money than that which has been provided and possibly providing things that they have not yet learned, then your conclusion would be don't permit any more satellite transfers?

Dr. SOKOLSKI. No.

Mr. MENENDEZ. Then why not?

Dr. SOKOLSKI. My conclusion would be, how do you prevent them from getting capabilities sooner than we have alternatives? How do we make sure that they don't get these capabilities militarized before our relations actually improve with the Chinese? Because, believe it not, I'm an optimist. I think that's very likely.

And there are a variety of ways to do it. You could delay certain kinds of satellites. I indicated that. You might OK the ones of the types that we've already had them launch or you could come to the conclusion, as I've indicated as well, that for the moment this is just not worth it. We'll wait out the 36 months and we'll have these alternative launch capabilities from Russia and the U.S. come on line and we'll just wait it out a little bit and see how things work.

Mr. MENENDEZ. But, ultimately, you're still standing by the statement that if the customer's intent on diverting technology, even the best controls do not prevent it.

Dr. SOKOLSKI. If the conclusion of this committee is that there's more value to get—and I think, by the way, I think there probably is. I think they're interested in getting satellite technology under their belts. I think they're interested in getting more maneuverability, not just independent targetability—and if you're convinced there's no way to control it, then, yes, I would think that a reasonable conclusion would be, you're going to have to cutoff at least the most advanced types of satellites.

Mr. MENENDEZ. Dr. Milhollin, you have spent a lot of time in your written statement with reference to saying, you know, that the transfer from State to Commerce for all of the jurisdictional issues is something, at least as it has been pursued, I think would be fair to characterize it as not a good thing. Is that a fair characterization?

Dr. MILHOLLIN. Yes, that's fair.

Mr. MENENDEZ. Now, so that means that you would be in disagreement with President Bush in 1992 when he said that in the future certain commercial telecommunications satellites, imaging technologies, and navigational technologies would be removed from the U.S. munitions list and added to the CCL. That wouldn't have been a good idea in your view.

Dr. MILHOLLIN. That's right.

Mr. MENENDEZ. And you would also be of the view that the vote of the House Foreign Affairs Committee in 1994, as well as that of its Subcommittee on Economic Policy and Trade, which was passed by voice unanimously, that the export of commercial communications satellites to be regulated only by the Secretary of Commerce was not a good idea?

Dr. MILHOLLIN. That's right.

Mr. MENENDEZ. Dr. Freese, let me just ask you—I've heard some of my colleagues today try to paint you in a certain way. I don't know you, so, you know, I'm trying to develop my own ideas.

Would you agree with Colin Powell in October 1988 when he was the national security advisor when he wrote that, "Legislation may be offered to prohibit or delay issuance of licenses authorizing the export of U.S. satellites for launch on Chinese vehicles. This would be a serious mistake. I request your assistance," this is a letter to Dante Fascell who was the Chair at the time, "in forestalling any last minute actions in Congress that could jeopardize the important commercial and national security interests we're seeking to advance in our approach."?

Dr. JOHNSON-FREESE. I would agree wholeheartedly that it seeks a balanced approach.

Mr. MENENDEZ. Would you agree with former Defense Secretary Frank Carlucci in the Reagan administration when he wrote to also Chairman Fascell that, "I remain concerned that another attempt may be made to block the export of these satellites. Your support for this important national security issue can make a difference. I strongly urge you to support the administration's initiative to license these satellites to the Chinese." Would that be in line with your view?

Dr. JOHNSON-FREESE. Absolutely.

Mr. MENENDEZ. So you share the views of those predecessors in the previous administration? And that's in essence the focus of your testimony today?

Dr. JOHNSON-FREESE. Absolutely.

Mr. MENENDEZ. Thank you, Mr. Chairman.

Chairman SPENCE. Mr. Ryun.

Mr. RYUN. Thank you, Mr. Chairman. My question is basically for Dr. Sokolski, but anyone else is welcome to respond. You generated a question earlier that I know that we will at some point have to respond to. How do you prevent the leakage of technology? Now we've talked about controls; we've talked about monitors. And, if you'll allow me this analogy, the little boy is standing at the dam. He has 10 fingers. He's put them all in the dam and at some point the dam breaks anyway.

You know, what—how do you—you've asked the question. Perhaps you've also thought about a solution as to how you might control that leakage of technology that could be detrimental.

Dr. SOKOLSKI. I want to see more investigation. I really do not want to get into this. And I'll tell you why. I really don't have the answers. I think, if I've done anything to try to give you some of the questions, I've done my job. The answers are the ones you're going to have to reach as you get the answers to those questions.

Everyone else says just change the jurisdiction. Four weeks after this summit, it will off the news pages. It'll be better than nothing. I don't know that. I've even heard someone who I respect argue that you can have just as tough controls under commodities jurisdiction. I find that hard to believe, but I think he ought to be called up here and explain that. Maybe you can instruct the Secretary of Commerce to do things very, very differently.

What worries me most, though, is having raised so many issues and having so many people here finally focused on something concerning national security, I think the first time since Iraq, that we not blow it. We now have folks on both sides of the aisle interested in what is a pretty important set of questions. It would be a shame just to do business as usual and try to get a fix in. So I think I'm going to beg off.

Mr. RYUN. Would there be anybody else on the panel that would dare to venture into that?

Dr. MILHOLLIN. Well, I think you could stop some of the leaks of technology just by applying the law, as I've said before. We do have laws on the books, passed by Congress, supported by these committees, that aren't being enforced. We could take that modest first step, which would take us a fair distance.

Second, I think you have a problem that the Commerce Department has a conflict of interest. The Commerce Department's main job is to promote commerce. It is not a good idea to have the same agency in charge of regulating it, in charge of actually denying exports for national security reasons. These decisions are made where security and trade intersect and what you have now is a referee who's biased in favor of trade. And so that's why we're seeing such a poor result over time.

Ms. KAN. The current agreements that are in place include not only the safeguards for the transfer of not only technology, but also, very importantly, technical assistance. And that's administered by State, not just Commerce. So State also oversees the technical assistance agreements. But there's also something that has been negotiated before that's underlying all of this and that other technology safeguards agreements. When Reagan first decided to approve satellites for exports he said that there would be conditions to safeguard the concerns that we have about national security. And so there was an agreement negotiated with the Chinese to safeguard technology. One question for you to ask, perhaps, would be is this agreement being enforced?

And, second, the agreement was amended in 1993. The agreement—the first agreement that was concluded in 1988, was in place, but for some reason, and you may want to get to the bottom of what the reason was, the agreement had to be amended in 1993.

Dr. SOKOLSKI. By the way, that is a very important starting point for your history. That agreement spelled out what each agency was going to do to implement the plan. That amendment in 1993 vitiated the active participation and assignment of duties, is my understanding. So that's certainly a good place to begin and I guess maybe I'm going back a little bit. Yes. Try to see how well the law's being enforced and implemented and then you do have to ask the secondary question: Is that ever going to be enough? And what is at risk? What more is it that the Chinese might gain from us? You have to answer those three questions before you come to your conclusion.

By the way, I repeat, I will hand off to the staff a list of contacts: people who were monitors, contractor counterparts, and officials that may help your investigations. You should at least be talking to all of these people.

Dr. JOHNSON-FREESE. I would wholeheartedly concur that focusing on whether or not the safeguards, the assurances in place, are being adhered to and carried out properly is certainly an important topic that you really need to get all the information in in this investigation. I would suggest again, however, that both State and Commerce are involved, as is DOD. The change in one area was to Commerce as the lead agency. This does not mean that they are the sole, that they are the veto. I think emphasis may have changed, but the players were not excluded.

I would simply encourage you that until you have this information, until all the information is in, not to change the policy, which has much broader national security implications.

Mr. RYUN. Thank you.

Chairman SPENCE. Thank you. Mr. Bereuter.

Mr. BEREUTER. Thank you, Mr. Chairman. I've been here for most of the hearing, but made the mistake of being 3 minutes late at the beginning. I want to thank the panel for your help to us today. There is already, unfortunately, underway a game of political "gotcha" in defense on those issues and we really ought to keep our focus on what happened, what has happened that shouldn't happen, and what is our national interest. I want to ask three questions, if I can, and I'll try to be succinct and I hope that you can respond that way.

Dr. Milhollin, I followed carefully, I believe, your oral and written comments about the difference in definition of missile technology as applied by Commerce versus State. Having noted your suggestion to us that we need to focus on implementation of existing law and accepting that, I want to ask you this question. As a bare minimum, as we look to make changes in the law, we should ensure that our missile sanction laws are applied to the export of all missile-related technology even if imbedded in satellites. Do you agree with that? Would that be a position you would support?

Dr. MILHOLLIN. Yes. Regardless of which agency is doing the regulating, I think that I would support that position.

Mr. BEREUTER. Thank you. Second—and this would be for you or for Dr. Sokolski, for example—the GAO has also testified that Commerce does not clearly control the transfer of technical information that might be militarily significant. In particular, they reported that the Commerce regulations do not have separate export commodity control categories for technical data, leaving it unclear how this information is licensed. Do you believe this is a problem? In this situation, how would DOD have the opportunity to review the need for monitors and safeguards or attend technical meetings to ensure that sensitive information is not inadvertently disclosed?

Dr. SOKOLSKI. That is one——

Mr. BEREUTER. Dr. Sokolski.

Dr. SOKOLSKI [continuing]. Of the key deficiencies you've pinpointed in the GAO. As it is now, and the way it effectively works, it's up to the contractor, essentially, to decide when and why he needs or wants to have a monitor and, as I understand it, they have to pay for it. So it's a pretty high threshold for some folks to demand that monitor whereas, before, they didn't have any choice. They had the monitor whether they like it or not.

Mr. BEREUTER. So, for example, at the independent review committee that met in Palo Alto and Beijing, it was not deemed necessary by Loral to have DOD there? Is that right? And, in fact, later they suggested—they admitted to a transfer of sensitive information.

Dr. SOKOLSKI. Yes and, as I raised before, there's an entire case, APSTAR II, where I'm sure anybody has any report of any sort about what was said about the accident.

Mr. BEREUTER. Dr. Milhollin, do you have a judgment about the initial question I raised regarding the transfer of technical data and the lack of, apparently, regulation in that area?

Dr. MILHOLLIN. Well, GAO seems to be correct, but I think it raises a broader question. Since the State Department has been set up to do this and has the expertise and has the experience, why the transfer to Commerce? That is, how does it make sense to abandon a system that seems to be at least in place, with experience behind it, to transfer the responsibility elsewhere where that doesn't exist?

Mr. BEREUTER. Well, that is a good question and we'll have to see what the administration says to justify their position to switch it.

I would ask this of any of the panelists now, my final question. First of all, I would observe that the U.S. is not the only source of technology and know-how or what somebody here said know-why for the improvement of Chinese missiles. There are, I guess, four categories of sources. One is the United States; one is indigenous or Chinese; third would be third countries, for example, Russia or Russian rogue military entrepreneurs; and fourth would be some combination of those things.

So how are we to determine—what would be your best suggestions about where new technology available for such things as MIRVing or improving missiles guidance systems and therefore accuracy and reliability, how are we to judge whether or not that was a transfer of American technology by our corporations or if in fact it came from Russia or other sources? How do we—who do we go to to get this information if possible?

Dr. MILHOLLIN. We depend on our intelligence agencies to provide that; that is, to figure out if the Chinese are getting things they shouldn't be getting, who's doing it. In my testimony earlier, I spent some time on the work of the intelligence community and putting together a picture of Chinese proliferation to Pakistan and Iran. I think that that system is in place. It works. Now the problem is we're not doing anything when the system produces the result we're paying it to produce. So I don't think, it's at least from what I've seen, it's not an intelligence problem. It's not a problem of figuring out who's getting what from whom.

The problem is that we need to have the fortitude in the United States to do something about it and the way we influence third countries is primarily through access to our own technology. We don't have perfect control over what third countries do, but we can use our assets, access to our market, access to our technology—those are the two things the rest of the world wants from us most is access to our market and access to our technology. I think we need in our own best interests to ration those to outsiders, to foreigners, to foreign companies and foreign governments who share our values and who play according to the rules. That's not a perfect solution to the problem, but it's one that is possible and that we have and that has worked in the past.

Dr. SOKOLSKI. In answer to your question as well, I wrote a detailed article which was used in a debate over the Harmon amendment which was defeated, I believe, in 1995, which essentially did what the President later decided to do without a public debate and that is to move much of the satellite—commercial satellites to commodities jurisdiction. That article prompted a phone call from the CIA. They wanted me to brief them on how U.S. technology may have helped the Chinese program. Which I was happy to do, but I was a little nervous about that. I don't want to oversell my expertise. Someone else ought to be a doing a better job than me.

I think, in addition, you should be able to ask these questions of your DIA and CIA—what's interesting is they will disagree and that will be a very useful hearing to have—you ask, how good was the satellite dispensing capability in 1981, that was Chinese, that's indigenous. What is it that was transferred by the Russians relating to the SS–25 technology that may be going into the DF–31 and 41? How good is it? And, finally, how would you rate the MIRV-related capabilities associated with the Motorola launches? You're going to get disputes, but you'll learn a great deal.

Mr. BEREUTER. Thank you. Thank you.

Chairman SPENCE. Thank you. Now, Dr. Sokolski, I know you have to leave and we appreciate your testimony and wish you well.

Mr. BURTON. Mr. Chairman.

Chairman SPENCE. Yes.

Mr. BURTON. I had a couple of questions. Does he have to leave right now?

Chairman SPENCE. Yes, he has to catch a——

Dr. SOKOLSKI. I'm afraid I have the responsibility with the military service schools of the country to go to Colorado Springs and I rebought a ticket at about $1,200 expense to be here, but it will go off without me if I don't go now. [Laughter.]

And so I'm torn. It is public service there too if it's any mitigating circumstance. But I will answer anything for the record and I'll be happy to visit with any of your staff.

Chairman SPENCE. We thank you.

Mr. BURTON. Well, if I could submit something to you for the record for later, I would appreciate it now.

Dr. SOKOLSKI. I would be honored to answer your questions.

Mr. BURTON. Thank you.

Chairman SPENCE. That would be a good way of doing it. And you can go ahead now and ask the rest of them.

Mr. HUNTER. Give us your home phone number. [Laughter.]

Dr. SOKOLSKI. Everything's on the top of my testimony. Fax numbers, everything else.

Chairman SPENCE. Thank you again, Doctor. Thank you. Thank you.

Dr. SOKOLSKI. Thank you very much.

Chairman SPENCE. Good luck to you.

Mr. BURTON. Thank you, Mr. Chairman. I only—am I next?

Chairman SPENCE. You're next, yes.

Mr. BURTON. OK. I only have about two or three questions, but I do have a statement I'd like to make that expresses some of my concerns. In May, 1997, the Department of Defense assessment which, according to press accounts, concluded that United States national security was harmed by the technology transfer and a similar conclusion was reached by the Air Force. A decision in February, 1998, by the President to waive existing sanctions law and allow export of another SSL satellite to China over the reported objections of the Justice Department, which believed the decision might compromise its criminal investigation, has also a higher profile in light of the controversy over campaign financing.

I'm concerned that the Chinese Communist government is getting, in all probability, the Long Beach Naval Facility which DIA and CIA is concerned about. They're building facilities at both ends of the Panama Canal, which is of concern to me. And now they're getting this kind of technology, which may jeopardize our national security.

China has rattled their sabres in the not-too-distant past regarding Taiwan and they had military exercises in the Taiwan Straits. And during that time, a General Xiong Guang-kai, an intelligence officer, said that he wasn't concerned about the United States intervention in that area because Americans care more about Los Angeles than they do about Taiwan. And it was a statement that a Mr. Freeman characterized as an indirect threat by China to use nuclear weapons against the United States at some point in the future.

Those are great concerns of mine. Because it appears as though China is moving toward checkmating the United States at some point in the future. Panama Canal, Long Beach Naval Station, saying things like they said about Los Angeles and the possibility of attacking us if we did anything in that part of the world regarding Taiwan.

Now, Dr. Sokolski said that by the year 2015, in all probability, things will change in China and there won't be a big problem. Or that's what he was—that was some conjecture that he was talking about. The fact of the matter is, we've got 17 years until the year 2015 and the Chinese Communist government is moving on a number of fronts that are of great concern to me and we are giving them directly, in my opinion, some technology that may assist them.

Now, Ms. Freese said that, without our support and without our assistance, China, in all probability, will advance, their technology will advance. My concern is that, without our assistance, they probably won't advance as rapidly as they would otherwise and, therefore, time might be on our side if we ever do have a confrontation with them, which we hope doesn't happen, but it could. And so for us to give them any assistance whatsoever that might speed up their ability to target American cities is something that we should not assist them in.

Now I have a couple of questions. First of all, Dr. Milhollin, at the end of the 20th century, we realize that technology has surpassed the old accepted mechanized standards of warfare, i.e., the infantry, et cetera. On the battlefield, the war will be won sometimes without firing a shot. I feel, at this time, that the war of the future is information warfare and we need to look no further than our own DOD unclassified publication I think. Have you seen this publication?

Dr. MILHOLLIN. I have.

Mr. BURTON. You have. Do you believe that the technology that we lost puts us at risk on the next battlefield and—you are familiar with this document, you said?

Dr. MILHOLLIN. I've seen it. I haven't said I was familiar with it. I've seen it.

Mr. BURTON. Yes.

Dr. MILHOLLIN. I think that the Gulf War taught the whole world, including us, an important lesson and that was that the person or the side who can process information faster is going to win. If you can see over the horizon and your opponent cannot; if you have smart weapons, your opponent doesn't; if you can pick him up on your radar and he can't pick you up on his; you have real-time battlefield information from remote sensors that he doesn't have; all of which can be processed—or must be processed by computers. Then you can see that what you've said is absolutely right, that the next battlefield, in fact, it's already here, the Gulf War was such a battlefield, is an information-technology battlefield.

Mr. BURTON. But technology is advancing in quantum leaps and we're one of the leaders in technology in the world. And it just seems to me that any technology transfer that has military significance should not take place, especially with a potential enemy like China. And what you've just said verifies my concern and that is that the technology leaders of the future are the ones that will control military battlefields and maybe even the world. And for us to give technology that may give a potential enemy an advantage is something that we should not be involved in. Would you agree with that?

Dr. MILHOLLIN. I agree completely and that's why I think the export of the supercomputers I alluded to previously, I suspect now pushing 100 supercomputers to China, is a big strategic shift in the world and one that we may well live to regret.

Mr. BURTON. Mr. Chairman, can I ask one more question?

Chairman SPENCE. We've got to move on. We have a rule. Can you submit it for the record?

Mr. BURTON. Well, some of the other members—some of the other members, Mr. Chairman, have been given some latitude, like Mr. Gejdenson, but if you can't allow me the extra time, I'll understand.

Chairman SPENCE. How about getting back to you in the second round, then?

Mr. BURTON. Well, I hope you'll have the same time constraints on Mr. Gejdenson and the others in the future. Thank you, Mr. Chairman.

Chairman SPENCE. I'll take that under advisement. [Laughter.]

Mr. REYES. Thank you, Mr. Chairman and I do have a statement for the record that I would like to submit.

[The prepared statement of Mr. Reyes can be found in the appendix on page 191.]

Mr. REYES. And I would like—I am curious because I have heard and read a lot about the issue with China and there are several areas that I'd like to focus in on. The first one is I'd like your opinion on the technical safeguards that are associated with satellite licensing. Are these procedures adequate? Do we have enough safeguards in the system? If, assuming that they're all followed and abided by, are these safeguards sufficient to protect the national security of this country? I'd like an opinion from each one of you.

Ms. KAN. I'd like to mention something that I said earlier in answer to another question. When President Reagan first issued the licenses, he said that there would be conditions and one of the conditions was a technical safeguard agreement which was concluded in 1988. At that time, the administration said that that was sufficient.

New problems probably arose because they had to—the administration, later on, Clinton administration, had to renegotiate that agreement and it was done in 1993. So one question for you to get at would be what problems arose that required another agreement to be signed and negotiated and concluded? And do we have any new problems since that agreement in 1993? Has the agreement been implemented and enforced, not only by American companies, but by the Chinese? Have the Chinese lived up to their part of that agreement since 1993? And a number of years have passed.

Dr. MILHOLLIN. Sir, I would say that one big data point we don't have in answering your question is the Air Force report and other information, possibly, that will tell us exactly what was transferred in the Loral case. And I think once we know what technologies or what equip—well, what information was actually transferred, then I think we can answer the question whether the safeguards regime is adequate. But we don't have that yet.

Dr. JOHNSON-FREESE. I guess I would start by saying that I would certainly want to look at what caused the renegotiation. Were the problems involved covered by those renegotiations? Second, it's very difficult to say what is enough because of course the nature of dual-use technology is such that at what point do you have enough safeguards, is there enough?

I actually would separate that into two questions that it is up to you as the astute body to decide what is enough and then are they being adequately carried out.

Mr. REYES. Well, in an area that is so complex and so technical and, as we see, so controversial, both politically and otherwise, sometimes it's difficult to separate the rhetoric from the reality of the consequences. One thing that I believe that we all have in common here is—at least I would hope so—that we're all interested in safeguarding the national security of our country. And that's why, for me, it's important to get the opinions of independent people that are looking even at us, that are in this arena, from a, I guess, nonpartisan perspective or a perspective that doesn't get tainted by politics or anything else.

And one other thing that comes to my mind is the question that I've asked before and that's are our companies, are U.S. companies the most constrained in terms of providing international commercial information and equipment of anyone in the world? Are we the only ones that are perhaps putting these constraints unfairly and constraining them from being able to compete internationally as has been stated here in testimony, if they don't get it from American companies who may have developed it to begin with, all they've got to do is get it from someplace else and, apparently, from what I understand, there are people lined up to provide that information, that equipment, that technology, throughout the world, at least, from three other different sources that I heard previously in testimony. Is that correct?

Dr. JOHNSON-FREESE. I would agree with you that, indeed, in—and it certainly varies country to country and on specific pieces of technology—but, overall, U.S. companies are the most controlled in these areas and I think you're absolutely correct that, if we step out, someone else will step in.

I would also add too that, in looking at this, I was very heartened to hear you say that looking at this as a broad picture issue, that this is more than a non-proliferation issue, this policy has very, very broad-reaching implications for our national security policy, which encompasses non-proliferation and economics, though that hasn't been particularly popular for us today, and human rights and stability in Asia. And this is all part of a greater whole. And to look at it in that regard, I think, would behoove all of us and the national security of the United States.

Dr. MILHOLLIN. Sir, in a general sense, I think it's been true since World War II that U.S. companies have been more tightly controlled in what they can sell than any other companies in the world. That's one of the reasons we won the Cold War. Because the United States took a leadership position with respect to technology transfers to the Warsaw Pact. It was a system known as CoCom. We took the leadership position in saying to the world, look, we're going to control our companies, thusly and we want you to do the same.

And it worked. Our leadership position paid off. We stayed ahead of the Warsaw Pact and, ultimately, we deny them so much technology that it's still difficult to make a phone call in some places. That was a case where we took the leadership position. Our companies were more restricted, but it worked and we won. And so I think, even though the cold war is over, we're still in basically the same posture and that putting more controls on our companies than other countries put on theirs is simply the price of leadership and I don't think the United States can avoid it.

Mr. REYES. Even in cases where technology is readily available from alternative sources, to the detriment of the companies in the U.S.?

Dr. MILHOLLIN. Yes. I think, as I've testified before, we don't compete with the Chinese in selling missile components to Syria or Iran. We could make those sales. We have better missiles than the Chinese. We have better missiles than the North Koreans. We could make those sales. Our companies are also controlled for the export of torture devices. Other countries don't control torture devices. I think it's a good thing for us to control torture devices.

Mr. REYES. Well, I was more referring to technology than your run-of-the-mill torture type device. [Laughter.]

Dr. MILHOLLIN. But the principle is the same.

Mr. REYES. Thanks. Did you have a comment?

Ms. KAN. I would just point out that the fact is the Chinese who have some capability of building rockets as well as satellites have still chosen to buy U.S. satellites for whatever reason. Maybe we're more advanced, which apparently is so. Loral says the latest satellite that China bought—China's had eight—which is a concern of the latest Presidential waiver, is the most powerful satellite that China has ever bought. So the fact still remains that China's choosing to buy U.S. satellites at this time, although they would like to have their own satellites, built by themselves, and export it in the future.

Mr. REYES. Thank you, Mr. Chairman.

Chairman SPENCE. It looks like we have some people who still want to debate this issue, instead of eating lunch. So, let's see, Mr. Hunter.

Mr. HUNTER. Thank you, Mr. Chairman. This has been a very interesting hearing. We really appreciate your participation. Let's get this straight. I think it's fairly clear from your testimony and from the information that we have that the satellite that was lost, the $200 million satellite, the Loral satellite, in 1996, went down when there was a malfunction with the rocket. Is that accurate? It was a malfunction with the rocket.

You've made it clear in your testimony, and of course all of our supporting documentation, I'm sure the witnesses tomorrow from the administration will reassert, that the policy of the U.S. has been to very strongly segregate the satellite package from the booster for the reason that the booster, of course, is the same booster that can boost nuclear warheads into our cities. So we don't want to make it more reliable. We don't want to improve it. So that our interaction with the Chinese government has to do solely with the satellite.

Now if that's the case—and it was clear early on when this rocket crashed in 1996 that it was a rocket malfunction, not a satellite malfunction. None of you have said there was any possibility that somehow the satellite caused this crash or the components of the satellite caused the crash—why did Hughes and Loral engage at all with China? If their job was to keep hands off the rocket section, the rocket aspect of this launch and to only be involved with the satellite aspect of the launch, when the rocket section failed, why did they engage at all with China?

Ms. Kan, any answer for that?

Ms. KAN. I think that's a good question and the answer that Loral has given is that they were asked to by the insurance companies and by the satellite owner. The other question is something I mentioned in my prepared statement——

Mr. HUNTER. Well, let me hold you up on that. So they did engage with the Chinese with respect to the capability or the malfunction of the rocket itself, not the satellite?

Ms. KAN. To definitively answer that question, you would have to look at the actual report which was done by the independent review committee and that report—I'm not sure it has been given to Members of Congress who have conducted an investigation, but that will tell you what has been transferred. According to news reports, they did talk about guidance of the rockets. There has been this continuing tension, as I mentioned, between the unreliability of the Chinese rockets and the interest of the American companies to have the rockets work. So this has been a continuing tension all along and this has characterized the policy.

Mr. HUNTER. OK. Dr. Milhollin.

Dr. MILHOLLIN. Your question literally is: What were the motives of the companies or why would they do that? And that's hard——

Mr. HUNTER. Well, first, is it legitimate at all for them to have engaged with respect to the rocket?

Dr. MILHOLLIN. No, I don't think it is. You simply can't make a satellite launcher better without making an ICBM better. It's impossible. As I think I said——

Mr. HUNTER. And wasn't our—hasn't our policy been to keep their hands off the launcher?

Dr. MILHOLLIN. Yes, I mean, that's—unless you want to actively help them make better ICBMs, you have to have that policy.

Mr. HUNTER. So their engagement with the Chinese with respect to what happened when the launcher failed was in itself illegitimate?

Dr. MILHOLLIN. It seems—it's hard for me to see how a company could in good conscience do that, knowing that the risk would be high in improving the ability of the rocket itself.

Mr. HUNTER. OK. Dr. Freese, do you have a comment on that, the appropriateness of addressing the problems with the Chinese launcher?

Dr. JOHNSON-FREESE. As long as the United States has a policy that we can launch satellites on foreign launchers, I would think it is reasonable for these satellite companies to want to look into the launcher that is going to put its very expensive piece of technology in orbit.

Mr. HUNTER. But they're burdened by regulations that say we—the effect of which is to say, do not make these launchers, because they also carry nuclear warheads, more reliable. You can bring the satellites to China, you can guard them, you can do all these other things, but do not interact with China with respect to making its launchers more reliable. So wouldn't you characterize the attempt to figure out what went wrong when the launcher failed—and when it's clear that it's a launcher problem, not a satellite problem—and our people are supposed to be—to regulate all of their activities solely to the area of the satellite itself, the payload—wouldn't you agree that that was an illegitimate engagement? Because the only thing that could make it relevant was to show them where they failed, where they messed up, and how they might make the launcher more reliable, which accrues to our detriment inasmuch as the satellite carrier is also a nuclear weapons carrier?

Dr. JOHNSON-FREESE. Well, I think Ms. Kan covered that in the aspect that the insurance companies asked for an outside review. The Chinese were very reluctant to allow it, had never allowed it before, saw their commercial market dwindling rapidly away or going away completely, and it was at that point that they reviewed the Chinese——

Mr. HUNTER. Well, yes, but I'm not talking about the motivation of the Chinese or the insurance companies. I'm just talking about from the standpoint of our direction to our companies, it's clear that our direction is keep your hands off the launcher. We don't want you to make the launchers more reliable. That's very critical military information. You have to do—your work should be confined solely to the satellite. The failed launch was a rocket problem; it wasn't a satellite problem, right?

Dr. JOHNSON-FREESE. Right.

Mr. HUNTER. Then what business did they have engaging with them with respect to the rocket?

Dr. JOHNSON-FREESE. I think part of that is being handled currently by the Grand Jury the specifics of the case, but I would suggest that, again, they are going to reasonably try and figure out if the Chinese have figured out what's wrong with the rocket.

Mr. HUNTER. OK. Thank you, Mr. Chairman.

Chairman SPENCE. Thank you. Thank you. Mr. Gejdenson.

Mr. GEJDENSON. Thank you, Mr. Chairman. I think certainly Mr. Hunter focuses on the real important element in this whole discussion which is in Loral's and the other company's actions, post-failure, what was their motivation? What was their responsibility? And I think that is an essential element here.

The other essential element, and I think there are honest disagreements here again—I'd like to ask Mr. Milhollin, would you sell the Chinese a 333 megahertz Pentium computer with Windows 95—or 98 when it finally comes out—would you sell them that?

Dr. MILHOLLIN. It depends on who's the buyer.

Mr. GEJDENSON. Would you sell it to the Chinese government?

Dr. MILHOLLIN. I wouldn't sell it to the Chinese Aerospace Corporation.

Mr. GEJDENSON. OK. So, now, if you decided not to sell them that 386—that 300 megahertz computer, that would deprive the government of that technology. Other countries don't make 300 megahertz computers that trade regularly with China?

Dr. MILHOLLIN. No. I mean, if I were a gun dealer and a guy came in my shop——

Mr. GEJDENSON. No, I think there's a difference here. And I know you like—I reclaim my time.

Dr. MILHOLLIN. I'm trying to answer your question.

Mr. GEJDENSON. I don't want to talk about guns. I don't want to talk about bombs. I want to talk about non-lethal technology that is generally available. Because what I saw after 18 years in the committee is the Defense Department with the State Department crippling the machine tool industry to a point where we started with people like, who hold your position, who said, we cannot sell American machine tools overseas because they're just too good. And then, a few years later, the Defense Department's saying, hey, we don't want to buy American machine tools, we want Japanese machine tools. They've been outcompeting in the international marketplace. They have the best technology.

And so I guess the question is: You would not sell a 386 computer to China, not based on a moral kind of argument we shouldn't do any trade with this country. You'd sell them grain, wouldn't you? Would you sell them grain?

Dr. MILHOLLIN. Yes. I'd sell them——

Mr. GEJDENSON. You'd sell them grain. OK, so they'll think——

Dr. MILHOLLIN. I would sell anything that's moral to sell.

Mr. GEJDENSON. OK, and grain is moral to sell, but computers that they can get over the entire planet is immoral to sell, apparently.

Dr. MILHOLLIN. No, I said I wouldn't sell a computer to China Aerospace Corporation because when they mount that nuclear warhead on that missile pointed at New York, I don't want to have contributed to the success of that project, even if they get the technology elsewhere.

Mr. GEJDENSON. But when you sell them the grain, sir—reclaiming my time. When you sell them the grain, sir—reclaiming my time. When you sell them the grain, you feed the soldiers that launch—that fill the missiles up. Using your logic, we should have a complete trade embargo with China.

Now I believe in embargoes in certain instances. In South Africa, I think this Congress led the effort and we finally got President Reagan to go along with us and we took an embargo based on a moral principle that we wanted to bring about a change in that racist government. Now, I think it's harder to do it with a country as big as China. I think that's probably been the problem with Republican and Democratic Presidents alike.

But I'd like to finish with Dr. Freese here. What did the Clinton administration do differently than the Bush administration did? And you might also comment on the Bush administration's sale of items that needed a nuclear license in exports to Iraq, just prior to the Iran-Iraq war, if you're familiar with them. You're not familiar with those?

Dr. JOHNSON-FREESE. No, I'm not.

Mr. GEJDENSON. Then just stick to the—I mean the Bush administration sold things to Iraq that, in hindsight, look pretty silly because they sold them things that had nuclear licenses, which I think is a much higher level. I think another important debate we have to have here is on the whole issue of Nunn-Lugar to make sure we have the funds to stop fissionable material.

But in the administration's actions in the licenses, what did they do, or did they do anything, in either the Bush or the Clinton administration, that endangered American national security?

Dr. JOHNSON-FREESE. I don't see any difference between the two administrations. The lead agency was changed from State to Commerce. No one was cut out of the process. And it goes to the heart of what you are saying: dual-use technology can be taken to a very long extreme. And satellites, specifically, telecommunications, is a global industry.

I'd like to also bring out—we've talked about, why are we sending these satellites to China? Why don't we launch them domestically? I'm all in favor of that. We have not the assets. The launch assets currently in the United States are not there to launch those systems which are going on-line. It's telecommunications that was to be the focus of Commerce because it was such an international agency. State is also involved, as is DOD. Everybody gets their say.

Mr. GEJDENSON. Let me just go over that process again. And the big fight in that process during the legislation was whether or not it would finally get kicked up to the President. And in all this discussion about the interagency process, the folks at DOD, they are in no hurry to export anything. I understand their instincts. If you say yes to anything and it's any kind of application to a military situation from trucks to satellites, of course, they are going to be nervous. The guys at State are somewhere in the middle, and the Commerce guys I think are a little more an advocate for exports.

But the end of the day and the process in the Clinton administration, in the Bush administration, in the Reagan administration, in every instance where there was any kind of serious debate it got kicked up to the President. And the President in the White House makes decisions and that's why we elect the President of the United States. Isn't that correct?

Dr. JOHNSON-FREESE. Yes.

Mr. GEJDENSON. So, the final arbitrators stayed the same. All the players stayed the same. Again, if you look at the history of the reason it got moved from State, is people believe like, Mr. Milhollin, were making decisions over there—frankly, didn't think we should sell anything even when it was generally available and wasn't a critical technology.

Now, I think the place where we're missing the boat today is we have failed to adequately control fissionable material out of the Soviet Union. I think that's probably the single biggest danger to American national security and security globally. I think, if we look at, even though I support unilateral export sanctions, they simply don't work. Especially living in the days of CoCom while we were sitting here trying to control technology—and CoCom is better than what we have today—everybody else was out there selling everything else to everybody.

Chairman SPENCE. Thank you.

Mr. Burton.

Mr. BURTON. Mr. Chairman, I hope you'll grant me the latitude you granted Mr. Gejdenson again.

Let me start of by saying there is a difference between this administration and the previous administration—excuse me just one second. I want to suspend while Mr. Gejdenson shuts up. There is a difference between this administration and the previous administration in that, the previous administration when the Justice Department was investigating criminal wrong-doing regarding a technology transfer did not, after the fact, grant a waiver. This President granted a waiver after the fact and impeded, according to people at the Justice Department, their investigation and may have made it very difficult to prosecute those who allowed this technology transfer to go forth.

Now, I have just a couple of questions. We've heard a lot of talk about an excess of sanctions laws having been passed by Congress, yet we may have to change some of the those laws as a result of these hearings. But there is a bill in Congress to create speed bumps to keep us from passing such bills quickly and to slow down the administration's enforcement of sanctions. Do you think it's a good idea to make it harder to pass or enforce sanctions a kind of legislation? Dr. Milhollin.

Dr. MILHOLLIN. No. I think we need strong sanctions legislation. As I said before, we need to ration access to our high-technology and to our market and one of the best ways to do that is through sanctions.

It's true that some sanctions are unilateral, but they work. I think the decision of Brazil and Argentina, to come into the fold was primarily because they were being denied high-technology from the rest of the world.

Mr. BURTON. Well, I appreciate that and I understand an amendment in the Senate is going to be put into the DOD authorization bill and I hope that Kan Freese will take what she just said into consideration because I don't believe it should be harder to impose sanctions. And I agree with what you had to say.

Representative Rohrabacher asked a question about explosive bolts, laser gyros and MIRVing, and not to mention U.S. command and control technology. Are these American technologies now being put on Chinese rockets?

Ms. KAN. As far as one of the most sensitive types of technology you mentioned is the MIRV technology. From what I understand the smart dispenser or multi-satellite dispenser that the Chinese used to launch to Iridium satellites was Chinese. However, you could argue that the technical assistance involved between Motorola or Iridium in talking to the Chinese Great Wall Company, in passing the technical assistance, that this may have involved some sort of expertise that was passed. And second, the launch of Iridium satellites on the Chinese using the smart dispenser allows the Chinese to tempt their system.

Mr. BURTON. And about the other technologies?

Dr. MILHOLLIN. I'm not aware. I don't know specifically the answers to your questions, but it would be easy for the committee to find out.

Mr. BURTON. Well, we'll check that out.

Dr. MILHOLLIN. These items are licensable items. If they went to China from us, they would have to be licensed and, you, the committee, have the jurisdiction to obtain the licensing records. I recommend you do it as soon as you can.

Mr. BURTON. We ought to check that out.

One last question: This 1993 amendment that we talked about earlier today that was put into place during the Clinton administration, can you elaborate on that any? I think Dr. Sokolski was going to comment on that and I'm sure I ever understood it. So, if any of you have knowledge on that regarding the export licensing, the 1993 amendment or change that was made: Are any of you familiar with that?

Ms. KAN. From what I understand, some of the technology for satellites, if they contain certain characteristics, if they do not contain certain sensitive technology were transferred over to Commerce before the transfer of the whole satellites from State to Commerce in 1996.

Mr. BURTON. So in 1993, we transferred some of the very sensitive decisions to the Commerce department?

Ms. KAN. So Commerce has argued that the process has started in 1992 and 1993.

Mr. BURTON. Did anybody object to that when that took place? Was there any concern about that?

Dr. MILHOLLIN. The Secretary of State objected in 1996, as I recall.

Mr. BURTON. Christopher, I think Warren Christopher did, yes. Nevertheless, the administration went ahead and changed it and transferred over to Ron Brown.

Dr. MILHOLLIN. That's my impression.

Mr. BURTON. OK.

Ms. KAN. It was also tied in with the implementation of the missile proliferation sanctions that we had in place in August 1993.

Mr. BURTON. OK.

Mr. Chairman, thank you very much. That's all the questions I have.

Chairman SPENCE. Thank you.

Mr. Berman.

Mr. BERMAN. Thank you, Mr. Chairman.

One point, I hate to do this because the gentleman from California, Mr. Rohrabacher, has left, but he wouldn't yield to me. I do think it's important to clarify and, I think, to correct the point that he made that was not responded to. I have listened to almost all of this hearing. Dr. Sokolski was not saying transferring satellites was fine. It was this transfer of Loral's that was the bad thing. He was pointing out and he was trying to make the case that all these transfers of satellites create a leakage of technology which could and it is our job to find out if it did, enhance Chinese missile capabilities.

Dr. Milhollin, as I understand it, says when you transfer those satellites you are eviscerating any sanction for Chinese proliferating conduct and therefore, we should not have transferred satellites for launches on China because, in effect, their proliferating and the one club we have on that, we're giving up on. In fact, rewarding them to the extent that we're helping them with their missile capabilities, we helping them with proliferation capabilities. So, I understand why Mr. Rohrabacher says this because he, as he acknowledged, came down on the side of transferring those satellites.

But, to now distort the fundamental testimony of the critics of the administration's policy into saying that it was something that it wasn't, I think is wrong. So I just wanted to make that point. I wish he'd have been here so I could have made it, but he wouldn't yield to me when he was.

Dr. Sokolski's also not here, but a person with much more knowledge about this than I have, wrote a note and I think it's important to get that in the record.

Mr. BERMAN. Dr. Sokolski's chart, that is the chart that shows the 75 percent effectiveness until 1996, and then 100 percent effectiveness, he says, "is circumstantial evidence. There is no hard data as yet that ties China's improved success rate since 1996 to assistance from U.S. firms. Among questions that this raises are: What data is available in the academic and trade literature that would give the Chinese the same data? What other foreign assistance is available to the Chinese? Are there other countries assisting the Chinese Space Missile program? And what other countries are Chinese space scientists studying or training? What other factors could account for the sudden improvement?"

In other words, I'm not sure that if he were here he would disagree with this. But part of the job is to find out the answers to these questions. We can't quite assume exactly, based on what we know so far, that that chart establishes causation.

Another issue which I was going to ask if any of the panelist have an idea, I'd be interested in their response: Is the technology associated with Iridium which allowed the multiple release of the satellites part of the payload adaptor, that is a physical device, which may not have been available to Chinese inspection and study under the license? In other words, was it part of what we put up there and did ourselves with our own team, and the Chinese never really had a chance to study it or was it software which could have been necessary to pass to the Chinese to ensure that the launch vehicle followed a very exact trajectory? This would be construed as improving the launch vehicle.

That's the question. In other words, does anybody have the answer to that one?

Dr. JOHNSON-FREESE. I'm not sure. If you're asking was it part of a hardware or a software package——

Mr. BERMAN. In other words, was the Iridium that we heard a lot about for this dual-satellite launch which creates this potential and this capability, perhaps, for missile MIRVing and therefore, has real implications on a security question, was this something that we gave to the Chinese or was is something that was part of the payload—what do they call it—the payload adaptor and therefore, not in the hands of the Chinese? They couldn't have inspected it. Does anybody have any thoughts on that?

Dr. JOHNSON-FREESE. I think you're going to get diverse opinions on that again. Certainly, the hardware was a part of——

Mr. BERMAN. That seems to me like "that's because I don't understand the question."

Dr. JOHNSON-FREESE. There is certainly hardware that was part of the satellite package that was not available to the Chinese. Now, I'm sure it can be argued back, traced back to, but if you had that, you'd have to have different weight qualification, et cetera, et cetera. Information will be shared on regard. But the technology was part of the satellite package which was not available to the Chinese.

It's important to go back to that technology was in the United States, first commercial. This was not a military technology.

The point you also raised about how much can the Chinese—Are there other areas that they have simply copied? One of the examples I raise in my book that I would like to suggest to you is, they are excellent at copying and have done it for years. One of their launch sites is sighted at approximately the same latitude as Kennedy Space Center so that they could read our publications and get trajectories and basically duplicate them. They are very astute at that, but that doesn't mean we should discuss whether or not we should totally retract the First Amendment. We put controls on it and we make sure that those controls are carried out.

I think that again, goes to your question of we put controls on the technology, the hardware and then, it's our responsibility to make sure they are adhered to.

Chairman SPENCE. Thank you.

Mr. Taylor.

Mr. TAYLOR. Thank you, Mr. Chairman. And I want to thank our panel. You all have been extremely patient to be here this long and I know all the members are grateful, and I'm certainly grateful for you staying this long.

Going back to my previous question about what happens with the United States and Russia teaming up on the actual platform and do these sanctions need to be universal or should we just pick and choose? And I thought your answer was, they have to be universal because there is absolutely no guarantee that a potential partner like the Russians or anybody would not, in turn, sell their technology or make it available to a potential foe. Is that correct?

Dr. MILHOLLIN. Yes, I think the universality is part a credibility, because otherwise, it just looks like you have a foreign policy of opportunism that allows you to cooperate with people with whom you want to cooperate and not cooperate with others. So, my response was, I think a universal regime is the most credible regime and it's the strongest regime.

Mr. TAYLOR. Dr. Freese.

Dr. JOHNSON-FREESE. Well, I would not pose it in terms of sanctions, but if we're actually saying that all satellites launched on foreign technology, inherently transferred technology that we don't want to let go, then basically what we're saying is we can't launch U.S. satellites on foreign launchers. And I don't think that's something we want to consider.

Mr. TAYLOR. Ms. Kan, would you like to comment?

Ms. KAN. Of course, multilateral sanctions and multilateral efforts are always going to be more effective. But we cannot always get that, and that's where U.S. leadership comes in. We have to make a judgment as to whether we go it alone or we wait and get everyone else to come on board. Sometimes the U.S. has more leverage that we can use.

Mr. TAYLOR. My follow-up would be, and this goes back to a statement you made, Dr. Freese, and please correct me if I'm wrong: You said, why not launch these satellites here? And then, your answer to your own question was: Because the assets aren't there.

I've heard this on the NAFTA debate; I heard this on the MFN debate, it seems like on every trade debate they say, well, we can't do it here that's why we do it someplace else. Well heck, if you can get it for a-third of the price someplace else, you will never establish a market here. That's why there are very few American made televisions anymore, if any, et cetera, et cetera.

So my question to you is: If we, as a nation, strictly enforced these sanctions would we not as a nation benefit in that those foreigners that are now spending money elsewhere would spend money and subsidize the cost of our space program?

Or, number two: Would an American firm not evolve to fill that need?

Dr. JOHNSON-FREESE. I would certainly hope so, over time, but I don't think that it would be within the next two years or within the next five years. I think, we are getting more systems on-line, but as many systems as we are putting on-line, we are having telecommunications systems with the need to launch satellites. I would wholly support the development of a U.S. commercial space industry which was not dependent on the government as its dominant customer or even telecommunications, but we haven't done it.

Mr. TAYLOR. But, getting back to my previous question: How in the heck can we do it, when low-cost options are available from potential foes using slave labor?

Dr. JOHNSON-FREESE. There are problems with the U.S. industry, too. I don't know that we want to get into that completely. But, the U.S. commercial space industry basically—we took, we did exactly what we're talking about with the Chinese—we initially took our ICBM fleet and turned it into commercial launchers. They were not developed for cheap operational costs; they were developed for other purposes. We haven't got a launcher that was specifically built for launching these kind of——

Mr. TAYLOR. Then why can't we continue to do that?

Dr. JOHNSON-FREESE. Why can't we do it? Because right now, the industry doesn't see the potential for getting the return that they need. It's very high-risk, very, very, high-risk field and unless they have a guaranteed customer, and as long as it's out there elsewhere, they are not going to do it.

Mr. TAYLOR. You just answered my question. As long as it's cheaper elsewhere.

Doctor, would you like to comment on that?

Dr. MILHOLLIN. I think the committee should investigate this question of unavailability of alternatives. I think that's very important. I'm not an expert on that by any means. I'm informed that there are other possibilities that could be taken advantage of in countries that we are friendly with for the European space launch consortium, for example. They launch satellites; they're expensive, but they're at least, not potential foes.

In general, in our export policy, we have distinguished between potential foes and countries we can count on. I think we should just do the same for satellites.

Mr. TAYLOR. Ms. Kan? Is it C–A–N, or Kan? I don't mean to be rude.

Ms. KAN. No, that's fine. Kan is fine.

I would decline to make a policy judgment on that.

Mr. TAYLOR. I'm sorry?

Ms. KAN. I decline to make a policy judgment on that.

Mr. TAYLOR. Okay.

Chairman SPENCE. That's it?

You've been very patient with us and I appreciate that and for staying so long. You have, believe or not, you've contributed greatly to our deliberations. As I've said in my few comments today, one of them was that I'm continually amazed that we can look at the same set of facts and arrive at conclusions under these degrees and part that's what it's all about. We find that kind of thing. I think you helped lay the groundwork for our hearing tomorrow. The background for the hearing tomorrow will be administration witnesses representing DOD, State, Commerce and all that. We'll get into all those kind of things. But, you've helped greatly to lay that ground work.

And Mr. Skelton has walked back in and wants to——

Mr. SKELTON. Yes, may I ask a clarifying question. I think Mr. Sokolski brought it up and maybe each or one of you could answer the question as I understood it: The "how" has been transferred, and the question is whether the "why" has been transferred? To those of us who are not technically oriented, please explain that—anyone, please?

Dr. MILHOLLIN. You mean the difference between the two concepts?

Mr. SKELTON. You bet. To someone who is not technically oriented.

Dr. MILHOLLIN. Maybe I should defer to my colleagues to go first. But it seems to me that the question he was getting at is why the launch failed. But, perhaps, I'm misunderstanding what he said.

Mr. SKELTON. That's not what I understood, no.

Dr. JOHNSON-FREESE. I would certainly say, you can look at what is done, how it's done, and why it's done. But I don't understand the why or I didn't understand——

Mr. SKELTON. May I submit that question to be answered, Mr. Chairman by Dr. Sokolski?

Chairman SPENCE. That would be fine.

Mr. SKELTON. All right and with that, I thank you so much for your indulgence.

Chairman SPENCE. Thank you. And again, thank you very much. Good luck to all of you. Thank you.

[Whereupon, at 2:18 p.m., the committees adjourned subject to the call of the Chair.]

UNITED STATES POLICY REGARDING THE EXPORTS OF SATELLITES TO CHINA

HOUSE OF REPRESENTATIVES, COMMITTEE ON NATIONAL SECURITY, MEETING JOINTLY WITH COMMITTEE ON INTERNATIONAL RELATIONS, *Washington, DC, Wednesday, June* 18, 1998.

The Committees met jointly, pursuant to notice, at 10:07 a.m., in room 2118, Rayburn House Office Building, Hon. Benjamin A. Gilman (Chairman of the Committee on International Relations) presiding.

OPENING STATEMENT OF HON. BENJAMIN A. GILMAN, A REPRESENTATIVE FROM NEW YORK, CHAIRMAN, COMMITTEE ON INTERNATIONAL RELATIONS

Chairman GILMAN. The committee will come to order.

We are pleased to open this second day of joint hearings between the Committee on International Relations and National Security and the United States policy regarding the export of satellites to China. We continue today with our examination of whether our nation's national interest may have been put at risk by the export of satellites for launch in China.

Yesterday's hearing clearly helped clarify and focus the issues before us. Today, we have the opportunity to further pursue them with administration witnesses. In my mind, the key to understanding why our national interest may be at risk are, first, the decision by the Administration to effectively immunize both the Chinese Government and U.S. satellite companies from the application of sanctions, designed to protect our national security, coupled, with their decision to change licensing jurisdiction from the State Department to the Commerce Department for the export of commercial satellites.

It is the combination of those decisions, whereby, our national security agencies' veto was weakened and our sanctions' leverage on the Chinese and our companies was removed that has brought us to where we are today. In that regard, let me make a couple of points about this Administration's implementation of our missile proliferation laws.

The Administration undercuts these laws in two ways. First, it has agreed upon a legal interpretation that allows satellites to be exported even if sanctions are imposed. Second, in those cases where it is clear that if sanctions were to be imposed that satellite exports would, as a consequence, be halted. It has refused to make the necessary determinations, often in the face of overwhelming evidence.

We want to salute those in the bureaucracy who have been fighting the good fight. They must have been saddened by the decisions of some of their higher-ups. If you are a mid-level official of the Defense Department would it not be hard to fight for an honest satellite decision if you saw missile sanctions were being effectively waived?

We hope our witnesses today understand the importance that members of our two committees place on determining the degree to which our national security interests might have been compromised by satellite transfers to China. If there should be any doubt to that, I should point out that we expect that sometime during this hearing we will go to the floor to vote on constituting a special committee dedicated to getting to the bottom of this matter.

Instead of trying to induce China to join the missile technology control regime, we should be strengthening the multilateral missile regime to protect against the "made-in-Beijing" proliferation threats being examined in this morning's hearings.

Before we introduce our witnesses and yield to Chairman Spence, let me say that as the President prepares to go to China and visit the very same square where protestors were killed some nine years ago, he must remain mindful of the debate in this nation and avoid any further efforts or activities that could give the Chinese a message that our policies on the spread of weapons and human rights abuses could be altered by commercial considerations.

As the President prepares for his summit meeting with the Chinese officials, President Clinton should leave his bag of carrots at home as far as our national security is concerned. There should be no concessions, no deals, no permanent waivers, no new technology or science agreements and, most importantly, no shoe-horning of China into a missile technology control regime they have been violating over the past decade.

And, with that said, let me introduce today's witnesses.

We have with us the Honorable John Holum, Acting Under Secretary of Political Affairs in the Department of State. He is accompanied by Mr. John Barker, Deputy Assistant Secretary for Export Controls of the Department of State. The Honorable William Reinsch, Under Secretary for Export Administration in the Department of Commerce. He is accompanied by Roger Majack, Assistant Secretary of Export Administration, Department of Commerce. The Honorable Jan Lodal, Principal Deputy Under Secretary of Defense for Policy, Department of Defense; accompanied by Mr. David Tarbell, Director of Defense Technology Security Administration.

We welcome all of our panelists and we appreciate your being here. But, before turning to our witnesses, let me briefly touch on several administrative matters and then I want to recognize my colleague, Chairman Spence, Mr. Hamilton, and Mr. Skelton for opening remarks.

Members should be advised that we are operating today under the rules of the Committee on International Relations. And for that purpose of recognizing members for questions, we will follow the same process that everyone is used to as our respective committee rules are similar. We may try to limit you to as short a discussion as possible since our time is going to be limited today.

Further, as a combined membership of our committees total over 100 members, it will be a challenge to conduct these hearings in a manner that ensures each and every member will be afforded the opportunity to be fully recognized.

Accordingly, we will diligently observe the five-minute rule. We urge the cooperation of all members in closely adhering to this limit. In order to accommodate as many members as possible, we ask the cooperation of our witnesses as well in observing a limitation on time, and, please, watch our lights that are before you on the witness table.

Finally, I want to state that I just learned yesterday, that Chairman Spence and I have just learned late yesterday that a witness is planning to leave to testify in the Senate at 2 p.m. We had no prior notice of your plans except for Secretary Reinsch's obligation. Given that we have a considerable number of members who may want to ask questions, and we expect at least one vote during the next few hours, I do not see how we are going to be able to finish today's testimony on time.

Accordingly, we strongly urge our witnesses to remain as long as they can until all members have asked questions. This issue is deserving of your attention and all members involved in this unprecedented joint meeting of our committees have the right to put questions before you. Should we not finish, we would like to announce—and this is after discussion with Chairman Spence—that we will reconvene this joint hearing at 10 a.m. on Tuesday morning.

Is there any objection to continuing the existing order of the Committee on International Relations with respect to this joint hearing in effect at this time? Are there any objections?

Mr. HAMILTON. Reserving the right to object.

Chairman GILMAN. Mr. Hamilton.

Mr. HAMILTON. Has this been worked out with the Administration witnesses?

Chairman GILMAN. We discussed that briefly with the Administration witnesses before they started.

Mr. HAMILTON. Is it acceptable to them?

Chairman GILMAN. I have not heard any objections.

Secretary HOLUM. We will have to check the schedule but, in principle, we would like to finish today if we can and we are committed to trying to accomplish that.

Mr. HAMILTON. Well, Mr. Chairman, the usual practice when you have Administration witnesses is to try to work out a mutually convenient time and not for a congressional committee to just dictate time and place of the next hearing.

I would hope that we would follow that usual practice.

Chairman GILMAN. Mr. Hamilton, we have tried to be cooperative. We did not learn about the shortage of time and the need for the panelists to go over to the Senate until late yesterday. And we responded to them, asking them if they could delay the Senate meeting, and we were told that they could not. And then we discussed this morning with them the need, if we did not complete our testimony today, to go over until next Tuesday. We will try to cooperate with the Administration every way possible. We would welcome their cooperating with our committee, as well.

Mr. HAMILTON. Well, Mr. Chairman, it does not appear to me that it is cooperative if you just tell the Administration the time and place of the next hearing. That is not the way we do things in this Congress. I do not know that I have ever heard of that before.

Chairman GILMAN. Well, Mr. Hamilton, if the gentleman would yield, we will try to be cooperative. We have two committees, we have over 100 members and I think the Administration could have been a little more cooperative with us if they had told us they were going to have a short time today. We will try——

Mr. HAMILTON. Mr. Chairman, I do not understand——

Chairman GILMAN. Mr. Hamilton——

Mr. HAMILTON [continuing]. Where do we stand at this moment? Are you telling the Administration they have got to be here at a certain time, at a certain hour——

Chairman GILMAN. We have——

Mr. HAMILTON [continuing]. Tuesday or are you saying you are going to be cooperative and work it out with them?

Chairman GILMAN. Well, if the gentleman would yield, I have stated we will try to be cooperative with the Administration and we are asking the Administration to try to be cooperative with our joint committees.

Mr. HAMILTON. Well, that is fine. If it is worked out on a cooperative basis, cooperation would appear to me to mean that both parties would have to agree. If the Administration wants to appear at 10 o'clock next Tuesday, that is fine with me.

Chairman GILMAN. Thank you, Mr. Hamilton.

Mr. HAMILTON. What I object to is the committees dictating to the Administration without their consent or agreement the time and place of the next meeting.

Chairman GILMAN. Mr. Hamilton, as I have stated, we are not trying to dictate to them, we will welcome their cooperation and we will try to be cooperative.

Mr. SKELTON. Reserving the right to object, if we abide by the 5-minute rule we will get along much further than we think. And I certainly urge—I know folks on this side of the aisle, because we have spoken about it, will do their best.

Chairman GILMAN. I thank the gentleman for his suggestion.

Is there an objection to temporarily adjourning until Tuesday, providing the Administration can agree to that adjourn time; and, if not, we will have to adjourn it to a later date.

Without objection, so be it.

Let me now recognize the Chairman of the Committee on National Security, Mr. Spence for any opening remarks.

STATEMENT OF HON. FLOYD D. SPENCE, A REPRESENTATIVE FROM SOUTH CAROLINA, CHAIRMAN, COMMITTEE ON NATIONAL SECURITY

Chairman SPENCE. Thank you, Chairman Gilman.

After yesterday's testimony one point seems abundantly clear. At issue is not just the relatively narrow question of whether or not an American satellite company divulged military-related information to China without U.S. Government authorization, rather, the broader issue to consider is the extent to which national security concerns are being subjugated to commercial interests in the Administration's granting of licenses and waivers for the export of commercial communications satellites to China.

Yesterday's hearings only reinforced my long-standing view that United States' policy with regard to high technology exports, in general and more specifically, as it relates to commercial satellites is fundamentally flawed. There are significant similarities between space-launched vehicles and ballistic missiles.

The Administration's 1996 change in export control policies with respect to commercial communication satellites has encouraged United States' industry to seek subsidized and lower-cost Chinese launch services. Accordingly, it is easy to understand both China's and the United States' industry desire to increase the reliability of Chinese boosters.

Unfortunately, these are the same type of boosters that carry Chinese nuclear weapons. The United States' export control policy should not directly or indirectly serve to facilitate China's capacity to target the United States for nuclear missiles, period. However, based on what we have learned to date, it is difficult not to conclude that improvement of Chinese ballistic missiles has been one practical effect of an export control policy that seems focused almost exclusively on profit margins.

Is allowing United States' satellites to be launched on Chinese launch vehicles in the United States' national security interests? Last month, in overwhelmingly bipartisan fashion, the House said, no. This morning we are likely to hear again and again that this controversy is not about the export of weapons or military technologies but, instead, that there really is no controversy at all since we are simply exporting commercial communications satellites for non-military uses.

In reality, however, commercial communications satellites contain sensitive military technology that can and do fall into the wrong hands. For example, press reports indicate that the Loral satellite that crashed after launch in 1996 contained a sophisticated encryption device. A device that was never recovered by the United States and may have been confiscated by the Chinese at the crash site. If true, I cannot imagine that anyone in this room would argue that this is a benign development.

Moreover, according to press reports just last weekend, China has increasingly been using United States-made commercial satellite communications to relay encrypted communications for the People's Liberation Army. Despite some of the inevitable smokescreens, we should not lose sight of the fact that what we are talking about involves a lot more than commercial satellites for non-military uses.

We are also likely to hear that allowing China to launch United States' commercial communication satellites is in the United States' national interests because it is good for American satellite manufacturers. While none of us wish to needlessly inhibit the global competitiveness of American industry, neither should any of us countenance the selling of American advanced technology, technology with significant military applications, to the highest bidder.

Moreover, in my opinion, the Administration's commercial satellite export control policy directly undermines the development of a competitive American space launch capability by removing the incentive for U.S. industry to make our own space launch services more cost-effective. Such a development surely is not in the interests of America's aerospace industry.

We may hear also that China has not learned anything of military significance from these satellite exports that it does not already know, as evidenced by the fact that China has had the ability to target the United States with ballistic missiles since the early 1980's. However, China's ballistic missile program has not stood still these past two decades. Beijing is working overtime and at great expense to develop newer, more lethal and more accurate ballistic missiles, including missiles with multiple warheads.

As we discussed yesterday, since 1996, China has somehow been able to turn what was previously a notoriously crash-prone launch capability into one that now has a success rate of 100 percent. To argue that China's ballistic missile program has not benefited from launching United States' satellites defies common sense and ignores reality.

We will hear that the Administration's loosened satellite export control policy provides important political leverage that encourages China to abide by its nonproliferation commitments. The premise of this argument, however, is that licenses will be denied or revoked if China misbehaves. In reality, this remains an unproven premise because the Administration has not yet said, no.

Not a single license or waiver request for American satellite launch in China has been denied by this Administration. So, where is the leverage?

Indeed, despite China's repeated violations of its nonproliferation commitments, the Administration's transfer of licensing jurisdiction over commercial communication satellite exports to the Commerce Department serves to avoid sanctions.

According to documents recently released by the White House, the President's former National Security Advisor, Anthony Lake, noted in July 1994, that "Commerce controlled satellites are not covered by the missile sanctions law and can, therefore, be processed for export to China."

As was pointed out during yesterday's hearing, a compelling case can be made that the Administration's satellite export control policy could, in fact, be encouraging China to continue proliferating.

And, finally, we will certainly be told that the Administration's satellite export control policy simply continues the policies initiated under President Reagan and continued under President Bush. My first point, as someone who is on the record as in strong favor of disagreement, at least, with the actions taken by the Bush Administration in export control, is that two wrongs do not make a right.

My second point, however, is that there are significant differences between the Bush and Clinton Administrations when it comes to both satellite export and sanctions. Under the Bush Administration, commercial satellites were handled under the more stringent State Department export control procedures, given the military utility of the technologies involved. These procedures are designed to give principal and priority consideration to national security factors in making an export decision.

Under the Clinton Administration, satellite exports are now reviewed under the Department of Commerce procedures, procedures that effectively downgrade national security considerations and elevate commercial and trade factors.

Further, under the Bush Administration, the Department of Defense representing the national security perspective effectively had a veto in the export license review process. Under the Clinton Administration procedures, DOD has been downgraded to one of four agencies that have a say, three of which must agree in order to block any proposed satellite export.

Additionally, under the Bush process, all satellite exports required intrusive DOD monitors to ensure that no improper exchanges were occurring during the Chinese and American companies, between the Chinese and American companies.

Under Clinton, the requirement for monitors has been all over the map from requiring virtually no monitors to now finally moving back in the direction of the Bush policy.

In sum on this point, the General Accounting Office has concluded that the process changes adopted during the Clinton Administration have led to a situation where "Defense's power to influence the decision making process has diminished." In my view, such a pattern of varied and important distinctions hardly qualifies as a continuation of past policy.

In conclusion, I want to step back and urge my colleagues to step back from the complexities and nuances of this debate because the debate really is not that complicated. This controversy, this debate is about whether as a matter of policy we should be making it easier for China to target the United States and other countries with lethal, lethal ballistic missiles. In my view, the answer is a no-brainer. The answer is, no.

Whether under the guise of diplomatic engagement or economic interests, the answer should always be an unequivocal, no. It is really that simple.

Thank you, Mr. Chairman.

Chairman GILMAN. Thank you, Mr. Spence.

Let me now recognize Mr. Hamilton, our ranking minority member of the committee and then ask Mr. Skelton for any opening remarks he may wish to make.

STATEMENT OF HON. LEE H. HAMILTON, A REPRESENTATIVE FROM INDIANA, RANKING MINORITY MEMBER, COMMITTEE ON INTERNATIONAL RELATIONS

Mr. HAMILTON. Thank you very much, Mr. Chairman.

Let me say to you and Chairman Spence that I think the hearing yesterday was a very good one and I think this will be a good hearing today. We have got a complex subject matter before us and the variety of opinion that we have had expressed will be helpful to all of us.

I want to thank each of our distinguished witnesses today for coming. I hope that they will address two of the major criticisms that were made yesterday. The most compelling criticism, I think, is that the cumulative impact of years, perhaps going back as long as a decade, of U.S. satellite launches has been to improve the accuracy and the reliability of the Chinese missiles and that stopping the launches would slow Chinese missile improvements.

Now, that is an exceedingly serious criticism of American policy and I hope we will get some solid answers to that.

A second criticism, of course, relates to this transfer of jurisdiction over commercial satellites from the State to the Commerce Department. And it had removed, thereby, any meaningful impact Category 2 missile technology control regime sanctions against China; that the Administration had failed to impose sanctions against China when the circumstances warranted.

I might just say that one of the experts frequently quoted here has been Gordon Ohler, who was formerly—I am not sure of his exact title, but kind of the nonproliferation chief at the CIA—and my understanding is that in Senate testimony he said that he preferred the present export licensing system because in that system the CIA now sees all of the Commerce licenses, and when it was under State jurisdiction, he did not see it.

So, that was a little surprising testimony, I think, from him. But we certainly want you to comment on the effectiveness of this export control regime and how well it has been implemented.

So, I think I look forward to hearing the testimony and appreciate, very much, this hearing.

Thank you.

Chairman GILMAN. Thank you, Mr. Hamilton.

Now, I will recognize Mr. Skelton, the ranking minority member of the Committee on National Security for any opening statement he may wish to give.

STATEMENT OF HON. IKE SKELTON, A REPRESENTATIVE FROM MISSOURI, RANKING MEMBER, COMMITTEE ON NATIONAL SECURITY

Mr. SKELTON. Thank you, Mr. Chairman.

I agree with Mr. Hamilton that yesterday was an excellent hearing. I know this will be, as well. And I commend both the Chairmen for forming this joint hearing because it is of tremendous help for both of these committees.

Yesterday, a panel of outside witnesses helped us in our efforts to gather information on two issues. Whether our 10-year old satellite export policy has resulted in the transfer of sensitive space and missile technology to China; and the reasons behind the decision to transfer licensing jurisdiction over commercial satellites from the State Department to the Department of Commerce in 1996.

During yesterday's hearings two other issues were touched upon. The specific case of Loral and the possible release of export control information in the review it had conducted subsequent to the February 15, 1996, crash of the Chinese space-launch vehicle which attempted to carry one of its satellites into space. It is a matter currently under investigation by the U.S. Attorney for the District of Columbia. As a result, it is an issue which I believe our witnesses will be unable to address today because of the ongoing nature of that investigation.

The other issue concerns the February waiver for the launch of another Loral satellite from China. The waiver was granted despite concerns expressed by Justice Department about the potential impact of the waiver on its investigation in the 1996 incident and its ability to persuade a jury to convict Loral in the event the incident would warrant prosecution. Speculation has centered on the fact that because the Chairman of Loral is a heavy contributor to the Democratic Party, Loral received favorable consideration in getting the waiver.

This, I believe, we can explore today and I am sure will be explored to greater extent by the Select Committee on China that will probably be formed later today.

Our witnesses today from the Departments of State, Commerce and Defense will provide the two committees further information to help us in our efforts to understand this current policy.

Thank you.

Chairman GILMAN. Thank you, Mr. Skelton.

We will now proceed with our panel of witnesses and I urge each of you to summarize your presentations and to try to limit them to around five minutes.

And, now, I ask unanimous consent that each of your prepared statements be entered into the record and, without objection, so ordered.

I would now like to ask all the witnesses to, please, rise to be sworn in.

[Witnesses sworn.]

Chairman GILMAN. We will start with Secretary Holum and then he will be followed by Secretary Reinsch, and then Secretary Lodal.

STATEMENT OF JOHN HOLUM, ACTING UNDER SECRETARY OF STATE FOR POLITICAL AFFAIRS, U.S. DEPARTMENT OF STATE, ACCOMPANIED BY JOHN BARKER, DEPUTY ASSISTANT SECRETARY OF STATE FOR EXPORT CONTROLS; WILLIAM REINSCH, UNDER SECRETARY OF COMMERCE FOR EXPORT ADMINISTRATION, UNITED STATES DEPARTMENT OF COMMERCE, ACCOMPANIED BY F. AMANDA DEBUSK, ASSISTANT SECRETARY OF COMMERCE FOR EXPORT ENFORCEMENT; JAN LODAL, PRINCIPAL DEPUTY UNDER SECRETARY OF DEFENSE FOR POLICY, DEPARTMENT OF DEFENSE, ACCOMPANIED BY DAVID TARBELL, DIRECTOR, DEFENSE TECHNOLOGY SECURITY ADMINISTRATION

Secretary HOLUM. Mr. Chairman and members of the committee, thank you for the opportunity to address U.S. policy regarding the launching of U.S. commercial satellites from China. I would like to make several fundamental points to place this issue in context.

The first is that nonproliferation of weapons, of weapons of mass destruction, is a cornerstone of U.S. foreign and national security policy. We have placed enormous emphasis on strengthening international standards, on improving detection, on incentives both positive and negative, and on all the tools of diplomacy at every level.

Trends, such as Iran's progress toward a medium range missile capability and the nuclear explosions in South Asia make clear that these are not theoretical concerns but looming threats. There is no disagreement between the Executive Branch and the Congress on the vital importance of these issues.

The second point is that China is indispensable to any effective nonproliferation policy. China is a nuclear weapons state, it is a potential supplier of weapons, materials and technology to countries of proliferation concern. China's approach can make the critical difference between success and failure across the broad range of our nonproliferation efforts.

Now, unquestionably, China has been part of the proliferation problem. Its relationship with Pakistan's nuclear weapons has been a major concern since the 1970's. We take sharp issue with China's chemical and missile cooperation with Iran. In 1991, the Bush Administration sanctioned two Chinese entities and in 1993, the Clinton Administration sanctioned 11 Chinese entities for transferring missile equipment and technology to Pakistan.

At the same time, my third basic point is that although we still have serious concerns, China's approach to nonproliferation has changed markedly in recent years. I recall that China once advocated proliferation of nuclear weapons as a means of breaking the hegemony of nuclear powers, of the super powers.

But, in 1992, China joined the Nuclear Nonproliferation Treaty and, thereby, accepted a legal not to assist others acquire nuclear arms. China has ratified the Chemical Weapons Convention and signed the Comprehensive Test Ban Treaty. In 1994, China committed not to export MTCR-class ground-to-ground missiles. China's exports of missile-related components and technology appear to reflect the narrower understanding of the MTC guidelines than ours, but we have no evidence that China has acted inconsistently with its basic 1994 commitment.

Similarly, we assess that China continues to abide by its 1996 agreement to end assistance to unsafeguarded nuclear facilities in Pakistan or anywhere else. China is taking steps to improve its export controls, including its recent announcement of new dual-use nuclear controls. And last year, very significantly, China agreed to conclude its nuclear cooperation with Iran and also to terminate its exports of cruise missiles to that country.

So, the picture is mixed. The progress is substantial, but not enough, especially given the stakes. Therefore, my final broad point is that we must use all the tools at our disposal to make China part of the nonproliferation solution, including intensive diplomacy, including technical collaboration such as on export controls, including sanctions, and including positive incentives.

Let me emphasize, however, that there are clear limits to incentives. Of particular relevance to the subject of this hearing, neither this Administration nor its predecessors have been willing to sell China arms or transfer sensitive technologies that could contribute to China's own weapons of mass destruction or missile programs.

Now, one aspect of our efforts to persuade China to adopt the more responsible nonproliferation policy, particularly regarding missile transfers, has been the basic policy of three administrations beginning in 1988 to allow U.S.-made satellites and foreign satellites with significant U.S. components and technology to be launched on Chinese rockets.

But, again, the incentive is clearly limited to exclude transfer of sensitive missile or satellite technology when satellites are licensed for launch from China. The U.S. has a very strict policy secured in a bilateral technologies safeguards agreement between the United States and China designed to prevent the transfer of sensitive missile technology to China that could assist in its space-launch vehicle program, let alone, in its missile program.

Licenses for the export of satellites are subject to protection of sensitive technologies. Under current regulation Department of Defense representatives attend meetings between manufacturers and Chinese launch service providers to ensure that there are no transfers of unauthorized technology or data.

We do not believe that the commercial space launch activities that have been authorized by licenses and monitored under these procedures have benefited China's missile or military satellite capabilities.

And let me comment very briefly on the transfer of jurisdiction over commercial satellite exports. In September 1995, as you know, Secretary Christopher received an approved recommendation from an interagency group narrowing, but not eliminating, U.S. munitions list controls.

Further, interagency review under the aegis of the National Security Council, however, led to a common recommendation from State, Commerce, Defense and the intelligence community with two important parts.

First, commercial communications satellites would be controlled by Commerce even if they had embedded in them individual munitions list components or technologies. In all other cases, munitions list technologies or components, themselves, would continue to be subject to State Department licensing.

However, the further shift in control was accompanied by new control procedures. State and Defense were given the right to review all Commerce export license applications. A new foreign policy and national security control was established whereby State and Defense could recommend denial of a satellite export to any destination on the basis of national security or foreign policy interests.

Commercial communication satellites further were made exempt from the foreign availability requirements of the Export Administration Act. Now, Secretary Christopher noted in a recent letter published in the Los Angeles Times, "These features made it possible for the State Department to support the March 1996 recommendation to the President."

The bottom line question, of course, is has this change resulted in a degradation of protection for U.S. national security? It was Secretary Christopher's conclusion and remains the judgment of the Department of State that the changes made in the Commerce export licensing system in 1996 were sufficient to deal with the national security sensitivities associated with foreign launches of communication satellites.

They provide a degree of protection for these items when under Commerce control that approximates the strict controls of the international traffic and arms regulations.

Finally, let me briefly report that the waiver of Tiananmen sanctions earlier this year for Loral's Chinasat–8 project was handled in a normal manner, in accordance with the procedures used in previous waiver requests.

Specifically, after all consulted agencies concurred in the issuance of the proposed technical assistance agreements, the State Department recommended to the President that he waive the Tiananmen sanctions in accordance with established procedures.

Senior Administration officials were aware at the time that Loral was under criminal investigation for alleged violations of the Arms Export Control Act. But the State Department's long-standing policy has been that provided the activity proposed for waiver is consistent with U.S. national security and foreign policy, we do not deny export privileges to U.S. firms that are under investigation but have not been indicted.

However, if a U.S. firm is indicted, the Department does adopt a denial policy on the basis of the indictment and does not wait for a conviction.

It is against this backdrop that the United States conducts commercial space cooperation with China. We strive to accommodate U.S. commercial and economic interests, including promotion of U.S. satellite exports but within our paramount nonproliferation and national security objectives.

We have a system involving the licensing and technical safeguards process to deny access to sensitive missile technology by China. At the same time, if any person violates our laws and regulations in this area then such violations need to be investigated fully and prosecuted accordingly.

The United States has engaged China at the highest levels regarding its nonproliferation policies and practices. We continually encourage China to strengthen its export controls and bring its nonproliferation policies and practices more in line with international norms.

The prospect of launching U.S. satellites under technology safeguards is an important inducement to a positive evolution in Chinese policy which, in turn, is indispensable to the containment of proliferation in a dangerous world.

Thank you, Mr. Chairman.

[The prepared statement of Secretary Holum can be found in the appendix on page 247.]

Chairman GILMAN. Thank you, Secretary Holum.

We will now turn to Under Secretary of Commerce, Mr. Reinsch.

Secretary REINSCH. Thank you very much, Mr. Chairman.

I am pleased to have the opportunity to appear before you. I believe this Administration's policy on the export of commercial communication satellites to China both protects our national security and facilitates our economic well-being.

In allowing China to launch U.S. satellites and transferring licensing jurisdiction for commercial communication satellites to Commerce, this Administration has continued to enhance the policy of the Reagan and Bush Administrations and has been consistent with Congress' expressed intent. Our current policy continues the decision by previous administrations to allow China to launch U.S. built satellites subject to bilateral agreements on price, number of launches, and technology safeguards.

Our view, like that of Presidents Reagan and Bush, is that under the appropriate safeguards these launches need not pose a risk to national security. In a moment I will describe these safeguards as they apply to Commerce-licensed commercial comsats.

Commerce licensing of commercial communication satellites grew out of a 1990 decision by President Bush to veto a revised Export Administration Act which would, among other things, have moved all communication satellites, all commercial communication satellites to Commerce jurisdiction.

President Bush's veto was not related to the satellite issue, but in his veto message he directed that State review its control list to determine if a range of items, including communication satellites, could be moved to Commerce jurisdiction in light of the strong interest expressed by members of both parties in the jurisdictional issue, and as the U.S. was the only country in the world to control communication satellites as munitions items, that is as weapons.

It is worth noting that in 1990 and in 1992, both Houses of Congress passed legislation that would have transferred jurisdiction over commercial communication satellites to Commerce and in 1994 committees in the House introduced and then the Senate reported bills with the same provision. These actions are in addition to the letters the administration received from a number of Members of Congress urging either the jurisdiction transfer or the export of satellites to China.

Now, Under Secretary Holum has gone into some detail on the history of this series of events, so, I will skip that part of my testimony in the interest of time and go right to the question of Commerce safeguards. Commerce licenses for communication satellites contain numerous conditions and provisos developed in conjunction with the Department of Defense and the Department of State. Under Commerce licenses, exporters are obliged to comply with the terms of the satellite technology safeguards agreement between the U.S. and China which requires them to do the following.

Develop a technology transfer control plan that identifies the level and extent of technical data to be released and which also includes plans for securing the satellite during its transportation to the launch site.

Have all technical data under the license reviewed by the Defense Department prior to its release to the launch service provider.

And have a Department of Defense monitor present at technical meetings and launch activities with the Chinese launch service provider.

Transport the satellite in a sealed container allowing no access to equipment or technical data and with U.S. monitors to accompany the satellite if it is transported on a non-U.S. aircraft.

Have a separate cryptographic equipment safeguard plan for communications security equipment and limit technology which can be released under the Commerce license to only form, fit and function data used to mate the satellite to the rocket. And require the exporter in the event of a launch failure to obtain a license from State before releasing any new technical data.

In light of these safeguards, I believe the existing Commerce licensing system fully protects our national security and foreign policy concerns. There have been no allegations regarding export control violations of Commerce satellite licenses since the transfer of jurisdiction.

I understand there have been questions raised about an analysis conducted of the 1995 ABSTAR–2 launch failure. After that failure the company involved conducted an analysis without the participation of the Chinese launch service provider. The analysis was written in order to satisfy insurance requirements. The analysis was reviewed by the Department of Commerce which determined that it contained only information already authorized for export under the original Commerce license issued in February 1994. The unclassified report was first provided to a consortium of western insurance companies and later to the Chinese launch service provider.

I would also like to correct some misunderstandings that have arisen regarding the Commerce licensing process as a result of an earlier review by the General Accounting Office, which was alluded to in the opening statements. In particular, GAO asserted that there are five differences in the treatment of satellite licenses at Commerce and at State. A closer look shows these differences do not affect national security.

With respect to them, GAO reported first, that on the Congressional notification matter, the Congressional notification of individual licenses is not required in the Commerce system. Commerce regularly briefs the Hill, issues annual reports, provides licensing documentation and answers inquiries upon request. We have provided briefings on satellite exports and we briefed on the transfer of jurisdiction in 1996.

We are not aware that the Congress has objected to any satellite export. And the message that Congress has consistently sent is that it wants satellites controlled as dual use items under the Export Administration Act, which does not generally provide for Congressional review of individual licenses for a lot of reasons that I can go into in questions, if you wish.

In the specific case of satellites, of course, there can be no exports to China without a Tiananmen Square waiver, which is notified to the Congress. GAO also said that sanctions for missile proliferation do not apply to Commerce licenses.

Sanctions do apply to Commerce, of course, in cases of Category 1 violations. And the President generally has a flexibility to include dual use export sanctions in other cases if he chooses to do so. Normally, however, Category 1 missile sanctions apply only to munitions and dual use items controlled under the Missile Technology Control Regime. Commercial communication satellites fall into neither category.

Congress, we believe, clearly intended Category 2 sanctions to be less onerous than Category 1 sanctions which, as I said, do cover dual use items. Defense's power to influence the decision making process has diminished was GAO's third point. We believe DoD's authority has not diminished in this regard.

Commerce has denied licenses when Defense has raised national security concerns, found credible by all of the reviewing or the majority of the reviewing agencies. But Executive Order 12981 does not give Defense or any other agency a veto over licensing, which would be contrary to legislative authorities and Congressional intent. It does, however, permit an agency including the Department of Defense to prevent approval of any license until the President has heard and decided on its objections.

The fourth GAO comment concerned technical information and the way we control it. Since Commerce technology conditions are almost identical to those used at State, it is hard for us to see how the level of control on technology has changed since the 1996 transfer.

Finally, GAO reported that the additional controls placed on communications satellites transferred in 1996 do not apply to those transferred in the Bush Administration. This assertion misses the point. In practice, all satellite applications subject to Commerce license after the transfer are subject to the same safeguards and the other agencies have the same review and escalation rights.

Let me also suggest, finally, as a matter of policy that there are several reasons why allowing Chinese launches of U.S. manufactured satellites is in our interest. First, this is a large and important industry that is growing rapidly. U.S. industry revenues were $23.1 billion, a 15 percent increase from the previous year. Employment in 1997 was over 100,000, a 10 percent increase from the previous year.

The industry indicates that it currently has $1.7 billion in launch contracts on Chinese rockets with 8,000 U.S. aerospace jobs directly supported by those contracts. The industry further indicates that over the next 5 years, they have approximately $8 billion involved in China launches with 16,000 jobs at stake.

With over 1,200 satellites expected to be launched over the next 10 years, it is clear that the U.S. industry will continue to need access to the full range of launch providers if it is to remain the world's leader. Not to be able to offer a competitive launch alternative puts our industry, our satellite manufacturers at a decided disadvantage vis-a-vis their competitors. That is a status, I think, that all of us would regret occurring because a strong civil communications satellite industry is not only good for our economy, it is good for our military and our national security, as well.

As the line between military and civilian technology becomes increasingly blurred, what remains clear is that a second-class commercial satellite industry means a second-class military satellite industry, as well. The same companies make both products and they depend on exports for their health and for the revenues that allow them to develop the next generation of products that we all rely on.

Second, some of these satellites bring telephone, television and Internet services to the Chinese people. I believe that such services are an integral part of any effort to bring democracy and freedom to China.

History has shown that it was the successful example of the West, not only in military strength but in standard of living and freedom of expression that brought the Cold War to an end. Our goal should be to bring not only our products but our ideas and our values to China. We cannot do that if they do not have the technological tools to receive them.

Thank you very much, Mr. Chairman.

[The prepared statement of Secretary Reinsch can be found in the appendix on page 256.]

Chairman GILMAN. Thank you, Mr. Reinsch.

We now will hear from Secretary Lodal.

Please, be brief, if you would.

Secretary LODAL. Thank you very much, Mr. Chairman.

I will summarize my remarks as briefly as possible. In that regard, I would say that I believe you have heard the history accurately portrayed by my colleagues here, beginning in 1988 with President Reagan's decision to permit the launch of U.S. commercial satellites and going on through the 1990 congressional bill to transfer jurisdiction, the 1992 decision of the Bush Administration to make this transfer and the Clinton Administration's 1996 revisions of this procedure.

Let me pick up at that point in 1996 when President Clinton decided to transfer additional jurisdiction for commercial communication satellites to Commerce. DoD supported this transfer because the transfer did not involve certain sensitive technologies associated with satellites and launch vehicles. And it was accompanied by several changes in procedures that protect DoD's ability to ensure that transfers are consistent with U.S. national security.

Under Secretary Reinsch has covered these in some depth. I will just highlight, once again, a couple of the most important ones.

Companies can export complete commercial communication satellites but they cannot export the individual military technologies under a Commerce license. Those still require State Department licenses. Commerce can license certain limited form, fit and function technical data to mate the satellite to the launch vehicle but State retains control over all launch vehicles, the technical data associated with the launch vehicles or the integration of the satellite payloads with the launch vehicles, and all design and manufacturing data for satellites, and all technical assistance that might be provided by U.S. companies to Chinese launch service providers, including any launch failure analysis.

In addition to this key point about what is transferred and what is not transferred to Commerce jurisdiction, I might emphasize a couple of the key changes that were made to strengthen the Commerce system.

First, licensing determinations, even in the first instance for these particular items, are made by a majority vote of the reviewing agencies, unlike the previous system in which the initial determination for dual use items was always made by the Commerce Department acting alone. So, this eliminates the need to escalate, if you will, in the first instance if there is a majority that would oppose it.

Licenses can be denied for broad national security reasons even though they are not on the State Department munitions list. Communications satellites are not subject to formal foreign availability determinations, unlike many other dual use items. And all communications satellite licenses must include strong safeguards including DoD monitoring and the payment of DoD monitoring expenses by the companies.

DoD currently reviews all communication satellite licenses to ensure that the proposed export would be consistent with U.S. national security interests. DoD's recommendations reflect inputs from several relevant DoD components such as the Air Force, the

National Security Agency, and DoD's recommendations to approve such satellite exports are conditioned on strong safeguards, as I mentioned.

These include a requirement that the satellite exporter prepare a technology control plan and that that plan be approved by the Department of Defense. This technology control plan must include a detailed transportation plan for shipping the satellite to ensure that only U.S. personnel have access to the satellite at all times, and a detailed physical and operational security plan, including procedures for the supervised mating of the satellite to the launch vehicle.

There is a requirement that technical data that the U.S. company wants to transfer to the Chinese launch service provider is approved in advance by DoD's Defense Technology Security Administration. And there is a requirement that a DoD monitor be present at technical meetings between the U.S. exporter and Chinese launch service personnel to ensure that no information is exchanged that would improve Chinese missile or satellite capabilities. This includes a requirement that DoD monitors be present at the launch site in China to oversee physical site security and launch operations.

With regard to monitoring. Monitoring by the U.S. Government was required in all licenses for launches of satellites that contain one or more of the identified military-related technologies or kick motors and any launch vehicle integration technical data or any technical assistance. In other words, all licenses issued by the Department of State.

After implementation of the 1992 Bush Administration decision to transfer to Commerce jurisdiction over commercial satellites and before the revision of the procedures in 1996, there were three launches that were not monitored.

These were launches of purely commercial satellites that were licensed by Commerce. Monitoring had always been associated with the licenses issued by the State Department, and the DoD license review procedures anticipated that there would be at least one State license required for the launch of even these commercial satellites that were now licensed by Commerce.

However, these launches did not require any State licenses. We are not aware of any transfer of technology from these unmonitored launches that contributed to China's missile or military satellite capabilities.

Nevertheless, DoD did conclude that full monitoring would be a strong safeguard at relatively low cost to the companies that should be applied to all license cases, even those that did not require Department of State licenses. This was agreed by all agencies and incorporated as a requirement in 1996 when jurisdiction was transferred to Commerce for all commercial communication satellites and when the other changes and improvements that I have discussed earlier were implemented.

Mr. Chairman, in September 1988, President Reagan decided to permit the launch of U.S. commercial communication satellites by China. This decision was motivated by a desire to allow commercial relations with China to expand in a more normal manner.

The Reagan Administration understood the potential risk that such a program could lead to the transfer of missile-related technology to China but also recognized that China had, for many years, had the basic technology necessary to develop and deploy effective ballistic missiles, including intercontinental missiles capable of hitting the United States.

To ensure that no significant missile or satellite technology is transferred to China, the U.S. negotiated a bilateral technology safeguards agreement with China that remains in force today. Since that time, we have implemented the many safeguards and procedures that I have discussed and that have been described to you by my colleagues. This overall policy was a wise policy when it was adopted in 1988, when it was expanded in 1992, and it remains a wise policy today.

DoD takes its overall role in the development and implementation of export control policies very seriously. The case of commercial communication satellites with China presents a significant challenge to the U.S. export control system as we seek to ensure that no technology is transferred that would improve Chinese indigenous missile or satellite capabilities.

We believe that the current system protects our national security and meets this goal.

Thank you very much, Mr. Chairman.

Chairman GILMAN. Thank you, Mr. Secretary. I now address all of the panelists—or Mr. Holum, rather.

Mr. Holum, on June 16, 1998, our Committee on International Relations was provided with a number of documents pursuant to our April 23, 1998 request to you as the director of the Arms Control and Disarmament Agency. And one of those documents is entitled "Comments to Chinasat-8 Waiver." I assume that it is referring to the waiver that is required under the Tiananmen Square sanctions, to allow satellites to be exported to China, and for which the State Department has to make a recommendation.

The document reads as follows:

> ACDA continues to believe that the United States should link its willingness to make Presidential national interest waivers for China's satellite launches to Beijing's taking concrete steps to resolve our concerns about its failure to implement the MTCR equipment and technology annex, and the ongoing missile related exports to Pakistan and Iran.

I go on to read:

> Although it is true that our nation has in the past seen approval of space launches and related exports as an incentive for Beijing to exercise restraint in its missile related exports, there has been no evidence to date that this policy is having any effect. Chinese transfers to Pakistan and Iran continue. China has refused to discuss missile non-proliferation in our most recent dialogs with them. It's time to reassess this policy. We have no other effective means to induce Chinese cooperation and linkage that might lead to a serious dialog to resolve our proliferation concerns. History has proven that the only time we have gotten movement from the Chinese on missile non-proliferation has been in the face of a penalty being imposed. Carrots have gotten us nothing.

Mr. Holum, I want to commend you for that position. And if this Administration is going to continue to make national interest waivers, then it should link those waivers to concrete improvements in Chinese missile non-proliferation behavior. Can I ask you, why did the Administration refuse to adopt this memo?

Secretary HOLUM. First of all, I think it is important to note that this recommendation was first offered in 1996. It had not been raised in the past. One of the Arms Control and Disarmament Agency's roles is to look for every possible avenue of leverage to try to advance non-proliferation. It is a single-minded, single-focus agency, as you know. The views that ACDA expressed at that time were considered carefully, in the context of examining whether we should try to use more direct leverage in the satellite waiver process to have a specific impact on missile behavior in China.

The reasons not to do that, the reasons why that idea was not adopted, were that we had in place, going back to 1988, as we have testified, an established pattern that would have been of relationships in communication satellite launches that would have been very disrupted to curtail at that stage. One of the particular problems is that a lot of those waivers involved packages with other countries involved—not just China, but allies, Germany and other countries that we have close relationships with. We would have had damage not only on our relationships with China, but on our relationships with other countries as well.

A broader and perhaps more salient, even more salient, concern is the fact that when we deny satellite launches on Chinese launchers, we deny China essentially nothing. There is a shortage internationally of launch capabilities. The very strong likelihood is that, as Secretary Reinsch has explained, among the 1,200 launches anticipated over the next decade, those launch opportunities will be immediately picked up by someone else. So the effect is to deny China essentially nothing, but to deny ourselves a great deal; to disrupt our leadership or threaten our leadership in satellite capabilities, and cause enormous economic consequences here at home.

So in any case, where sanctions or denial of commercial opportunities is considered, we obviously have to balance the potential effect, if any, on the target, with the effect on our own national interest. And that is the basis for taking a different approach.

Chairman GILMAN. Well, Mr. Holum, the paragraph I read that you wrote back in, I guess, 1996—is that the date of it?

Secretary HOLUM. I am not sure of the date, Mr. Chairman.

Chairman GILMAN. Well, at any rate—I think it was 1997, if I am not mistaken, or possibly 1998. At any rate, that paragraph admits that your policy was not working. So do you stand by that statement, or not stand by it?

Secretary HOLUM. The statement that the policy is not working?

Chairman GILMAN. Yes. "We have no other effective means to induce Chinese cooperation and linkage that might lead to a serious dialog to resolve our proliferation concerns." I am reading it again. "History has proven that the only time we have gotten movement from the Chinese on missile non-proliferation has been in the face of a penalty being imposed. Carrots have gotten us nothing." Do you stand by that statement?

Secretary HOLUM. I stand by it in terms of where we have made and not made progress in the past. At the same time, I think there are opportunities using satellite launches and, in particular, expansions of launch quotas, as a means of inducing improvements in Chinese behavior.

55-784 99 - 4

And keep in mind that we are now focusing with China on missile controls in particular, not on their transfers, or at least their commitment to transfer full-up missile, but on their interpretation, versus ours, of the controls on technology and other forms of assistance. It is a narrower question. I think there are other means by which we can achieve those ends without disrupting our national interests in space launch activities.

Chairman GILMAN. Well, the perception is that you may have changed your position from Acting Secretary of ACDA to Under Secretary of International Security. Am I correct?

Secretary HOLUM. It is certainly the case that the Under Secretary for International Security Affairs has a requirement to consider a broader range of concerns. ACDA is a specifically focused industry. As ACDA director, I had a narrower focus.

Chairman GILMAN. Well, it sounds like the facts are still the same, but your perception, I guess, is a bit different. Thank you very much.

Mr. Hamilton.

Mr. HAMILTON. Under Secretary Holum, your statement at the bottom of page 4 of your testimony is that, "We do not believe that the commercial space launch activities that have been authorized by licenses and monitored under these procedures have benefited China's missile or military satellite capabilities."

If you will look at this chart at the end of the room here, which I presume is prepared by the majority, their opening sentence is precisely the opposite of your conclusion. And I quote it: "The Administration's technology transfer policies have allowed China to obtain sensitive rocket and satellite technology and apply it to enhancing the reliability, accuracy, and potency of its ballistic missiles." The next sentence is, "The following are some of the ways that the Administration's technology transfer policies could aid Chinese ballistic missile development."

[See chart F in the appendix on page 371.]

Now, it is worth noting the verb there. It says "could." It does not say "did aid." And with that verb, I presume all of us can agree with the sentence. But then, they go on to list several things that could aid: payload dispersal technology; kick motor technology; radiation-hardened electronics; encryption devices; launcher attitude control; stage separation systems; stage coupling load analysis; and business development.

What is your response—And not just directing this to you, but perhaps to the whole panel. What is your response here? We have your statement, which is directly contrary to this statement; and in the statement prepared by the majority, a number of very specific things that they say could—they do not say "did," but they say could—aid the Chinese ballistic missile development. Now, let me have your point of view on that.

Secretary HOLUM. Well, first of all, we have to separate out—and I am not sure whether the chart does—any unauthorized transactions that may have occurred, any transfers which are now under investigation and are something we cannot examine here.

Mr. HAMILTON. Well, that is only one transfer, right, that might be unauthorized? You are talking about the Loral?

Secretary HOLUM. I am talking about the case that is under investigation, that is right.

Mr. HAMILTON. Yes.

Secretary HOLUM. That is why I made it clear, we are satisfied that no authorized transfers contributed to their military missile or even to their space launch vehicle program. In terms of specific capabilities, you have to go one by one. And I will ask for my colleagues from the Department of Defense to help here.

But for example, in dispensing capabilities dealing with the Iridium launch and the notion that by launching a number of different satellites from one booster you somehow acquire a MIRV capability, I am told by the experts that there is a dramatic difference—indeed, a difference in kind, rather than quantity—between a smart dispenser that can dispatch multiple satellites to different orbits, as distinct from a reentry vehicle that would be able to dispense multiple warheads to different targets on land. The technologies are dramatically different. And so our judgment on the lack of a contribution through authorized launches of commercial satellites is based on a considered evaluation of the various technologies that might be involved.

Remember that, aside from basic form, fit, and function, non-sophisticated technology dealing with mating the commercial satellite to the booster, any other technology involving the items on the list here that would be ITAR controlled is subject to monitoring on a continuous basis by the Department of Defense.

Jan, do you want to add anything?

Secretary LODAL. No, that is correct, Congressman. I think, picking up on this MIRV example—and there are others up there—it is, by very rough analogy, kind of like the difference between getting your car in the right state of the Union, and getting it in the garage in a particular location; what you have to do in one case, versus the other. There is a very big difference between getting into a parking or transition orbit, and getting a ballistic missile reentry vehicle right on the target, and quite different technologies involved. And we did review these in some detail in making these determinations.

I would just emphasize once again the key point, that all of those—well, I have not actually read this document, but to the extent that I can see it from here—and I will ask Mr. Tarbell if he has any contrary view—it looks to me like those are all things that are not permitted to be included in what we call commercial satellite transfers, unless they are completely embedded in the satellite, monitored at all times. Some of those technologies I guess can be included, such as the kick motors, if they are associated entirely with this launch and monitored and end up in space.

Mr. HAMILTON. Well, let us go after the kick motor, because one of the witnesses—maybe two of them—yesterday made quite a point of that. The kick motor, as that chart says up there, is used to put the satellite precisely into the desired orbit.

Secretary LODAL. Right.

Mr. HAMILTON. Why would that not help Chinese military capabilities, if the kick motor technology is conveyed?

Secretary LODAL. It would, if they could use the kick motor independently, or reverse-engineer it, or have access to it. But what is approved, what is defined as commercial satellite launch does not include any of that. They can only use the kick motor for that particular satellite for that launch, associated with that launch, and it has to be monitored and tracked all the way.

So they have a device which has technology embedded in it that could be useful. And if they could take it apart, or figure out how to build one just like it, or learn how to build more just like that, that would be useful. But we do not allow that. We allow them only to take that device, attach it to that particular satellite, and put it in that orbit. So they get the end use out of it, but they do not get any of the technology that is embedded in it, under the policy. Is that right?

Mr. TARBELL. That is correct.

Mr. HAMILTON. Mr. Chairman, I know my time is expired. It might be helpful if our witnesses would just respond item by item to these various technologies, and let us know what the facts are, how you would respond to that chart. Could we have that in writing, please?

[The information referred to can be found in the appendix beginning on page 381.]

Chairman GILMAN. Without objection, it will be.

Mr. TARBELL. Could you give us a copy of the chart?

Mr. HAMILTON. It might be helpful for you to have the chart.

Chairman GILMAN. Will the staff make certain that copies of the chart are made available to the witnesses, so that Mr. Hamilton's request can be complied with?

Chairman GILMAN. Chairman Spence.

Chairman SPENCE. Thank you, Mr. Chairman.

You know, we talked a lot about export control policies. We talked about the launching of satellites in China. But when it gets right down to it, the main question that we are concerned about is whether or not we have given either directly or indirectly, on purpose or otherwise, sensitive information, equipment, to China. It is difficult to define the answer. I have to commend all of you. You are doing a good job of getting your stories together, and I know it took a lot of preparation to do it.

Let me just cut right back down to it. We heard yesterday—and it is undisputed, I think—that over a period of time, the Chinese went from a 75 percent success rate in their launches to 100 percent. Now, is it the position of DOD, Mr. Lodal and Mr. Tarbell, that allowing the launch of our satellites on Chinese rockets improves China's ballistic missile capabilities, or not?

[See chart C in the appendix on page 365.]

Secretary LODAL. Well, Mr. Chairman, I am not familiar with the study that I heard about this morning that you saw yesterday, so it is a little hard to comment on either its statistical validity or exactly what it means in particular. And there are a lot of changes in technology, of course, that have occurred in this decade, from a lot of different sources, that China has obtained. So it is probably somewhat difficult to figure out all the causes that might be behind improvements, if they do exist on a consistent basis, in the reliability of their space launch vehicles.

There is also, I think, an important distinction between the space launch vehicles and the military ICBMs, which are different items.

And the ICBMs that are deployed today have technologies in them that are considerably different from those technologies that are in the space launch vehicles. They have some overlapping technologies, of course.

We, as I concluded in my statement, do not believe that there has been any transfer of these sensitive technologies. We do not have any evidence that there has been transfer of these sensitive technologies to their military programs that have made an impact on the Chinese military programs.

Chairman SPENCE. Mr. Tarbell, do you want to comment on that?

Mr. TARBELL. I would agree with everything that Mr. Lodal has said.

Chairman SPENCE. I thought you would, yes. [Laughter.]

That is a good try, gentlemen. I appreciate it. I know you have put in long hours on that, and I appreciate it. The answer is quite evident.

Chairman GILMAN. Mr Skelton.

Mr. SKELTON. Thank you, Mr. Chairman.

Yesterday, our witnesses mentioned a change in the bilateral technology safeguards agreement between the United States and China that was signed on February the 11th, 1993. Am I correct, Mr. Holum, that this agreement was negotiated during the Bush Administration and signed approximately 3 weeks into the Clinton Administration?

Secretary HOLUM. That is correct. It was signed in February 1993, February 11th.

Mr. SKELTON. And negotiated in the prior Administration?

Secretary HOLUM. That is correct.

Mr. SKELTON. What did the agreement change from the original 1988 Reagan Administration agreement with China?

Secretary HOLUM. It is my understanding that the agreement basically embraced what had been in the previous agreement and strengthened it in some specific respects. For example, in the purpose, the new language established under the purpose of the agreement that, except as otherwise specified, there will be no unmonitored or unescorted access to any of the equipment and technical data described below, nor will such equipment——

Mr. SKELTON. In other words, if DOD officials and others were present. Is that correct?

Secretary HOLUM. That is correct.

Mr. SKELTON. In other words, it was a tightening up of the previous launches?

Secretary HOLUM. That is right. This replaced language. This escort requirement replaced language that simply provided that no technical data would be transferred.

Mr. SKELTON. Now, let me ask, the State Department was opposed to the shift in exporting licensing jurisdiction for commercial communications satellites to the Commerce Department; is that correct?

Secretary HOLUM. The State Department in 1995 opposed the specific proposal that had been raised for a transfer. That is correct.

Mr. SKELTON. Now, however, I understand that former Secretary of State Warren Christopher agreed, and that he did have a significant role, and the State Department did have a significant role, in commercial satellite licensing decisions. Is that correct?

Secretary HOLUM. That is correct.

Mr. SKELTON. Would you please explain the difference?

Secretary HOLUM. Well, when the original idea was recommended in 1995—And this goes back, incidentally, as we have discussed in our testimony, to the 1990 legislation that had approved such a transfer. So the transfer occurred in several tranches. But in 1995, there was a proposal resulting from an inter-agency study that basically had a split recommendation. One recommendation was that all commercial space launches should go to the Department of Commerce; and another was that some components should go, but most would remain with the State Department. Secretary Christopher supported the option of keeping the majority of the control in the Department of State.

Subsequently, there was a further inter-agency review, in which the State Department, the Defense Department, the Commerce Department, and others participated, which arrived at a consensus recommendation. Secretary Christopher supported that recommendation to the President. He supported it on the grounds that there was a new national security and foreign policy control established in the Commerce controls that would allow denial of exports to any destination in the world. The national security controls now do not apply to every destination. This went to every destination in the world.

Foreign availability was ruled out as a consideration. And most importantly, the Secretary of State and Secretary of Defense had an active role in reviewing and access to every license that was proposed for commercial space launch activity, a satellite launch. Therefore, the protection, the right to denial and the protection of access to the process, the right to appeal up to the President of the United States by any agency, protected our national concerns. So Secretary Christopher supported that recommendation.

I would underscore, as well, that as a practical matter, in the applications that have been approved for commercial space launch vehicles, no agency—State nor Defense—has dissented. These have all been recommendations approved by all of those agencies.

Mr. SKELTON. Thank you, Mr. Chairman. I will reserve my further questions until the second round.

Chairman GILMAN. Mr. Hunter.

Mr. HUNTER. Thank you, Mr. Chairman.

Good morning, gentlemen. Mr. Tarbell, let me direct a few questions to you. There have been some press stories that have indicated that you directed subordinates within DTSA to delete any information, information including memos and draft documents, that opposed the granting of a license to Loral, or a waiver for the export of Loral's Chinasat-8 satellite to China. Is that true?

Mr. TARBELL. I am aware of those allegations, and they are absolutely and totally false.

Mr. HUNTER. OK. Were any DTSA analysts opposed to the granting of a license or waiver for Loral's Chinasat-8 satellite?

Mr. TARBELL. No. All of the recommendations that came in Chinasat-8 recommended approval.

Mr. HUNTER. Did you shred any documents related to either the May 1997 DTSA analysis or the Chinasat-8 license or waiver?

Mr. TARBELL. No, I did not shred any documents.

Mr. HUNTER. Did you direct any of your subordinates to shred documents?

Mr. TARBELL. I did not direct anyone to shred any documents.

Mr. HUNTER. Did you direct any subordinates not to draft memos opposing the license or waiver decision for Chinasat-8?

Mr. TARBELL. Not that I can recollect.

Mr. HUNTER. And if you recollect anything in the near future, could you submit that for the record?

Mr. TARBELL. Absolutely.

Mr. HUNTER. Did you receive any orders from superiors directing you to take any action against DTSA subordinates critical of a license or waiver?

Mr. TARBELL. No, I did not.

Mr. HUNTER. OK. So to your understanding, there were no DTSA officials or employees who opposed this waiver?

Mr. TARBELL. There were no organizations who provided recommendations on the license.

Mr. HUNTER. OK, now, I am talking about people. Were there any personnel, to your knowledge?

Mr. TARBELL. I am not aware of all of the personnel's views that went into the formulation of organizations' submission of recommendations on that license.

Mr. HUNTER. So to your knowledge, the answer is "No"?

Mr. TARBELL. I do not know. I do not know whether there might have been one employee in the Defense Department who thought this was a bad idea.

Mr. HUNTER. Have you heard that there was?

Mr. TARBELL. No.

Mr. HUNTER. Did you ever direct any member of your staff not to speak to Congress or congressional staff about either the 1997 analysis or the 1998 waiver decision?

Mr. TARBELL. I do not recall directing my staff to do that. I have general policies and procedures. Let me separate a couple of things out here. First of all, on the analysis that we performed on the 1996 launch failure and launch by Loral and Hughes, the Justice Department has instructed me not to comment on that subject. As far as Chinasat-8 goes, I can speak to that question.

Mr. HUNTER. Have you preserved all internal memos, documents, and computer discs relating to these issues?

Mr. TARBELL. Absolutely. And we are making them available to all of the inquiries associated with this matter.

Mr. HUNTER. How many copies of the May 1997 DTSA analysis were produced, and how many exist today?

Mr. TARBELL. I cannot tell you how many exist today, because I do not have right off the top of my head a count on that. I would have to get that for you.

Mr. HUNTER. Did you ever personally raise objections to the license or waiver that was approved in February 1998?

Mr. TARBELL. No, I did not.

Mr. HUNTER. Did you ever question the adequacy of satellite technology transfer control plans?

Mr. TARBELL. I do not review, personally, technology transfer control plans. That is done by members of my staff in the formulation of recommendations that go to the State Department.

I should note for you, Mr. Congressman, that our agency does 21,000 license application reviews a year. It is impossible for me personally to review every single one of those license recommendations.

Mr. HUNTER. It has been reported that you received a subpoena from the Justice Department. Can you describe the basis for that subpoena?

Mr. TARBELL. The subpoena refers to merchant communications satellites, and is the subject of an ongoing criminal investigation by Justice.

Mr. HUNTER. OK. Let me ask all of you a question. It is clear that the 1996 crash was a launch failure. Is that right?

Secretary HOLUM. Yes.

Mr. HUNTER. It was not a satellite failure; it was a launch failure. The direction of all your testimony has been to the effect that you very carefully segregate the satellite package—that is what our companies supply—with any interaction with the Chinese with respect to the launch system. Because the launch system, of course, also propels nuclear weapons, some of which are aimed at U.S. cities; and therein lies the security problem.

If that is the case, and the 1996 failure was indeed a launch failure, and was not a satellite failure—the Loral satellite did not fail; it was just the package that was carried on the failed launcher—why was it in order for the American companies to engage in any way with the Chinese with respect to that launch failure?

Secretary HOLUM. Congressman, that is something we can only discuss in a very limited way, because of the ongoing investigation by the Department of Justice. But let me say and confirm for you that the conditions of license and the technology safeguards agreement with China provide that no U.S. person can assist China in any way in the design, development, operation, maintenance, modification, or repair of launch vehicles.

Mr. HUNTER. Mr. Holum, I understand that. You understand it. That is why I am asking my question. The failure was not a satellite failure.

Secretary HOLUM. Right.

Mr. HUNTER. The essence of our direction to our companies, both in terms of policy and in terms of law, is, "You can interact with the Chinese with respect to your satellite; you cannot interact with the Chinese with respect to making your missiles more reliable, your launchers."

Now, this failure was a launch failure. It was not a satellite failure. So why did the Administration countenance any engagement by Hughes and Loral scientists with the Chinese following launch failure?

Secretary HOLUM. Congressman, when information of such an exchange came to the attention of the Department of State, that matter was referred promptly to the Department of Justice for investigation.

Mr. HUNTER. Thank you, Mr. Chairman.

Chairman GILMAN. Thank you.

Mr. Sisisky.

Mr. SISISKY. Thank you, Mr. Chairman.

There is something that is confusing me, and maybe you can explain it to me. I had a series of questions, but did we find China guilty of proliferation, of selling to Pakistan, missiles? Is that a fair assessment?

Secretary HOLUM. I am sorry, I did not hear the question.

Mr. SISISKY. Have we found China guilty of selling missiles to other countries? Let us put it that way.

Secretary HOLUM. Well, in 1991, the Bush Administration, and again in 1993 the Clinton Administration sanctioned China for transfers of missile-related technology to Pakistan. There has been further continuous review, still underway, constantly monitored, of the possibility that China may have transferred missiles, M–11 missiles, to Pakistan. We have not reached a judgment that such a transaction has occurred. If we did reach that judgment, we would apply category one sanctions under the Missile Technology Control Regime legislation.

Mr. SISISKY. Which is what? What are category one sanctions?

Secretary HOLUM. It basically prohibits for a period of 2 years all export licenses to the sanctioned entities, and all contracts with the sanctioned entities. China, under the Helms Amendment, is treated as a non-market economy. And as a consequence of those limits on specific entities, the reach would be very broad. It would reach other governmental entities.

Mr. SISISKY. Did we do that in 1991 or 1993?

Secretary HOLUM. 1991 and 1993 were category two sanctions that related to technology, as opposed to full-up missiles or major components.

Mr. SISISKY. And I also understand we have incentives, if they do not proliferate. What are the incentives?

Secretary HOLUM. Well, we have been proposing, we have been discussing with China the concept of enhanced quotas under the space launch agreement if they improve their behavior. This relates specifically to our concern that China appears to have a different interpretation of what their 1994 undertaking means; not as to complete missiles, but as to components and related technologies. We want to nail down that they have the same interpretation that we do.

Mr. SISISKY. And how long has this been going on, the difference in interpretation?

Secretary HOLUM. We have been pursuing this since 1994.

Mr. SISISKY. You are kidding me. Since 1994 you have come to no agreement on a written contract, I guess, I mean, or a rule?

Secretary HOLUM. We have not concluded that effort. The effort is still underway. There have been a lot of steps forward. It is difficult in an open session to characterize the specific give-and-take.

Mr. SISISKY. I guess that is why I am so confused. Let me just ask you another question. Yesterday, the witnesses testified that there is not much of a difference between the price in China for launching a missile as it is in the United States. There is only about a 10 percent difference now? I think that is what they said.

Secretary HOLUM. I would defer to Under Secretary Reinsch.

Secretary REINSCH. If I can comment on that, Mr. Sisisky, I would say two things. There has been sufficient concern in the past on the part of the American rocket industry, if you will—the launch industry, as opposed to the satellite industry—about price differentials, that they have approached the Government, primarily in the Reagan and Bush Administrations, to urge them—successfully, as it turned out—to negotiate a pricing agreement with the Chinese in order to avoid price undercutting. The numbers I have heard recently would be probably more in the 15 percent range.

But I would add that there is another issue, and that is the launch window, or the availability. The economics of this thing generally are that the satellite user, the communications company or whoever it is, generally pays for the satellite up front, and its revenue arrives only after the satellite is up and operating. Therefore, a short launch window—a launch at an earlier point in time in the future, as opposed to several years in the future—has a big effect on their revenue stream arriving, and has a big effect on the health of the company.

One of the competitive issues in deciding where to go for a launch and which satellite to buy is who can launch within the appropriate time. So it is not only a question of dollars. If the Chinese are available in 6 months, and the French are available in 2 years, that also makes a difference.

Mr. SISISKY. I have got you now. I am glad you cleared that up. One other question. I noticed in both of your testimonies—Mr. Hamilton said the State Department is satisfied with what we have. We now believe the commercial space launch activities that have been authorized and monitored under these procedures have benefited China's missile military satellite capabilities. The Defense Department says we believe that the current system protects the national security.

Just a question. Would you make any changes, if you had the opportunity to do it now, in the agreement that we have?

Secretary HOLUM. In the procedures?

Mr. SISISKY. Yes.

Secretary HOLUM. No. I think the procedures work.

Secretary LODAL. I concur.

Mr. SISISKY. Thank you, Mr. Chairman.

Chairman GILMAN. Mr. Bereuter.

Mr. BEREUTER. Thank you, Mr. Chairman.

Gentlemen, I want to focus on the missile launch failure investigations. An independent review committee was established after the 1996 missile launch chaired by Space Systems Loral officers and consisting of that company, their participants, along with representatives of Hughes Electronics, Great Britain, and Germany. They held two meetings, one in Palo Alto, and one in Beijing.

The presence of a Department of Defense official at these meetings to monitor technical discussions in order to avoid inadvertent releases of export controlled information was deemed not necessary by Loral, and the minutes of the meeting and the draft preliminary independent review committee's report were transmitted to officials of the China Great Wall Industry Corporation, the manufacturer of the Long March booster.

I would have this question initially, and I am going to hold it for you, Mr. Holum, or perhaps Mr. Lodal. Who appoints the IRC, and on what and whose authority is it appointed?

Then, prior to this failure, there was a failure in December 1992 and a failure in January 1995. It is clear that both the original 1988 and then revised 1993 bilateral agreements between the U.S. and the PRC require the U.S. Government to monitor all aspects of U.S. satellite launches in China.

Regarding the December 1992 crash, an August 17, 1993 article in "Defense Daily" reports that a joint statement was released that said there were 7 months of vigorous and cooperative investigation. Further, the joint statement said that Hughes is making design and process improvements on the—and it lists the type of satellite—while the Great Wall Industry will make design and process improvements in the Long March 2E.

Regarding the January 1995 crash, a February 9, 1995 article in "The Wall Street Journal" refers to a Hughes official who said that a thorough investigation of the explosion would take months. And afterwards, another July 26, 1995 article in "Aerospace Daily" regarding the same failed launch indicates that the joint statement released by Hughes and China Great Wall said that they "would work together to eliminate the two mentioned possible causes of failure."

I think these articles are important because they point to a cooperative effort between the U.S. companies and the Chinese government to investigate failed launches long before the charges of unauthorized technology transfer by Loral came about as a result of the failed 1996 failed launch.

Now, exactly what kind of monitoring took place with respect to these crashes? Can you give me anything to assure me that there was no unauthorized transfer of technology involved in these cooperative efforts between the U.S. companies and China?

So I will go back to ask you about the independent review commission that convened in Palo Alto and Beijing; but then where and how were we involved in monitoring the two previous ones, and how do we know that there was not unauthorized transfer of technology with those two very extensive investigations that apparently took place?

Secretary HOLUM. I will have to defer to my Defense Department colleagues who do the monitoring.

Secretary LODAL. Congressman, the 1992 launch failure analysis was conducted by the company and the Chinese, and that was monitored by DOD. And my understanding is, we did not have problems with that particular launch failure analysis.

Mr. BEREUTER. And can we have a copy of that monitoring report? I make that request.

[The information referred to can be found in the appendix beginning on page 381.]

Secretary LODAL. The January 1995 ABSTAR–2—This is the ABSTAR–2, I believe, you are speaking of.

Mr. BEREUTER. Yes.

Secretary LODAL. It was not monitored by DOD. If there was any analysis there, I do not——

Mr. BEREUTER. Why was it not monitored? And why was the 1995 investigation, the one that most of the focus has been upon at this moment—Why were those not monitored by the U.S. Government at all? And certainly, why not by DOD?

Secretary LODAL. In 1995, you are speaking of?

Mr. BEREUTER. Well, both. The 1995 and 1996.

Secretary LODAL. Right. In 1995, the license—This is one of the three that did not require DOD monitoring in the license that I mentioned in my opening statement, in this interim period after the Bush decision to transfer to Commerce the commercial launches. The monitoring had been always tied to the licenses on the munitions list. And the full-up requirement for monitoring on all licenses did not come into play until 1996. And there were three where there was no requirement in the license to include monitoring.

Mr. BEREUTER. So in conclusion, then, since my time has expired, of the three launch failures that I have mentioned, two were not monitored. One was monitored, and you can provide us with a report of that monitoring. But when they are not monitored, are we saying that there is nothing there that technologically can be transferred, or is that an impossible conclusion to make at this point?

Secretary LODAL. Right, it is, of course, impossible to conclude what might have actually happened, since we were not there at the time, on the ones that we did not monitor.

Secretary HOLUM. Can I add that the 1996 launch was a monitored launch.

Secretary LODAL. Yes, that is a good point.

Secretary HOLUM. And the 1995 launch is a case where, given the license condition, the Government's understanding based on the Commerce license was that this did not incorporate accessible ITAR-controlled technology. We do not know the outcome. We have no reason to believe that any sensitive technology was transferred, in substantial part because of the character of that particular satellite.

Mr. BEREUTER. Well, I have different information with respect to the 1996 report. My information is that the IRC, the review committee, did not deem it necessary for DOD to monitor; and therefore, the meetings in Beijing and Palo Alto were not monitored by DOD.

Secretary LODAL. Perhaps Mr. Tarbell can clarify this for you just a little bit.

Mr. TARBELL. This is very difficult, because Justice has asked us not to talk about the activities that you are describing. But let me say that the launch itself in 1996 of the Intelsat–708 in February 1996 was monitored by DOD. Activities subsequent to that launch, any such activities, were not monitored by DOD.

Mr. BEREUTER. Thank you.

Chairman SPENCE. Mr. Lantos.

Mr. LANTOS. Thank you very much, Mr. Chairman.

There is a certain theater-of-the-absurd quality about this hearing, because we are intertwining highly technical issues with policy issues. And since most Members of Congress—and indeed, most of the American people—do not understand the technical issues, I think it would be helpful if we would focus on policy issues. And my comments will be directed at policy issues, and they are directed at policy issues from the vantage point of one who has opposed the China policy of both the previous and the current Administrations.

There is an attempt to portray the policy differences—and most recently, we see ads signed by our three previous Presidents and the who's-who of the defense and foreign policy establishments—which sets up the phony dichotomy that we are either for engagement or isolation. And that, of course, is absurd, because nobody is for isolation. It is the kind of engagement that the proper debate should be.

Now, one of the problems that disturbs me about this policy, which in my judgment was mistaken under Mr. Bush and is mistaken under the current Administration, is the labored but hopeless attempt to differentiate between civilian uses and military uses. As a matter of fact, on the front page of today's "New York Times," a Chinese official is quoted as saying—and this is with a straight face—"It would be up to the Chinese government regulators, Mr. Liu said, to decide if China's military could use the satellites." Well, let me advise everybody that in a totalitarian society, there is no distinction between military and civilian uses; and whatever civilian uses are available, those are also available to the military.

[The information referred to can be found in the appendix beginning on page 381.]

I suspect we are now at a stage of this particular inquiry where we are all reminded of the British ditty which says, "What shall we do with the drunken sailor?" Well, the proper question is not "What shall we do with the drunken sailor?" but "Why did he take to the bottle?" And I will tell you why he took to the bottle. Because under the previous Administration, and under this Administration, the horrendous pressure of business interests determined to make these deals with China succeeded in persuading sufficient numbers of members of this body and the Administration to move along this line. That is why we are dealing with this.

And Congress has now reached the point of saying, "We made a mistake. And let us rethink it." I believe it was the distinguished chairman, my friend and neighbor, who proposed a resolution last week, which we passed by a vote of 417-to-4. Now, this was not a mistake. This was a recognition that for a long period of time under Mr. Bush, with minor changes, the policy has been continued that many people felt—and I read in "The New York Times" today even the Pentagon and the State Department are raising questions—was clearly the wrong policy.

So if we do not get bogged down with the issue of whether this was a Democratic failure or a Republican failure, because I am convinced it was a bipartisan failure—Mr. Bush and his Administration made mistakes, as did this Administration—and we recognize that very serious policy mistakes led us to this series of hearings, we will be able to correct them, and deal with them. But I think it is a very serious mistake to be bogged down in an attempt at blame-fixing and details.

The whole policy is wrong. It is perfectly obvious that the Chinese have benefited in enhancing their technological capabilities as a result of this bipartisan policy. The time has come to call a halt. The time has come to admit mistakes. And the time has come in the future to be less starry-eyed about dealing with totalitarian governments.

I would be grateful, Mr. Holum and others, if you could react in some way.

Secretary HOLUM. Well, Mr. Lantos, I think it is very useful to bring this to a policy level, to a policy perspective. The technical aspects are complex, and beyond most of us. But I would urge consideration of one fundamental reality that I attempted to articulate in my statement. We have a very large non-proliferation policy. If we cutoff satellite transfers, or satellite launches, if we cutoff a wide range of our relationships with China, it may be satisfying to us, but——

Mr. LANTOS. Some of us find our satisfactions in other areas, so we are not recommending policies for satisfaction.

Secretary HOLUM. But we would wake up the next morning still having to deal with a non-proliferation problem, still requiring China's cooperation in order to deal with this. If we do not have their cooperation, they have the capacity to transfer missile and nuclear and other weapons of mass destruction technology to a wide variety of places that would be of concern to us.

Now, our approach to them, as you point out, has to include a variety of elements. As you say, it is not a matter of engagements; it is a matter of what kind of engagement. We have pursued a variety of approaches. Diplomacy at every level. And this is not just this Administration; it is previous administrations as well. Diplomacy at every level. The front-line, day-to-day hard work of non-proliferation—which basically consists of experts reading, sifting through intelligence and crafting approaches, crafting demarches, to try to interrupt shipments of dangerous materials—goes on all the time out of public view, and it is fundamentally important.

It also includes sanctions. And there is no question but what the sanctions on missiles that were imposed in 1991 and in 1993 had an impact on China's willingness to undertake a commitment not to export MTCR-class missiles anywhere in the world. In fact, that commitment goes beyond the Missile Technology Control Regime requirements, which is just a presumption of denial. So it is a very strong commitment. We have no evidence that they have not lived up to it. Sanctions were part of our accomplishing that.

Mr. LANTOS. Mr. Chairman, if I may just follow up one small item, I would like to ask Mr. Holum, or anybody else on the panel, to expand on what "The New York Times" means in its headline, "The Pentagon and the State Department Are Raising New Questions." What new questions are the State Department and the Department of Defense raising, in view of all that has happened in recent times?

Chairman GILMAN. Could we break right there for this vote, and get that answer when we come back?

Mr. LANTOS. Surely.

[Recess.]

Chairman SPENCE. The committee will please be in order.

I think the pending business is the answer to Mr. Lantos' question.

Mr. LANTOS. Thank you very much, Mr. Chairman.

The reference I had, to sort of recapture the continuity, The New York Times today says in its headline, "Pentagon And State Department Raising New Questions." In the body of the article, "Government officials said the Pentagon and the State Department were raising new questions about whether a Chinese controlled company with close ties to China's military should be allowed to buy the satellites which contain some of the United States' most sophisticated communications equipment."

I would like to ask any member of the panel, or several, to respond.

Secretary HOLUM. Essentially, Congressman Lantos, this relates to a proposal to modify a Comsat export that had previously been authorized. I think the license was actually just granted fairly recently, although the waiver was provided in 1996.

The elements of that transfer have changed. The capability of the satellite has changed. As a consequence of that, the exporter has to come in and acquire a new license. We will look at that license again, de novo, examining the impact on national security and foreign policy interests of any changes, and respond accordingly.

This is being reviewed now at the working level. I am aware of it but it hasn't reached me yet. It is going through the normal process.

Mr. LANTOS. Do you gentlemen have any comment about the 417 to 4 vote in the House of Representatives? It is a pretty basic policy change, overwhelming, bipartisan, and we would like to get a substantive response as to what your reading is.

Secretary HOLUM. Do you mean on ending satellite launches, commercial satellite launches?

Mr. LANTOS. Yes.

Secretary HOLUM. As someone who has principal responsibility for U.S. nonproliferation efforts, I would hope that wouldn't be approved, and the reason is that we need both the negative and positive——

Mr. LANTOS. What wouldn't be approved, Mr. Holum?

Secretary HOLUM. The legislation.

Mr. LANTOS. It has been approved, 417 to 4. That is a vote which we took the other day.

Secretary HOLUM. I understand. I hope it doesn't become law that we would no longer be allowed to use this inducement as part of our nonproliferation efforts. We need the full range of tools. The problems are serious. We need to able to provide positive as well as negative inducements if we are going to succeed in defeating proliferation of weapons of mass destruction.

Mr. LANTOS. Let me just say, and I don't want to use more time, Mr. Chairman, this was a 2 by 4, and I think it is important not to minimize its importance. The body will need to be persuaded, presumably by having the questions the Pentagon and the State Department themselves are raising, to our satisfaction.

And I think it would be a more useful policy in general to admit past mistakes under both administrations, than to try to justify every single procedure, policy, approval, gesture, move that was taken since Mr. Bush authorized this whole thing, because you don't get a 417 to 4 vote on a high-profile item without large numbers of Members having very serious questions about ongoing policy.

And I would be grateful if Department of Defense and Commerce would also comment on my question, Secretary Lodal.

Secretary LODAL. Thank you, Congressman Lantos. I would subscribe to what Mr. Holum said. I think that there are important national security implications of these types of exports in general, dual-use technologies, and we have established over a number of years our export control system which, for all of its complexity and difficulties and challenges, is the best in the world. I think it serves us quite well and does give us the ability to make these distinctions and allow commercial interactions to develop.

I agree with your statement at the beginning that at its core this is a fundamental policy issue. If we have a policy toward China which encompasses commercial relations across the board, then it is going to encompass dual-use kinds of items, communications equipment of various kinds, phone systems and things like that, airplanes, those kinds of things which we sell to China.

And in some sense these commercial things are used by all militaries. They are used by our military. Something like 90 percent of our communications take place on commercial communication systems. We buy commercial aircraft to do things with. They are not our main military aircraft, of course, but we do make use of these commercial items. The Chinese military make use of commercial items, some of which they import, some of which they build themselves.

And so it is a fundamental policy question about whether we want to try to isolate a country and not be involved with——

Mr. LANTOS. Now don't use that word "isolate" again, because you are beating a dead horse.

Secretary LODAL. Fair enough. Fair enough.

Mr. LANTOS. And those of us who have criticism of the modalities of engagement——

Secretary LODAL. I will take back the word.

Mr. BATEMAN. Mr. Chairman.

Secretary LODAL. I will take back the word.

Mr. BATEMAN. Mr. Chairman, the 5-minute rule I thought was applicable today.

Mr. LANTOS. Those of us who are opposed to the particular engagement favor many other kinds of engagement. None of us advocate isolation.

Secretary LODAL. I take your criticism, and I should have used a more precise word because I was referring only to this type of commercial activity. But I am trying to agree with you that this is a fundamental policy issue, whether you want to try to cutoff this kind of activity or not cut it off.

Mr. LANTOS. Thank you, Mr. Chairman.

Chairman SPENCE. Mr. Bateman.

Mr. BATEMAN. Thank you, Mr. Chairman.

There is much that the gentleman from California, Mr. Lantos, has said with which I agree, including the bottom line, that this is essentially a matter of what should the policy of the U.S. Government be in the interest of the security and vital interests of the United States and its people.

Quite frankly, I have heard more about, “Well, this was the policy of the Bush Administration. We’re just following the policy of the Bush Administration.” Since that has been injected into this, I don’t want a conclusion, I want the specifics of the bilateral agreements, the regulations, the standards, everything produced side-by-side to this committee, in order that we can determine whether those witnesses who appeared yesterday are correct or whether what we are hearing today is correct, that there is really no difference between the one administration’s policy and the other. If they are the same, they are both wrong, in my judgment.

Quite frankly, I am glad you are here today, appreciate your being here today, but what I have heard today raises more questions in my mind than before you came. Let me indicate to you that to say it is in our interest to utilize, virtually to promote a Chinese space launch capability, as opposed to developing that capability on economic and competitive terms within the United States, because there is $1.7 billion short range of business to be done with China and 8,000 American jobs, how many jobs and how many billions of revenue to American companies, were we promoting and encouraging that industry domestically?

Yet I am not aware that this administration has done a thing to encourage that, and I find that very, very unacceptable as a matter of policy, or justifying a line of conduct vis-a-vis China.

We have no disagreement that nonproliferation is a sound policy, but my gosh, how ineffective we have been in the implementation of that policy. And isn’t it curious that we have licensed technology to go to the Chinese which enhances their capability to develop missile systems improved over what they already had, which were not very accurate, and thereby perhaps to encourage the very thing that we say we least want, and that is proliferation as exercised by the Indians and the Pakistanis.

It is fine to say, oh, nonproliferation, the things we do vis-a-vis China, we are going to make some progress. Nine times or 11 times you have sanctioned the Chinese for violations of their policy. You have had from 1994, by your statement, until now to resolve a difference of interpretation that is incredibly important, and depending upon the interpretation, rips the heart out of any meaningful Chinese commitment to nonproliferation. It doesn’t make any sense.

Secretary Lodal, you flabbergast me when you speak in terms of someone who is a part of the decisionmaking processes as to granting or not granting licenses, when you say in response to a question, “I don’t know the success or failure rate of the Chinese space launch capability.” The success or failure rate of the Chinese space launching capability is a vital matter of interest and not something that is very difficult to ascertain. I don’t understand how you advise without knowing things like that.

Now getting into The New York Times edition of today, it is said—and I don’t know whether it is accurate, I hope you all will know—“The staff memorandum that the President relied on to approve the deal made no mention of the Loral-Hughes help for China’s rocket program. Three weeks before the memorandum to the President, the State Department had alleged in a letter to Hughes that there had been a violation of the arms export control law during the rocket accident review."

My question is this: If the staff memorandum that went to the President made no reference to it, did he ever get informed that there was a very interesting question of whether there was indictable breach of our national security constraints? Did any of you tell him? Did he make the decision without knowledge?

And if he made it with knowledge of an ongoing investigation of a potentially indictable offense and against, at least what is alleged, the advice of the Department of Justice, is there any precedent for any President ever granting a license under such circumstances?

Secretary HOLUM. I think that was question to me on the question of whether the fact of an investigation was included in the decisionmaking, and it was. The position of the Department of State made reference to that ongoing investigation when the recommendation for the waiver last December went forward from the Department of State.

But, as I said earlier, our policy, continuous, going back in history—and in fact we have looked for precedents of this the other way—is that we do not begin denying licenses on the basis of an investigation. We deny licenses if there is an indictment. A lot of investigations don't produce an indictment. If there is an indictment and it is a case where national security could be endangered, we have the option of suspending a license.

Mr. BATEMAN. Was the President informed of all of this in meaningful detail?

Secretary BARKER. Mr. Chairman, maybe I can help out in a little bit of detail on this, and perhaps it is important to separate two different items here.

In 1998, the Loral waiver, the President was informed in quite a bit of detail. You have of course seen the memorandum which has gone from the NSC staff to the President, which talked about it in a great amount of detail.

The 1996 one to which you first referred at the opening part of your statement, we will check back through our files, but to the best of my knowledge at present, we did not, and here is why:

Right at the beginning of that we, as you noted, State Department sent a letter to Hughes and we had some concerns. We send a lot of these letters to a lot of companies when we initially have concerns, and we ask for more information. All of the Hughes information at that point had not come back in.

When we received the quite considerable volume of information, at that point we thought it was necessary to start a much more formal review process. Remember, initially we hear things, we hear concerns, and the first thing we do is try to clarify them and get the information.

In 1996 when that first memorandum went to the White House, again, we will check to see if there was any referral on that, but that was in a quite preliminary part of the investigation. It was actually prior to us making the formal recommendation to the Justice Department that we thought that there was sufficient basis in order for there to be a violation of law, for the Justice Department to take up this particular issue.

Mr. BATEMAN. Mr. Chairman, if I could have unanimous consent to just make one brief comment, in 1996 the concern first is raised. It is now June 17th, 18th, of 1998, the matter still under investigation in the Department of Justice. It stretches my credibility why we don't have some resolution one way or another on this. It raises a question as to maybe this administration has overtaxed the capabilities of the Department of Justice to investigate.

Chairman SPENCE. Mr. Berman.

Mr. BERMAN. I assume this investigation hasn't cost $40 million.

Yesterday's hearing, we had four witnesses. One, from the Congressional Research Service, I think, gave an analysis of the law and administrative actions over the last 10 years. One witness was a strong supporter of the decision to transfer satellites to the Chinese. There were two critics, for very different reasons, and I would like—some of this has been touched on, particularly by—well, by other questioning, but I want to come back to three points that seem in criticism and hear your answers to them.

The point by Gary Milhollin, forget the question for a second of what was transferred to the Chinese that might help their missile program, the Chinese want to launch U.S. satellites. The Chinese, notwithstanding their assurances and their reassurances to Scowcroft and Eagelberger, to Christopher and others in 1993, their pledges regarding adherence to the MTCR, both by virtue of what we know they did and their, as you pointed out earlier, their strained and narrow interpretation of what their earlier commitments had been respecting items on the annex, particularly in the area of missiles, were proliferators.

We, Gary Milhollin says, should not be giving them something they want like this until they stop doing something that we want them to stop doing. So I am curious about your answer to this question. And I see a copy of a memo which indicates this was on at least John Holum's mind, as well.

And by the way, if Ben Gilman were here, I would say, "Ben, when you see what ACTA sent as part of this interagency process, one, the interagency process is sort of working, in the sense that issues are being raised." And by the way, the same holds true with the article in The New York Times today. You are seeing different agencies coming to different thoughts about a particular issue. That is good.

But my only question to Chairman Gilman would be, why are you trying to push the consolidation? Why are you in league with the State Department to push a consolidation which eliminates ACTA as an independent agency, when you are getting the kind of feedback that you—you are champing because it is an independent agency. But that gets off the point. So I would like to know the answer to that.

The second question is the Sokolski, two issues. One, almost inevitably in the transfer of rocket technology—I mean satellites—you are going to be transferring, not illegally, forget the Loral issue, but you are going to be transferring "know why," he says, as opposed to "know how"—"know why" about capabilities which are pretty much transferrable to the missile range, which raise national security issues, and you have got to be very careful when you do this.

He talks about the MIRVing capabilities that can come from the dual satellites, an effort to speak to that, although since I don't understand the—that is beyond my ability to understand, although what I am not sure is whether it is also beyond the ability of everybody in this room to understand. There are just a very few people around the world who understand this, why they are not related technologies, dual satellite launches versus MIRV capabilities.

And then the final issue is, in the transfer here from State to Commerce there has been an erosion, particularly in the area of monitoring, and he says specifically in the area of monitoring prelaunch conversations between the Chinese company, launch companies, and the American satellite manufacturers; that in the old days before that transfer, that was mandatory, now it is optional at the request of the American manufacturer, not at the request of the governmental agency.

That is it.

Secretary HOLUM. I will answer the first two parts and ask Jan Lodal to comment on the third, because he is more directly knowledgeable, and Bill Reinsch as well.

The question of how we use the various kinds of incentives and approaches at our disposal inevitably has to be a matter of judgment in terms of influencing Chinese behavior on missile controls. As I said earlier, we have used sanctions in 1993, leading to the 1994 agreement.

I think sanctions were in the air, if you will, and had some effect when the Chinese agreed to stop selling cruise missiles to Iran in 1997. Sanctions were certainly a possibility in 1996 when China agreed not to assist unsafeguarded nuclear facilities in the context of the ring magnet case, so those are very much a part of our set of inducements.

In fact, sanctions are still a possibility, and the Chinese are aware of this, in connection with our continuing scrutiny of the question of M–11 missiles that may have been transferred to Pakistan. We haven't made a judgment yet, but that is an ongoing review and it is continuous. So sanctions are part of the mix.

I have dealt with this for a number of years. A number of people in the department that I work with who are expert in this agree that in the case of trying to resolve the missile annex question, the scope of control under MTCR, that sanctions are less effective than positive inducements, but it is a mix. We have tried to resolve this through connections with our missile—in the future, with expansion in our missile launch capabilities, but it is always a matter of judgment. That is my judgment. Somebody else might have a different one.

In terms of knowing the "know why" question, the technical capability, sort of the experience gained by virtue of the fact that the launches take place, I take it, not that there is any specific information or technical data or technology transferred, but simply they gain something by the fact that there is a launch——

Mr. BERMAN. Particularly in the area of the dual satellite launches.

Secretary HOLUM. Right. The first thing that I can comment on is that the launches would likely take place whether or not we are there. We are not denying them anything if we don't provide satellites, if we don't hire launch services, because someone else will. There is a shortage of capability in the world. So they would gain that "know why" through their own resources, whether or not we were putting the bird on top of the launcher.

Mr. BERMAN. Also, "know why" means know why something went wrong. The interaction with American engineers and specialists in the analysis of what went wrong gives them a "know why."

Secretary HOLUM. They don't have access, under the license limitations, they don't have access to that information. That is precluded. If something went wrong and information is shared, you get a Justice Department investigation, but the licenses and monitoring preclude that.

Jan, you might want to comment on the monitoring process and the effects it has on that, on those possible transfers.

Mr. BERMAN. I mean, just to pin it down, Sokolski said yesterday that prior to the shift from State to Commerce there was a requirement for the monitoring of pre-launch conversations between contractor and Chinese launch company, or whomever in China you are talking to about this; afterwards, that was only required at the option of the American company provider of the satellites, it was not automatic.

Secretary LODAL. I am going to let Mr. Tarbell fill in any details, but my understanding is that that is inaccurate, that it is required today. Also, all of this, it is not optional.

But let me just reiterate perhaps the point about the monitoring that I made in my statement. There is a historical point, not a point about what the current policy is or has been since 1996, where full monitoring has been required on all of these launches.

After the Bush decision, which was accepted and implemented by the Clinton Administration, and there was for the first time a distinction between commercial and noncommercial satellites, there was a period of time until 1996 where in principle the monitoring was not always required. If it was purely commercial and no State Department licenses were required, the monitoring was not always required, and therefore there were three launches which took place. Our analysis was that the monitoring was——

Mr. BERMAN. This is what you said in response to Mr. Bereuter earlier.

Secretary LODAL. Right, and our analysis was—and it is in my statement also—our analysis was that the monitoring was relatively inexpensive, and that even in these pure commercial cases—which, by the way, in those days it was very strictly defined.

Not only did the satellite—did there have to be no transfer of the military technologies involved, but the military technologies could not even be incorporated inside the box, inside the satellite at all, but the definition that applied at that time for what constituted a commercial satellite. I mean, it clearly had to be quite a pure thing.

So if that is all you needed a license for, and the only kind of additional information you needed to transfer was form, fit and function—you know, what is its size and shape, how do you plug this thing onto the launch vehicle, in oversimplified terms—but if that is all there was, the monitoring was not required at that point in time.

Our analysis and recommendation was that we should nonetheless require the monitoring in all cases, to eliminate any ambiguity. It was not terribly expensive. It was not that difficult.

That was accepted by all agencies when the Clinton Administration made its decision to adjust the procedures and policies in 1996. So that could have been—I don't know what was being referred to yesterday, but that could have been what was referred to in the change after the Bush decision, because it was true for a period of time.

Mr. BERMAN. So you are saying——

Chairman GILMAN. The gentleman's time has expired.

Mr. Weldon.

Mr. WELDON. Thank you, Mr. Chairman.

First of all, let me make some comments about some of the discussion by my colleagues. Mr. Lantos mentioned that this should be a discussion of policy versus technical issues. I would just remind our friends on the International Affairs Committee that National Security puts through the largest bill of Federal spending each year in the defense authorization bill, and much of that is technical. My subcommittee, which oversees $37 billion a year, has to focus on technical issues.

So by that very nature, this is going to be a technical discussion. Whether they like it or not, we are going to ask technical questions because that is the basis of what has occurred here, and whether or not there has been an increase in the ability of China to cause problems for us that we then have to respond to and spend more money on.

Second, let me say to the panel here, I have never in my 12 years in Congress seen less confidence in the Congress on our ability to enforce arms control agreements. At yesterday's hearing and 3 weeks ago in the Congressional Record I put an analysis done by Congressional Research Service of 38 specific potential violations of arms control agreements since the early 1990's. In the case of China there were 21. Only one had sanctions that were carried through. In the case of Russia there were two where the sanctions were kept intact.

Now I have offered as recently as 3 years ago to the administration to work with people like Henry Sokolski, and his idea is to give more flexibility on enforcing arms control agreements. The administration said we don't need it. This Congress has lost confidence in your ability to monitor proliferation.

Now you say we don't always have evidence. I have held up repeatedly accelerometers and gyroscopes that were taken from Russian long-range missiles, that we have in our hands, that we captured on at least three occasions going to Iraq. Now the administration never imposed sanctions. In fact, we were told, "Well, Russia is doing a criminal investigation. Don't worry, they will take care of it." You all know that criminal investigation was ended last fall with no criminal actions taken against anyone inside of Russia.

The Congress has lost confidence. We just don't believe that you all want to do anything that might violate some predetermined policy position between the White House and the President of Russia or the White House and China.

And that is not just evidenced by the votes on China on our defense bill. Look back in November when we had the Iran missile sanctions bill, and the most recent vote on that, when the administration lobbied hard through their Minority Leader Gephardt to oppose that bill. Only 22 members of the Democrat Party opposed that conference report on the floor last week.

We have lost confidence in your ability to monitor proliferation. That is at the heart of this whole issue, along with the Loral issue.

Now let me ask some specific questions. Mr. Tarbell, one of the nine militarily sensitive technologies that could be included in commercial communication satellites is encryption hardware. Was there an encryption device on the 1995 failure, satellite launch failure, and the 1996 failure?

Mr. TARBELL. That is information that I would like to address with you in a closed hearing.

Mr. WELDON. Mr. Chairman, I would like to make this request that we go classified. I do not want the witness to say they can't answer that question. That is at the heart of this issue. Was there an encrypted device on either of those two satellites?

Secretary REINSCH. Mr. Weldon, I can respond for 1995. There was no device in the 1995 one that blew up.

Mr. WELDON. Was there one in the 1996 incident?

Mr. TARBELL. I don't have that information.

Secretary REINSCH. That wasn't our license. I couldn't say.

Mr. WELDON. Can you say definitely there was no encrypted device on that satellite?

Mr. TARBELL. I don't have that information available to me right now.

Mr. WELDON. Do you know the answer?

Mr. TARBELL. I don't know the answer right now.

Mr. WELDON. You don't know the answer? You don't know whether there was an encrypted device on that satellite?

Mr. TARBELL. I would have to—excuse me just for a moment and let refer to some material.

Mr. WELDON. OK, on your time you can refer to your material. Let me go on to some other questions.

And if there was an encrypted device—let me ask this question, Mr. Tarbell—in the investigation of the failure, is it true, as the press has reported, that our team was held back for 5 hours before they were allowed to go out and look at what evidence was left from the failure?

Mr. TARBELL. It is true that our team was held back for a substantial period of time because there was a substantial safety risk, and the Chinese were worried about trying to clean up the crash site because it crashed into a village, and they were worried about taking care of their problem before they let any of the——

Mr. WELDON. Let me ask the question: So we were held back for 5 hours?

Mr. TARBELL. I can't say that it was 5 hours. It was a substantial period of time, and it was along those lines.

Mr. WELDON. Where the Chinese had access to this site and we did not?

Mr. TARBELL. That is true, sir.

Mr. WELDON. Now, I used to be a volunteer fire chief, so I know something about life safety, and I chair that here, and I understand the life safety. But how can you not have our people at a site where you have the encrypted technology that is at the heart of our ability to control our satellites, and say that, well, for safety purposes we are not going to go to this site and we are not going to press this issue?

Mr. TARBELL. Congressman, we are on—in these circumstances we are on Chinese soil, and we are at the whim of the Chinese in this particular——

Mr. WELDON. OK. Let me ask this question: If that encrypted capability were in fact taken by the Chinese, without our support, in your opinion would that add a security problem to the U.S., an enhanced security problem?

Mr. TARBELL. This is a matter that NSA is——

Mr. WELDON. In your opinion.

Mr. TARBELL. In my opinion, I am not an expert on encryption and I can't tell you——

Mr. WELDON. Do you know what NSA's opinion is?

Secretary REINSCH. Mr. Weldon, I would like to respond to that if I could.

Mr. WELDON. Yes.

Secretary REINSCH. I have had some experience with encryption, unfortunately. The answer in my judgment is no, it wouldn't have had any effect on our national security. The encryption that you are talking about that—leaving aside the question of what was or was not in the 1996 one, because that wasn't one of ours so I don't know anything about it—but the encryption that is in these satellites is encryption that is designed to encrypt the signal that tells the satellite what to do, which way to point, what orbit to go into, whatever.

That encryption is encryption—and the reason why it is there is obvious. You don't want somebody else hacking into the signal and moving your satellite around, or you don't want somebody else's signal inadvertently bumping into yours. That encryption, as I understand it, is unique to each individual satellite.

Mr. WELDON. But does it have common features important to the NSA?

Secretary REINSCH. I don't think that there is anything that could be done with——

Mr. WELDON. So you are telling me on the record that the NSA would not think that if they got that encrypted device, that there would be a security risk enhancement?

Secretary REINSCH. Well——

Mr. WELDON. Are you telling me that on the record?

Secretary REINSCH. I am telling you what I think, and I also think——

Mr. WELDON. I am telling you officially what you think.

Secretary REINSCH. Well, if you ask them, I think they will tell you the same thing, but——

Mr. WELDON. Well, we will ask them. OK. We will ask them. But the bottom line is, for the record, you will tell us whether or not there was an encrypted device in that satellite?

Secretary REINSCH. Yes.

Mr. WELDON. That will be today, I assume, you will tell us that.

Secretary REINSCH. I'm sorry, sir?

Mr. WELDON. That will be today, you will tell us that?

Mr. TARBELL. We will get you an answer.

Chairman GILMAN. Mr. Tarbell, will you provide that response for the record?

Mr. TARBELL. Yes, sir.

Mr. WELDON. One final point, Mr. Chairman, and I know my time is running over as other members have done. I would like to pursue one additional question, and that regards, I understand, Mr. Lodal, even though it is not in your testimony, that you commented on the issue of MIRV capability. It is a very critical issue for us. And I don't know what your background is—I do know what it is. I do have your resume here. It is computers.

You have said on the record, I understand from staff, that there is no way that the capability provided by the launch of two satellites, namely the Motorola and Lockheed satellites, and the small communications satellites known as Iridium, that there is multiple satellite dispenser provided there which could give the Chinese the ability to understand the problem of proper mounting, release of the satellites, the coupling load analysis and attitude control. You are saying that there is no possible way that would benefit their ability to get MIRV technology.

And the reason why I ask that question, you know full well and Mr. Holum knows the Chinese have been trying to acquire MIRV technology for most of this decade. They tried to buy the Russian technology, the Ukrainian technology. We caught them trying to buy the Ukrainian technology through the SS–18. So it is something that we know they have wanted.

So let me get your point on the record, Mr. Lodal. Are you saying that the ability to have these multiple satellite dispensers in no way could benefit the MIRV capability of the Chinese nuclear long-range ICBMs?

Secretary LODAL. What I said, Congressman, was that there is quite a difference between the tasks that are required in the two cases, the multiple satellite launch and MIRVs, and that while the tasks are similar, the standards and requirements, both the technical standards and the procedural standards and requirements for satellite release are much lower than those required for MIRVs, that it is quite different.

I used a very rough analogy——

Mr. WELDON. The analogy you made, from what I understand, was a very dramatic analogy. I think you have overstated the case.

Secretary LODAL. Excuse me?

Mr. WELDON. My understanding was, it was a very dramatic analysis.

Secretary LODAL. Analogy, my analogy? Well, I was trying to say that, you know, it is a question of in one case you have to get the satellite in a so-called transition orbit, and it—I can't tell you quantitatively what the allowable variation is there, but it is a lot. I mean, it can be off by, I don't know, what is it, feet, miles? A lot.

In a MIRV situation you have got to get the warhead right on that track, or else by the time it flies another several thousand miles, a little bit of error—It is just like pointing, you know, a gun——

Mr. WELDON. It might hit Chicago instead of New York. It still does the damage.

Secretary LODAL. Well, not necessarily Chicago as opposed to New York.

Mr. WELDON. It still does the damage.

Secretary LODAL. It can certainly be——

Chairman GILMAN. The gentleman's time has expired.

Secretary LODAL [continuing]. It can certainly be quite a bit off, even if it is much, much more——

Mr. WELDON. Are you aware there are scientists who disagree with you?

Chairman GILMAN. The gentleman's time has expired, Mr. Weldon.

Mr. WELDON. I understand that, Mr. Chairman. I would just say that two of our Democrat colleagues had significantly longer periods of time. Mr. Lantos went on for half an hour, during which time there was a vote, and so did Mr. Berman.

Thank you.

Chairman GILMAN. Thank you, Mr. Weldon.

Mr. Snyder.

Mr. SNYDER. Thank you, Mr. Chairman.

I wanted to ask Mr. Hamilton's question about the cumulative effect of having the rockets. You know, gentlemen, I appreciate you being here. I think there are serious questions that Congress has, and also the American people, and unfortunately in the last few weeks we have had some serious rhetoric going on, too, in this town.

But behind that all are some questions that the American people have, and I hope that you can see past sometimes our statements that we may make on the House floor and simply can come up with some answers, because I think this China relationship with the United States is going to shape the first half of the 21st century. I would hate to see us do something incorrectly here in the next few months.

The specific chart that Mr. Sokolski showed yesterday, showed that in 1993 and 1994 there were six successful launches of satellites. Out of the next six, there were three failures, and then they have had a record of 10 successes in a row. In Mr. Hamilton's referring to that chart, the point was made that through the trial and error of working with the U.S. contractors, the satellite folks, that somehow they have gotten information that has now improved their success rate.

I have two questions on that. First, what is your opinion on that? And also, you all have talked specifically about the processes you go through for approving individual waiver processes. Do you have a process for sitting down and looking at the cumulative effect? Have you all had a serious process at some point in the last year or two or three and said, OK, we look at these individually; let's look at the whole program since 1988. What has the cumulative effect been?

[See chart C in the appendix on page 365.]

Secretary REINSCH. Let me comment on that, if I may, Mr. Snyder. With respect to Mr. Sokolski's chart, let me—I hate to get into the weeds, but I think we have to, to deal with this. My understanding is that the chart that he provided included only five of the ten different rocket models that the Chinese have developed over the past 28 years.

Three of the five versions on that chart have had 100 percent success rates in launching from their maiden voyages, which were in 1974, 1984, and 1992. Which means that of the 11 launches that I think he mentions, seven of them represent rockets that have never failed from the first fight onward.

With respect to the other two, one of them started—other two models—one of them started launching in 1984. It has had 3 failures out of 12 launches. Since its last failure, there has been only one further launch of it, which I would suggest is a little bit too soon to assume that that particular one has been proven perfect or whatever the right word is.

With respect to the other one, I think in total it has had only four launches, one of which was a failure, the first one of which was a failure. And I would suggest there, also, that that is not a sufficient record to really demonstrate what is going on.

What Mr. Sokolski is arguing I suppose is an elaborate version of "practice makes perfect," and the more launches they have, you know, the better they get.

Mr. SNYDER. And that the practice sessions may have occurred with U.S.——

Secretary REINSCH. Well, I think first of all if you get into the data, you can see that with respect to some of the models, they have been perfect from the beginning and it was well before we started, under any President, launching these things.

The other point I would make about that is the same point that Under Secretary Holum made a little while ago, which is that there are projected to be 1,200 launches in the next 10 years, and right now there are four launch providers. Launch space in the near future or for that period of time is going to be at a premium. There is no question in my mind that if we are not going to use Chinese launchers in the future, they are going to be taken up very quickly by others.

Mr. SNYDER. And how about an ongoing look at this cumulative effect?

Secretary HOLUM. I think the best way to consider that is that when we went through the transfers of jurisdiction in 1992, following up on, basically implementing the orders that the Bush Administration had initiated, and then again in 1996, you inevitably have as part of that an evaluation of the overall program, how it is working. So I think that has been a continuous part of our deliberations.

In addition, obviously we are looking at it very closely now as part of this process, and the judgment that we have reached is along the lines of what Under Secretary Reinsch is proposing: This has not, as my statement states, and Jan Lodal's as well, this has not provided significant assistance to their missile or space launch vehicle program.

Secretary REINSCH. If I could add one word to that, one additional fact I point out, which relates also to something that Mr. Weldon said, the Chinese first launched a rocket that had multiple satellites on it in September 1981. The first rocket they launched, period, was in 1964. There is a very long history to the Chinese rocket program and satellite program. There is no question that, you know, there have been American launches, but there are an awful lot of other things going on as well.

Mr. SNYDER. I wanted to ask something on the issue that you mentioned, the fact that if we cut off commercial satellite launchings by the Chinese, they will find other satellites to do their practice with. But in order for us to cutoff sending satellites to the Chinese, we would have to develop further launch facilities, abilities, in the United States, or someone else in the world would have to do that.

And I would assume that, given some of the failure rates of the Chinese, that in fact some of these other folks trying to launch satellites may in fact come to whatever new facilities are out there for launching satellites. I mean, it is a little bit simplistic, isn't it, to say if they don't do it, if we don't send them ours, they are going to find others to do, at least in the long run? Because we are going to find someone else, maybe, perhaps, ourselves, to do that.

Secretary REINSCH. If I could comment on that, I think the current market situation is that our share of the satellite market globally is about two-thirds, and the U.S. share of the launches is about 40 percent, so there is a gap there that means that we are searching for other sites. The choices right now are the French and the Russians, in addition to the Chinese.

There are other launch facilities under development, some of them coming on line sooner rather than later. How successful—one of them is an ocean platform that is a consortium of both American and other country entities. How successful they will be and how much market share they will be able to capture in the short term is uncertain.

Chairman GILMAN. The gentleman's time has expired. Thank you.

Mr. Buyer.

Mr. BUYER. Thank you, Mr. Chairman. I have got one quick comment off of, Mr. Reinsch, your opening statement. I went to check a little history on this.

Seems like anytime you get into trouble or you get allegations against you, everybody wants to reach out and find somebody, so——

Secretary REINSCH. You are going to make allegations?

Mr. BUYER [continuing]. Reaching out, well, the Bush and Reagan Administrations did it, and that is all we are doing. Oh, by the way, Congress, you even also did this. You said it was "worth noting" that in 1990 and 1992 both houses of Congress passed legislation to do this, and in 1994 even committees in the House introduced this measure, as if that is okay.

Well, then I went and did my research on it. I don't know who wrote that for you, but they should have done their research before they wrote it for you, because there were Members here—I wasn't even in Congress at the time. In 1990 there was a generic move for dual-use technologies, pocket vetoed by the President. In 1992 the commodity jurisdiction language was struck by the Kyl amendment on the floor. It was put back in in conference, and Mr. Hunter of this committee blocked it, and it never made it to the Floor.

So I wanted you to know that even in 1994, when it came out of the International Relations Committee, the National Security Committee took sequential referral and jerked that section out. So I want you to know that whoever wrote that for you, Mr. Reinsch, maybe was a little loose there with it.

So there are some of us here in Congress that are concerned about these transfers of technology, and I—one other thing I want to pick up off of is where Dr. Snyder was making his very good comments here. I am not an expert in this area at all. I will share that with the gentleman.

So I am at square one in trying to learn this, so I look at this very basic. I just look at this very simply and say, okay, I understand now the difference between satellites and the launchers.

And I will even use your definitions and call it a dual-use technology because I concur with you. If it is the same launcher that will launch a satellite, it is the same launcher that would launch a nuclear warhead, I suppose it is dual-use technology. So I will use your definition of dual-use.

But I don't understand, if in fact Secretary Holum said that China is part of the proliferation problem in the world, that they are part of the problem in the world in proliferation, the transfer of technology to Pakistan, so in this pursuit of satisfying commercial interests and bragging about $23.1 billion in space dollars last year, and it is jobs, what are we doing? I mean, I just look at this very simply and say, what are we doing?

And then some people will say, well, it is just a commercial satellite, a commercial satellite. Gosh. Satellites are also used in command and control. Command and control. And as we monitor China as an emerging nation and their desires for power projection capabilities beyond their own border, and how it upsets the balance of power in the Pacific Rim, we ought to ask these very plain and simple questions.

So I am trying to take a step back from this. I don't care if you reach out and say, well, we are only following policies of Bush or Reagan, whatever.

I think what Dr. Snyder is trying to get to, and some others have brought up, is a very good question on why, Commerce Department, why aren't we doing more to say, develop, do this launching here, so we can have the greater controls here? We wouldn't have Mr. Tarbell saying, "Well, we couldn't run in there because we were on somebody else's territory."

What by our commercial sector, just share with me, what are we doing to give incentives to do greater launches here, to control our own technology instead of sharing it with someone who is perhaps, in your own words, part of the problem in proliferation of these technologies to bad actors?

Secretary REINSCH. Let me make a general comment first, if I may, Congressman. I think that you have hit on one of the central difficulties of dealing with all these issues, namely, that making a decision involves balance. There is no decision that any of us make in this area where there are no dissenters, and your comments about the Congress are well taken.

In fact, I wrote those provisions, because I was here at the time, and am familiar with the history of them. And I am aware of the then-Armed Services Committee's and now National Security Committee's point of view. There was dissent then, just as there is now.

There is, every time we deal with one of these questions, a difference, you know, a weighing and balancing. The $23.1 billion that I cited is not bragging. It is simply stating a reality: These things matter. They matter economically. They matter to a lot of people. And that is one of the issues that we believe needs to be taken into account.

Now you did not make the point that we have been trying to make all morning, which is that we believe these things have gone forward and these exports have gone forward without any risk to U.S. national security, with the possible exception of the single case that is under investigation by the Justice Department. We would not have allowed these to go forward, had we felt that there was a risk to national security. That is paramount in our consideration.

Mr. BUYER. Let me interrupt. Let me ask you this: Am I to accept the assumption that it is just a mere coincidence that only through the passage of time, that they continue to have a greater success rate in their launches? Is that the assumption I am supposed to pass on to my constituents?

Secretary REINSCH. Their current success rate, their overall historically, is about 84 percent, which is comparable to their competitors. It is a little lower than their competitors, but comparable.

The point that I was trying to make with respect to Congressman Snyder's point is that there are a lot of factors that go into that. Does practice make perfect? Sure. The more they practice, the more they do, the more their rate is likely to go up. There is a lot of factors besides U.S. participation that relate to that.

Mr. BUYER. Let me throw this out: If in fact we had, after the '96 crash of the Loral satellite lift-off, with the post-crash investigation conducted by the——

Chairman GILMAN. The gentleman's time has expired, and I understand we are coming up on a vote, and I just want to be fair to all of the members.

Mr. Taylor.

Mr. TAYLOR. Thank you, Mr. Chairman.

To follow along what I thought was the line of thought of my colleagues, I have heard on several occasions several different agencies of the government say, "We're only doing what the Reagan and Bush Administration did." I also heard very clearly three of the four panelists yesterday say, very clearly, that what has happened has increased the capability of the People's Republic of China to launch satellites, to launch missiles against this country.

My question is, with all the people that we are paying to look into these things, why hasn't anyone stepped forward to say, "This is a bad policy. We don't care what our predecessors did. We got elected to make things better, and doing what they did does not make things better."

Secretary HOLUM. Let me suggest that we have certainly put this policy in the context of its historical precedents. However, I think everyone here at the table is defending the policy. We are saying that it is the right policy for national security and that it is the right policy for the United States.

From my perspective, as someone who deals with nonproliferation for a living and with a great deal of responsibility in that area, I am making the argument that along with all the tools in our arsenal for dealing with proliferation risks and the problem that China poses in this context, this is one we need. If we were to deny the ability to launch commercial satellites on Chinese launchers, we would gain nothing in terms of national security because nothing is being given away, in our view. We disagree——

Mr. TAYLOR. Mr. Holum, I disagree. I think the statistics—I mean, the people who keep pointing to the statistics, I think they are right. When you go from a high level of failure to almost no failure in the Chinese program, in the same amount of time that this information was given to them by Loral—or alleged to have been given to them by Loral, since there is an ongoing investigation—yes, something has changed.

Secretary HOLUM. Congressman——

Mr. TAYLOR. Now I do see person after person talking about the $23 billion in industry revenues. But, doggone it, there is a right way and a wrong way to make money, and the wrong way is making money by making your enemy stronger.

Secretary HOLUM. Congressman, I think there is a logical fallacy in the Sokolski chart that is glaring, and that is the assumption that because events coincide, one is caused by the other. What he assumes is that the only thing that is happening in the Chinese effort to make their space launch vehicles and their missiles more reliable is this commercial launch activity.

This commercial launch activity, as we have been describing all morning, has detailed provisions to prevent the transfer of any sensitive technologies. Second——

Mr. TAYLOR. Did it prevent the 200 pages of information from being faxed by Loral to the Chinese?

Secretary HOLUM. You are dealing with——

Mr. TAYLOR. Question. Yes or no answer.

Secretary HOLUM. You are dealing with a criminal investigation that I can't talk about.

Mr. TAYLOR. Did it prevent that from being transferred?

Secretary HOLUM. If I could finish my——

Mr. TAYLOR. So obviously it isn't working.

Secretary HOLUM. If I could finish my——

Mr. TAYLOR. So you are here to defend something that isn't working.

Secretary HOLUM. If I could finish my answer, the other part of the missing piece of information is that the Chinese are doing thousands of things, totally apart from this effort, to improve their missiles. They have an advanced capability in that area going back to the 1960's that is totally apart from anything that is happening in this area.

So I think it is impossible to—it is a huge stretch to make a cause-and-effect relationship between commercial satellite launches with all these technical safeguards, and the fact that they have managed to improve the reliability of their space launch vehicles.

Mr. TAYLOR. Tell me how the technical safeguards that you are so proud of prevented those 200 pages from being faxed to the Chinese without anyone in the Department of Defense, anyone in the Department of State, anyone in the Department of Commerce either signing off on it, A, or even being aware of it? Because it is one or the other. You have got to tell me you were aware if it and it is OK, or you weren't aware of it.

Secretary HOLUM. I am telling you——

Mr. TAYLOR. And if it is neither, then you are telling me, then you have just told me that what we are doing doesn't work.

Secretary HOLUM. No, what I am telling you is that you can't guarantee that there won't be a violation of law. When there was a question that came to our attention——

Mr. TAYLOR. But you told me earlier what we were doing is working. Those are your words, verbatim. How can you tell me what we are doing is working, when people are allowed to fax information of a highly sensitive nature to a potential foe of the United States of America, when there are supposed to be things in place to keep that from happening?

Secretary HOLUM. Congressman, I can't talk about a criminal investigation, as you know. What I can say is that our approach to deny, when the law is complied with, our denial of technical information that would help China in its space launch vehicles or its missile programs is working.

Chairman GILMAN. The gentleman's time has expired.

Mrs. Fowler.

Mrs. FOWLER. Thank you, Mr. Chairman. I didn't realize I was up next. Let me get my questions here.

Thank you gentlemen for being here, and I thank the Chairman for having this hearing today. As one who has been tracking the transfer of technology with military applications to the People's Republic of China for several years, this transmission of classified data by Loral and the administration's subsequent reaction I think has just been one piece of a very disturbing trend. And today I want to follow up on a couple of my questions that I had yesterday for outside witnesses.

On the PLA's use of satellite communications, it has been reported that the PLA has obtained dedicated access to transponders on the U.S.-built satellites that have been sold to China. Now the White House subsequently issued a statement which, you know, rather than denying the uses, tried to trivialize it by stating, and I quote, that "Militaries around the world use commercial satellites," and another quote, "We do not believe it provides any particular military benefit."

Do any of you, whoever wants to answer this, and I have three questions in relationship, do you agree with the statement made by the White House that providing the PLA with the ability to conduct military communications does not provide them with a military benefit?

Second question, was there any Defense, State, or Commerce Department analysis of the possibility that the PLA would obtain dedicated access to transponders on these satellites for its military communications, prior to the licensing of these satellite deals?

And, third, has the U.S. Government licensed the sale of any ground station technology to any Chinese entity that would support the satellite user's ability to encrypt its transmissions prior to sending them to the satellite or decrypt them once they have been transmitted? And, if so, who would oversee the issuance of such a license, and do we know with certainty that the supposed end users of these ground stations were not front organizations for the PLA?

Whoever would like to take that issue on, I would appreciate some answers.

Secretary HOLUM. First of all in terms of the commercial satellite access by military entities, we are well aware that a number of Chinese Government companies that may use the telecommunications services provided by these satellites could have ties to the military, even if they are not controlled directly by the military.

That is why commercial communications satellites exported to China are structured so that no sensitive technology is authorized to be provided to China which could compromise U.S. national security. We prohibit providing any information or technology that would improve the performance or design of the launch vehicle, as we have explained, and any design information relating to the manufacturer——

Mrs. FOWLER. As Mr. Taylor pointed out, that might not be working.

Secretary HOLUM. Pardon me?

Mrs. FOWLER. As Mr. Taylor just pointed out, that might not be working, because from what I understand, the PLA currently is using our U.S.-made satellites to transcript communications that are encrypted from the ground to the satellite and then back to their other units.

And was there knowledge, my question is, was there knowledge that they would obtain dedicated access to these transponders prior to the licensing of these satellite deals. That is my question, and if so, who oversaw that?

Secretary HOLUM. The communications satellites that we do authorize for sale, commercial communications satellites, we are aware, can be used by the Chinese military for communications purposes in the same way that the U.S. military frequently uses commercial telephone lines for its communications. We do not authorize for sale encryption capabilities, whether ground-based or elsewhere, under the ITARs, and they would be——

Mrs. FOWLER. OK, so your answer to my question, that has the U.S. Government—now listen carefully—licensed the sale of any ground station technology to any Chinese entity that would support the satellite user's ability to encrypt its transmissions prior to sending them to the satellite or decrypting once they have been transmitted, your answer is no, the U.S. Government has not licensed any Chinese entity to that? And I would be interested in any other person's answer to that, too.

Secretary HOLUM. You are getting beyond my level of technical——

55-784 99 - 5

Mrs. FOWLER. Mr. Tarbell, do you have any knowledge on that, probably, since that comes under your area? You are the entity that——

Mr. TARBELL. I mean, that is a very specific question about a very particular area, and I would have to do some research on that.

Mrs. FOWLER. Your agency oversees this area——

Mr. TARBELL. It does, so——

Mrs. FOWLER [continuing]. So I would appreciate if you could send me information for the record in answer to that.

Secretary REINSCH. Mrs. Fowler, if I could comment——

Mrs. FOWLER. Yes.

Secretary REINSCH [continuing]. From the Department of Commerce perspective, I would say, No. 1, we would have to go back and review individual licenses. Second of all, I believe the answer to the question is one that we would want to address in closed session.

Mrs. FOWLER. OK. Well, I will pursue that in closed session, but I would like——

Secretary LODAL. Mrs. Fowler, just to clarify, there are some types of encryption that are, I believe—Bill Reinsch can clarify this if I'm wrong—now exported under general licenses and so forth. So it is possible to get commercial encryption that is not controlled anymore, so it is possible the Chinese military could use some of that. Anybody can use it.

There is generally available encryption now, available in the commercial marketplace, and so that is one possibility, that that kind of thing could happen. And you can do that independent of what kind of communication system you are using. In other words, you put it on the end of the pipe. You put it on your phone call. The phone call goes on the satellite, comes down. It is just a bit-stream, and if it is encrypted or not encrypted, the satellite doesn't know, and of course we don't control that. So in that sense, if you have access to the commercial phone line, you have access potentially to some kind of encrypted information.

If I could also clarify just one point to make sure we didn't mislead people about what we consider dual-use in all of this, these kind of things are not considered dual-use and the missile—I mean the launch vehicle—is not dual use. Only the actual communication satellite itself is in the dual use category, and its narrowly related launch services. The rest of it is still all controlled under the munitions program. It is not considered dual-use or treated as dual-use or handled in that way.

Chairman GILMAN. The gentlelady's time has expired.

Mr. Bartlett.

Mrs. FOWLER. Mr. Chairman, I would just like to submit my question for the record so that I could get a better answer or a more clear one from the——

Chairman GILMAN. Without objection, the gentlelady's question shall be made part of the record.

Mr. Bartlett.

Mr. BARTLETT. Thank you very much.

Remembering that you are presumably still under oath, were any of your opening statements or answers approved or influenced by White House personnel?

Secretary HOLUM. We have discussed our preparations with the National Security Council staff, yes.

Mr. BARTLETT. So the answer to that question is yes?

Secretary HOLUM. Yes.

Mr. BARTLETT. Thank you. Considering the mission and objectives of the Commerce Department, putting them, giving them primary responsibility for licensing commercial satellite launches by China would appear to me to invoke Khrushchev's analogy that this is equivalent to putting the goat to guard the cabbage patch. If at any time you ask the goat, while he is guarding the cabbage patch, how the cabbage is doing, he will tell you, "Very well, thank you."

I agree with Mr. Lantos that the problem here is fundamentally a policy problem, but I would submit, contrary to what he feels, that there has been a policy change. I think now that the Commerce Department has far more authority and responsibility than it previously had for licensing these satellites, and therein I think is the root of the problem that we are facing.

Let me ask you a question: Considering the objectives of these companies that have these satellites launched, that their primary responsibility is to their stockholders, are we not subjecting them to temptations we ought not subject them to? They do not want to launch their satellites on vehicles that are not reliable, and if the vehicle is not reliable, they therefore have every motivation to improve the reliability of that launch vehicle because they want their satellite in orbit.

Considering this, don't you think that in our desire to get more satellites launched, that we have in fact, as Ben Franklin concluded as a young man, "paid too much for our whistle"? I just think that we inevitably invite the kind of problems that we now have. Don't you agree?

Secretary REINSCH. Let me just say if I can, on your particular point, Mr. Bartlett, about the satellite companies, the way the—as I understand the way the marketplace works here, the launch vehicle, whether it is the Chinese, the Americans, the Russians, is chosen by the buyer of the satellite, not the manufacturer of a satellite. So it would be a question for whoever is going to own and operate the satellite to decide who the launcher is going to be.

Mr. BARTLETT. The same commercial profit, loss incentives are involved, though, correct?

Secretary REINSCH. Well, commercial motives are involved, certainly. That is clear from the beginning. They are the motives of different parties, though, not always American parties.

Mr. BARTLETT. Yes, sir. I understand.

Secretary HOLUM. In that context, it seems to me it is important to recognize that there are a variety of ways the shareholders' interests can be served or damaged, and we don't rely on the companies and the profit motive to protect our national security interests. That is why we have these controls in place, and why the regulations and the license terms and the technical safeguards agreements are all very clear on what responsibilities are.

Leaving aside reference to any specific case, an indictment under these rules can lead to denial of any further license possibilities for a company. That is——

Mr. BARTLETT. I understand the sanctions involved, but if you are launching a satellite on a launcher that fails, and you know why it fails or think you know what it fails, don't you think that it is totally unreasonable for me to expect that you will not tell the chap who is launching your satellite why you think his rocket failed, so it won't fail next time when he is launching another one of your satellites?

Secretary HOLUM. If you have a legal obligation not to provide that kind of information, and you are being monitored, and the consequence, the potential consequence of that mistake, that violation, is to go out of business, to go out of this business completely, I think that is a pretty strong disincentive.

Mr. BARTLETT. But then I think you do it surreptitiously so that you minimize that risk, and I think that is exactly what has happened here. I think that the policy from the beginning predisposed that we were ultimately going to have just this kind of a problem, and I think that we have, unwittingly perhaps, contributed to the development of better boosters by the Chinese. They now have 13 of them pointed at our cities with nuclear weapons aboard them. And I think that, you know, perhaps totally unwitting, we have aided and abetted a potential enemy in doing this.

My question is, wouldn't we have been better off not to have proceeded down this road at all, but simply encouraged development of more launch facilities, either here or among our friends, not among our potential enemies.

Secretary HOLUM. One of the impacts of that, as Under Secretary Reinsch has pointed out, is the likelihood that our leadership in the satellite business would falter. If all other providers of communication satellites and related services in the world have an open market and go anywhere, to all of the potential launch sources, and ours are constrained, our satellite launchers, our satellite industry is at risk. And that satellite industry is something we depend on for national security, as well, so it is not a simple outcome.

Mr. BARTLETT. It is a judgment call that I think we made the wrong judgment. Thank you.

Chairman GILMAN. The gentleman's time has expired. The committee will now stand in recess until 10 a.m. Tuesday, unless otherwise notified by the Chair. We will continue with the recognition list established for today's hearing.

[Whereupon, at 1:35 p.m., the committee recessed, to reconvene on Tuesday, June 23, 1998 at 10 a.m.]

UNITED STATES POLICIES REGARDING THE EXPORT OF SATELLITES TO CHINA

HOUSE OF REPRESENTATIVES, COMMITTEE ON NATIONAL SECURITY, MEETING JOINTLY WITH COMMITTEE ON INTERNATIONAL RELATIONS, *Washington, DC, Tuesday, June 23, 1998.*

The committees met, pursuant to call, at 10:07 a.m. in room 2118, Rayburn House Office Building, Hon. Benjamin A. Gilman (Chairman of the Committee on International Relations) presiding.

STATEMENT OF HON. BENJAMIN A. GILMAN, A REPRESENTATIVE FROM NEW YORK, CHAIRMAN, COMMITTEE ON INTERNATIONAL RELATIONS

Chairman GILMAN. The committee will come to order.

The Committees on International Relations and National Security will continue our joint hearings, that were recessed to this date and time last Wednesday.

Pursuant to the unanimous consent agreement entered into at our last meeting, we will continue to be operating under the rules of the Committee on International Relations, which are similar to the rules of our National Security Committee.

The same witnesses that appeared previously are here now, except that Under Secretary for Policy, Mr. Slocombe, is present in lieu of Deputy Under Secretary Lodal.

Mr. Slocombe, will you please rise and be sworn in?

[Witness sworn.]

Chairman GILMAN. The Chair will note for the record that the witness has responded in the affirmative, and we will proceed with our hearing. The other witnesses have previously been sworn in and remain under oath.

Mr. Holum, I do want to correct an oversight, and I apologize for introducing you last Thursday as Under Secretary for Political Affairs, rather than your correct title, which I understand is Under Secretary for Arms Control and International Security Affairs.

Our procedure on recognition will be to recognize members in the order in which they would have been recognized at the last session of the committee. For the sake of convenience, this is our first list. Any member who was present at the time the gavel fell today but who is not on the first list will be placed on a second list in order of seniority. Any member who appeared after the gavel fell today and who was not on either of the first two lists will be placed on a third list in the order in which they appear, and we will be alternating by party.

If there are questions on this procedure, please inquire to the staff of the Chair, and we will try to iron it out.

Our last questioner was Mr. Taylor. We will now begin with Mr. Hefley. The Chair will be enforcing the 5-minute rule, and we will appreciate the cooperation of all of our members.

Mr. Hefley.

Mr. HEFLEY. Well, thank you very much. And I apologize, and I suppose all of us are kind of in the same boat, that not everybody could be here for all of it, so I may be covering ground that has been thoroughly covered.

One message that seemed to be coming through the other day when we were meeting was the excuse that, well, everybody is doing it. Well, I tell you, that excuse didn't work on my mama and it didn't work with me when my kids came to me and said, daddy, everybody is doing it.

So even if everybody was doing it, if Reagan and Bush had done it, I think that raises the question—I mean, the question still exists as to whether or not we should have been releasing this kind of technology at this time.

But the truth of the matter is, it is not the same. Under President Reagan, the United States and China negotiated an agreement to allow China to launch U.S. satellites and the agreement continued under Bush. But almost immediately it was suspended because of the crackdown after Tiananmen Square. And the requirement that the presidential waiver be granted for export of U.S. satellites to China, President Bush waived those sanctions four times. President Clinton has waived the sanctions 11 times. No satellite export licenses have been denied by the Clinton administration. And you might respond to that. If I am incorrect, I would like to know that.

Also, under the old way of doing it, under Reagan and Bush, DOD had an effective national security veto, the licensing was done under the Department of State, and the licensing requirements for technical data was clear and unambiguous. That is very different from what we seem to have now. National security considerations seem to be the sole determinant of whether or not an export license should be granted.

Under Clinton, licensing jurisdiction for the commercial communications satellites was transferred to Commerce, DOD's influence in the licensing process was diminished, veto power eliminated, technical data is not clearly controlled, and economic and commercial interests are the important factors in determining whether an export license should be granted.

So I would like to ask whichever one of you can answer from the standpoint of DOD whether or not you believe the influence of the Department of Defense has been strengthened by this transfer. And, if so, how can this be, if DOD has lost its veto power over an export and if national security considerations are no longer the sole determinants of whether or not export is in the national interest? And whoever wants to speak for DOD, I would appreciate you responding to that.

Secretary SLOCOMBE. I will be happy to respond for the Defense Department.

We feel that the procedures which are in place now allow the Department of Defense to have a fully effective voice in all of these matters. They provide that a Commerce license can be denied for any destination on broad national security or foreign policy reasons. They do not permit what was permitted in the past, which is a so-called foreign availability petition, which is, in a sense, another aspect of the everybody is doing it argument. That argument is not admissible under the new regulations.

They provided a procedure internally which makes it quite easy for us to raise concerns when we have them, and it guarantees DOD monitoring with reimbursement by the companies that are involved when all licenses are issued.

The specific safeguards that have been put in place include that—and these are for Commerce licenses—every satellite exporter has to produce a technology control plan that has to be approved by the Department of Defense. That technology control plan has to provide the plan to ensure that there will not be access by anybody but U.S. personnel to the satellite during the time that it is being transported and a physical and operational security plan for supervised meeting of the satellite to the launch vehicle.

It requires the Department of Defense monitor be present at every technical meeting to ensure that no information or technical data is exchanged that would improve Chinese missile capabilities. It requires that all technical data and information the U.S. company wants to transfer to the Chinese must be approved in advance by the Defense Department, and it requires that DOD monitors be present at the launch site in China to oversee physical site security and to ensure that the launch operations are conducted without any transfer of technology that would improve China's missile capabilities or give Chinese officials access to the satellite.

So I think that the procedure, while different in some details from various procedures that have been in place at various times in the past, works very well to fully protect our interests. That is the interest of the Department of Defense in having our voice heard.

If we had a situation—and I may say it has not arisen—in which the Department of Defense felt that a license should not be granted and other agencies didn't agree, we have an appeal through a series of steps which are easy to put in place right up to the President, and we would not in any way hesitate to take the appeal if we thought it was necessary to do so.

Mr. HEFLEY. I see the red light has gone on, so my time is up.

Let me ask one very brief follow-up question. Are you completely comfortable that this does protect our national interest in terms of defense when it comes to exchanging information that could be detrimental to our defense capability?

Secretary SLOCOMBE. Yes.

Mr. HEFLEY. OK. Thank you very much.

Chairman GILMAN. The gentleman's time has expired.

Mr. Pickett.

Mr. PICKETT. Thank you, Mr. Chairman.

I would like to know if, in light of all of the discussion and publicity that has been given the issue about the improper transfer of technology to China, if the Department of Defense has conducted any kind of internal review of its own. And if it has conducted such a review, what was the conclusion or determinations as a result of that review?

Secretary SLOCOMBE. Well, Mr. Pickett, as I think the committee is familiar from their earlier testimony in the transcript, the Justice Department is looking into allegations with respect to the specific Loral case.

Mr. PICKETT. I am not asking about Justice. Did the Department of Defense conduct its own internal review? And, if so, what was disclosed?

Secretary SLOCOMBE. It is a matter of public record that the Department of Defense did do a review of that incident. We have been asked by the Justice Department not to discuss its conclusions in public because of concern about an impact on the investigation, although my understanding is that a procedure has been set up so that the intelligence committees will have access to the document.

Mr. PICKETT. Were any changes in the DOD procedures for approving these kinds of launches made as a result of the investigation—the internal investigation conducted by DOD?

Secretary SLOCOMBE. Not at this point. And bear in mind that, without getting into details, the thrust of the investigation is over whether or not something happened which was not licensed and should have been licensed. It is not a question of giving a license which shouldn't have been given.

Mr. PICKETT. I see.

Thank you, Mr. Chairman.

Secretary SLOCOMBE. I am sorry. I have been advised the document in question has also been provided to the two Armed Services Committees in the two Houses of Congress.

Mr. PICKETT. Thank you.

Thank you, Mr. Chairman.

Chairman GILMAN. Thank you, Mr. Pickett.

Mr. Ryun.

Mr. RYUN. Thank you, Mr. Chairman.

I would like to direct my question to the panel as a whole.

Since 1994, there have been 20 various reports of transfers of missiles or missile technology from China to Pakistan and Iran. The administration has imposed sanctions only twice, only twice in those roughly 20 different reports of transfers and has refused to impose category one sanctions. Now category two is nice, but category one really sends a very strong message.

The administration has, however, imposed sanctions on North Korea four times, most recently in April for the transfer of missile technology to Pakistan. Clearly, the threat of proliferation by China is much more dangerous than that of the proliferation of North Korea. Could you please comment on why you believe the President has chosen to treat North Korea differently than it does China?

Secretary HOLUM. I will respond to that, and maybe others will have a comment.

First, I think I dispute the proposition that compared China and the DPRK. We think DPRK is also a very significant proliferation risk and one that we continue to pursue actively, including the recent imposition of sanctions for technology transfers.

In the case of both China and the DPRK and, in fact, all countries, we pursue cases of missile proliferation aggressively and follow them closely.

The case you are specifically referring to, I believe, relates to the possible transfer of full up M–11 missiles from China to Pakistan. We have twice sanctioned China for M–11-related transactions, once out of the Bush administration, more recently in 1993 under the Clinton administration. We have not yet reached a conclusion that category one sanctions are warranted by the evidence. This is something that is under continuous review. We simply have not yet reached that conclusion.

Now, given that the consequences of imposing sanctions are very far reaching, it is the case that we impose a very high standard of proof in order to impose these sanctions. But we are monitoring the situation closely. China knows we are. Pakistan knows we are. It is an open case, and we are continuing to evaluate.

Mr. RYUN. However, in light of what I have just said, you have only sanctioned twice when there have been over 20 various incidents. That doesn't, to me, seem like you are sending a very strong message. It indicates to me, perhaps, that you are putting our economic interests against national security interests. In fact, Mr. Slocombe or Mr. Tarbell, would you care to comment on that in terms of DOD policy in that regard?

Secretary SLOCOMBE. DOD's concern is with the national security aspect; and for the reasons I stated in response to the questions from Mr. Hefley, we are satisfied with the procedures that allow the national security concern to be articulated, if that is your question.

Mr. RYUN. Let me add something for the record and that is, during a Senate investigation, the Under Secretary of Commerce for Export testified before the Senate Governmental Affairs Committee that the U.S. satellite companies have contracts worth $1.7 billion with Chinese rocket firms and that this work would help support 16,000 U.S. jobs. To me, it seems that what we are doing is putting economics in front of national security, and it just seems this is a pretty obvious case.

Secretary SLOCOMBE. It obviously is an important economic activity, but I don't agree that we have put economics before national security.

Mr. RYUN. Well, I would like to submit as a matter of record, if I may, Mr. Chairman, a chart provided by the CRS which details the missile transfers from China and Pakistan as well as the sanctions. You will note on here that there are 21 different violations. Only twice has China been sanctioned.

Thank you, Mr. Chairman.

Secretary HOLUM. If I could just respond a bit further on that. We impose sanctions when we have a conclusion—when we reach a conclusion that the sanctions laws have been triggered. So I would contest the notion that these are cases where there is unambiguous demonstrated proof that there was a sanctionable event.

There are a lot of cases under continuous review involving missile-related transfers and other transfers from China and a number of other countries, and we follow those closely. We raise them continuously and repeatedly with our interlocutors in China and warn them of the possibility of sanctions and use the sanctions laws as leverage to try to obtain improvements in their behavior. But we don't sanction on the basis of some intelligence. We sanction on the basis of unambiguous evidence that there is a sanctionable event. Because the implications on our foreign policy, in some cases on our ability to work these issues, can be diminished as a result of sanctions.

Chairman GILMAN. Without objection, the material will be made a part of the record.

[See chart I in the appendix on page 375.]

Chairman GILMAN. The gentleman's time has expired. Thank you, Mr. Ryun.

Mr. Reyes.

Mr. REYES. Thank you, Mr. Chairman.

I have a couple questions. The first one is, while I know there are significant similarities between space-launched vehicles and ballistic missiles, are there significant differences as well?

Secretary SLOCOMBE. There are certainly significant differences in the terms that are relevant here to what you have to do to put a satellite into orbit and what you have to do to deliver an RV to a target on the ground, yes.

Mr. REYES. In the context, again, carrying that a little further, the capabilities to crossover from ballistic missiles to space-launched systems, how easy is that or how complex would that be in order to—what I am trying to get at here is there has been a lot of criticism about making available space-launched vehicles, and I am just trying to get a sense for how difficult or how easy it would be to convert them to ballistic missile threats for this country.

Secretary SLOCOMBE. Let me just give you an example of an important distinction that has come up, and some have been lost sight of.

That is the difference between what is necessary to disperse more than one satellite into an orbit so that it can function as a communications satellite and what is necessary for a MIRV, that is, to deliver a number of RVs to targets from a single launcher. The ultimate objectives are significantly different and what has to be done technically is significantly different.

The objective of a satellite release into a parking or transition orbit does not require exquisite accuracy and allows for a quite wide margin of error. That is, the satellite will work even if it is not in a mathematically exact orbit. The range of error which is permissible is a matter of miles.

By contrast, the objective of a MIRV system is extremely accurate reentry, with released timing and orientation, providing the only mechanism to ensure the warhead reaches the target. And, indeed, it is not simply a question of whether it hits the target. It is a question of whether it survives reentry through the atmosphere, rather than burning up because it is being released improperly.

So the tasks are in some very broad sense similar. That is, you are putting a single booster with more than one object that you want to separate from the booster, but the technical requirements and what you are in fact trying to do is quite different. And, in particular, the standards on the requirements for satellite release are far lower in the satellite case than those required for MIRV, so that is an important example of the difference between the two kinds of activities.

But I don't dispute that there is some connection between running a civil space launch program and running a military space program. But the idea that there is a one for one correlation and that you could use a civil space program as the driving force for a missile program is simply not right.

Mr. REYES. Does anybody else have any other opinion?

Secretary HOLUM. Well, just to emphasize one point. Here we are not talking about the boosters at all—or only tangentially. What we are talking about is the transfer of satellites to be mounted on the rockets, on the space-launched vehicles to be launched into space.

As Under Secretary Slocombe has testified, in the context of doing that there are a whole range of safeguards that prevent any transfer of technology related to either the booster, which is Chinese, or the satellite, which belongs—is provided by the United States.

So even given the remoteness or the differences between space-launched vehicles and missiles—and there are a great many similarities as well—it is sort of separate from the point that we are launching satellites, not designing rockets or missiles.

Secretary SLOCOMBE. It is very important to bear in mind what actually happens here.

The U.S. manufacturer builds a satellite. It is shipped under controlled conditions, which limit access during the transportation to China, and is mated up with the Chinese booster with minimum information necessary to do the connections. And the Chinese are now given the satellite and have a lot of time to take it apart and figure out how it works and put it back together again and then launch it.

It is a—the analogy is sometimes made that it is like delivering a package. The package stays in the box, so to speak, at least until it is taken out and actually made it onto the booster and then put into orbit.

Chairman GILMAN. The gentleman's time has expired.

Mr. Gibbons.

Mr. GIBBONS. Thank you very much, Mr. Chairman.

Prior to March 1996, the State Department essentially had control of communication satellite waivers, et cetera, in deals with China. What I want to know is, from the Commerce Department, at what point did the Commerce Department become involved in seeking national interest waivers for communication satellites?

Secretary REINSCH. Well, let me respond in two ways, Mr. Gibbons.

The Commerce Department first became involved in the issue of licensing commercial communication satellites in 1993 as a product of the Bush administration rationalization exercise that we described in some detail at the previous hearing. The decision to move those, some of the satellites, to the Commerce Department was made in 1992 and then reaffirmed without change in 1993.

The question of a waiver is, from our point of view, fairly mechanical. That is, one or the other agencies, meaning State or Commerce, would initiate the waiver request after an interagency decision has been made and, at least in our case, after an interagency decision has been made that it is appropriate to grant the license.

I believe, of the various waiver requests, I think there were four that were initiated by the Commerce Department and the rest were initiated by the Department of State, even post 1996. And the reason for that is also a point we made in previous testimony and that is the Department of Commerce is permitted, pursuant to the President's decision, to license only that technology which relates directly to the mating of the satellite to the rocket. Any other technology that goes beyond that point must be licensed separately by the Department of State.

That is why many of the licenses in question, including the one that received the 1998 waiver, the Loral one, had both a Commerce license for that technology and hardware that we could license and a State license for the additional technology that was required. In those cases, whether it was two licenses, I think it is a matter of indifference as to who initiates the waiver request.

In the case I referred to, it was the State Department that initiated the request, but it is really on the part of the agencies kind of a ministerial act in terms of turning it over to the White House.

Mr. GIBBONS. Mr. Reinsch, following the 1996 accident, this committee, you do—and I am sure each one of you on this panel knows that there were certain secured classified pieces of information that were not recovered from the 1996 accident. Do any of you know where that equipment is today?

Secretary REINSCH. All I can say, on behalf of the Department of Commerce, Mr. Gibbons, is that was a State Department license. That was prior to the President's decision to transfer jurisdiction. We were not involved in that in any way or in the aftermath of the investigation, and I have no information about it.

I defer to my colleague.

Mr. GIBBONS. Let me go back to that, Mr. Reinsch. You say you had nothing to do with the granting of the license for the 1996 accident?

Secretary REINSCH. The Loral one that blew up. That is correct.

Mr. GIBBONS. Well, Mr. Reinsch, why don't you provide for us three memos that came out of your office: one dated October 24, 1995, consisting of 10 pages, a memo to Betty Bauka from you regarding the national interest waiver for—and it is redacted, I presume that is Loral Communications, satellite project with attachment. That is one.

Then provide for me, if you will, another memo dated December 8, 1995, consisting of 17 pages, again, to Betty Bauka from you regarding a national interest waiver for—and it is redacted, and I presume it is Loral Communications, again, satellite project with attachments.

And, again, provide the third document dated December 4, 1995, a memo from you to Betty Bauka regarding a national interest waiver for—and it is redacted again, and I presume that is Loral Communications satellite projects with attachments, just so that we can see what your involvement was with that national waiver for that satellite to confirm that you had nothing to do with it.

And if you would provide that to this committee, I presume, Mr. Chairman, within the next week to 10 working days, if you could.

[The information referred to is classified and is retained in the committee files.]

But let me go back. Once you knew in 1998 that the equipment hadn't been recovered, that the classified equipment hadn't been recovered and that that is a critical component to a waiver, you still went forward in balancing national interest on that waiver by agreeing to the Chinasat–8 license, did you not?

Secretary REINSCH. All the agencies did. That is correct, Mr. Gibbons.

Mr. GIBBONS. Did you ever talk to DOJ about their investigation and the impact that investigation would have or this licensing waiver for the Chinasat–8 would have on that investigation?

Secretary REINSCH. Did I? No, I did not.

Mr. GIBBONS. Did anybody in your office?

Secretary REINSCH. Well, I would have to ask. I don't think so. That would have been an unusual event.

Normally, the Justice Department guards these investigations closely and wouldn't provide us information, even if we had sought it.

Mr. GIBBONS. Mr. Chairman, my time is up, but there is a line of communication in this regard that I would like to pursue once you have a second round.

Chairman SPENCE. Mr. Allen.

Mr. ALLEN. Thank you, Mr. Chairman.

This weekend the Washington Post had an article which quoted a Pentagon report from last year which noted that China is using the 21 satellite Global Positioning System to improve the accuracy of its weapons and the situational awareness of its operating forces. And the article in the paper quoted John Pike, the Director of the Space Policy Project of the Federation of American Scientists as saying, quote, on a scale of one to ten, GPS is a six and U.S. satellite sales and launches are less than one. Do you agree or disagree with Mr. Pike's evaluation?

That is for anyone who wants to comment.

Secretary SLOCOMBE. I am not sure I can do it in exactly the same one to ten basis. We do not believe that any of the satellite activities that have been properly under license have had any adverse impact on national security.

The question of GPS—obviously, anybody can go into Radio Shack and buy a GPS receiver. They are made all over the world, and GPS is a remarkable, remarkable system.

And the question—I think in an open session I don't want to say anything more than that we are very conscious of the question of how to ensure that, in a combat situation, our adversaries would not be able to rely on GPS and we would, and that is a subject which we give a fair amount of attention. In open session, I don't want to go beyond saying that.

Mr. ALLEN. The same article—and I don't want you to go any further than you feel comfortable, but in the same article, the senior congressional intelligence expert was quoted as saying, we can turn the system off, in the event of hostilities, the GPS system.

Secretary SLOCOMBE. That is indeed what the article says.

Mr. ALLEN. Secretary Holum, in your testimony on Thursday, you discussed the February 1998 Loral waiver. And we have been told that licenses and waivers permitting commercial satellites to be launched by Chinese rockets worked their way from the bottom up through the interagency process.

I have a couple questions. Number 1, did the waiver sought this past February follow standard procedures used in previous waiver requests? And, second, and I apologize if you have covered this before, but can you explain the reasons why each of your departments supported the waiver despite the concerns of the Justice Department?

Secretary HOLUM. Yes, as to the first question. The waiver followed our standard procedure.

And in response to the second question, it did, in specific terms, as related to the Justice Department investigation.

Our practice has been, going back many years, that when there is an investigation, an investigation under way concerning possible violation of license condition or other activity related to the license, we do not deny licenses, we do not put a company on the denial list, short of an indictment.

There are good reasons for that. All investigations don't lead to indictments. We operate under the—in broad terms, under the innocent-until-proven-guilty rubric, except that if there is sufficient probable cause to justify an indictment, then we think it is justified to deny licenses.

So in the context of this situation in 1998, we were aware of the—late 1997, early 1998, we were aware of the investigation. In fact, the information had come to the Department of State in the first instance, and we referred that information to the Department of Justice, which then led to the investigation. But that has been our standard practice.

Mr. ALLEN. Would members of the other departments wish to comment?

Secretary SLOCOMBE. That accurately summarizes the reasoning.

I should say that, just to emphasize, the normal practice was followed. What Defense and, specifically, DTSA does is to ensure that the proposed waiver is consistent with the underlying license, and we did that in the normal course in this proceeding.

Secretary REINSCH. We would adhere to the same reasoning that Mr. Holum described in our own practices.

Mr. ALLEN. Thank you.

One last question. In testimony before the Senate Select Committee on Intelligence earlier this month, Katherine Schinasi——

Secretary REINSCH. Schinasi.

Mr. ALLEN [continuing]. Schinasi of the GAO referred to the transfer of satellites from the jurisdiction of the State Department to the Commerce Department. She said that Defense's power to influence the decisionmaking process has diminished since the transfer, and I would be particularly interested in what the Department of Defense feels about that particular comment.

Secretary SLOCOMBE. For the reasons I stated in some detail in response to Congressman Hefley, I just don't think that is true. I think our ability to influence the process continues to be everything we need in order to ensure that we look at it from a defense perspective. If we have a concern, we have a full opportunity to raise that; and, if necessary, we will take it all the way to the President, if we have a concern that we think is not being properly reflected. Often, it is a question—and this goes to the issue of whether there have been denials or not. Often, it is a question of denying the license or not denying the license, but what conditions would be imposed.

Mr. ALLEN. Thank you very much.

Chairman SPENCE. Mr. Burr.

Mr. BURR. Thank you, Mr. Chairman.

Mr. Holum, does the State Department still object to the decision to transfer the jurisdiction on satellites to Commerce?

Secretary HOLUM. No. Secretary Christopher, in 1995, agreed to an interagency recommendation against the full transfer, under the conditions that that recommendation was presented to him.

Mr. BURR. What was the President's objection to State's belief that they could narrow but not eliminate State's jurisdiction?

Secretary HOLUM. Actually, it wasn't the President's objection. If I—and this is from my perspective, since I wasn't a direct participant, based on information provided to be me, but this led to an inter—the original rejection by Mr. Christopher led to further interagency review under the auspices of the National Security Council. That review led to a new decision that produced a consensus recommendation from all of the concerned agencies, including the Department of State and, specifically, Secretary Christopher.

Because of the additional access that had been included, the fact that—or additional terms, the fact that forward availability would no longer be a consideration, the fact that there was a new national security control that allowed denial of licenses to any destination in the world on these particular kinds of transfers, the fact that Defense and State and other agencies had regular, routine access to all of these processes and the right to appeal to the President. So Secretary Christopher felt the foreign policy and national security interests were fully protected.

Mr. BURR. Since we made the change, how many times has the State objected to waivers?

Secretary HOLUM. We have not objected to waivers.

Mr. BURR. Mr. Slocombe, how many times has DOD objected to waivers?

Secretary SLOCOMBE. To be clear, the issue would be objecting to a license.

Mr. BURR. Objected to a license?

Secretary SLOCOMBE. We have raised concerns about a number, but they have always been satisfactorily resolved, and we have not had to exercise our appeal rights.

I should say there is a series of appeal rights. You don't have to take everything straight from the desk officer to the President. We have been able to work things out so that we have not had to exercise our appeal rights, but we have no hesitation in doing it if we had to.

Mr. BURR. Mr. Tarbell, let me ask you about your testimony in front of the Senate Government Affairs Committee. You said that you often overruled technical experts when they urged that something not be approved. Define for me "often", would you?

Mr. TARBELL. My—the way that I would characterize overruling is making decisions. Recommendations are brought before me when individual elements within my organization or within DOD have conflicting recommendations and, in certain circumstances, we then have a discussion about that, all views are taken into account, and then I make a decision about whether or not——

Mr. BURR. Let me read you some of your testimony.

It says, the third kind of situation is the situation where this matter comes to my attention and we often get in touch with the Commerce Department and I make it known to them that there is something that we are very serious about. We reconcile on a consensus basis and try and come up with some type of compromise so that we don't have to take up the time of busy people up the line. Would you elaborate a little bit on the reluctance to take up the time of busy people up the line?

Mr. TARBELL. This is a matter of some judgment, and all of us are paid in the bureaucracy to make judgments about when matters are of such significance that they deserve the attention of people who are paid to deal with higher policy issues, paid to deal with a variety of different matters that take up their time. And when we are able to work out compromises with the Commerce Department and with other agencies——

Mr. BURR. We are talking about questions that have been raised about national security interest.

Mr. TARBELL. Yes, sir, and those security interests are foremost in my mind when I am having discussions with the Commerce Department. And usually what happens is that the Commerce Department comes in our direction, and I take great pride in the fact that we have had pretty good success in ensuring that that is the case.

Mr. BURR. Do you agree or disagree with the decision to move DTSA from the Office of Policy to the Office of Acquisition and Technology?

Mr. TARBELL. I think that this move will be something that will actually provide a circumstance where we can strengthen export controls within the Defense Department and benefit from a great deal of technical synergy that we expect from our partners in this new agency. And, as a consequence, I think that we are going to have a particular role that is no more—that is not diminished from the role we have today.

Mr. BURR. Let me move to Mr. Slocombe for one last question.

Secretary Lodal testified last week that DOD was responsible for security of satellites, and I think you went on to expand today that that is DOD's role from the standpoint of the transportation, all the way up to the launch. Am I correct?

Secretary SLOCOMBE. We are responsible for approving the plan for doing that. It isn't always done by DOD personnel, but we are responsible for approving the plan.

Mr. BURR. Let me ask you, as it relates to the failed launch in 1996, was DOD involved or was a representative of DOD involved in the actual assessment of—and review of the failed launch?

Secretary SLOCOMBE. Well, as I say, there was a DOD—were we involved in the assessment——

Mr. BURR. You have told me that the Department of Defense or somebody designated by the Department of Defense is in charge of security through the launch. My question is, does DOD believe and were they involved in the security of the satellite after the failed launch?

Secretary SLOCOMBE. You mean, after the failure, were we involved in the—David?

Mr. TARBELL. Let me try.

If you think about three stages of a particular launch program, the first stage is the preparations for the launch, shipping the satellite there, getting it all ready to ship, things of that nature. In that particular launch activity, we monitored that.

The next thing that happens is the launch occurs. We are at the launch site, and there can be a failure. In that particular circumstance, there was a failure. There was a crash. We were there at the launch site; and, as I testified before, we were on the ground and, in the immediate aftermath of the crash, were there to assist in the recovery of debris, which is the responsibility of the U.S. company involved.

Subsequent to that, there can be, in many circumstances, launch failure analyses that are conducted as to, you know, what caused all this and what happened. That matter is now under investigation by the Justice Department. We have been asked not to comment on that.

Mr. BURR. My question deals specifically with the security of the downed satellite. You were onsite at the crash site?

Secretary SLOCOMBE. My understanding is that the satellite, as I understand it, fell into a village near the launch base, and there was some delay in getting any Americans out to the site. I think that is a fact.

Mr. BURR. How long a delay?

Secretary SLOCOMBE. My understanding is it was several hours, 5 hours, something like that.

Chairman SPENCE. The time of the gentleman has expired.

Mr. BURR. Was that a logistics problem or was that restraint by the Chinese government?

Secretary SLOCOMBE. I will have to take that question. I don't know.

Mr. BURR. I thank the Chair for its indulgence. I yield back.

Chairman SPENCE. Ms. Sanchez.

Ms. SANCHEZ. Thank you, Mr. Chairman.

I want to begin by acknowledging Mr. Tarbell. I heard you being questioned by Mr. Burr earlier and talking about making judgment decisions about taking things further up, and it is nice to hear that bureaucrats are not—that they can be decisionmakers, so I applaud your ability to do that and would encourage it more in the Federal Government.

I really want to ask Mr. Slocombe, because you made some comments with respect to the technologies involved here. You know, I was in the corporate sector before, and one of the things that I know companies do is they buy a competitor's product and take it apart and figure out the innovation of the technology, cutting-edge technology, put it back together. And, of course, there are limitations to what they can do. They either have time constraints——

For example, in a large mainframe computer company, as I can recall from a few years back, we had a 30-day visiting period by the competition to come in and make sure the machine was actually put together and in working condition.

Can you tell me, either between the time constraints involved or who a company is, the satellites, et cetera, what we have as safeguards through the process when we go to use the—have the Chinese launched some of this equipment?

Secretary SLOCOMBE. Yes. It is designed specifically to make sure that the Chinese or any other foreign launcher can't do the dismantling and attempting to—first of all, understanding how it is built and then attempting, at least in principle, to reverse engineering.

The satellite—the physical object is accompanied by DOD monitors or by—I am sorry. There is a transportation plan which requires that there be U.S. guards. My understanding—David, correct me—is they are not normally DOD personnel, but they are contract guards. But Americans who accompany the satellite when it is being shipped to China, who accompany it while it is in China, and then there are DOD monitors who supervise—they don't supervise the operation of mating the satellite, but they are present to make sure that the satellite is, in fact, being handled the way it is supposed to be handled and simply being attached onto the Chinese booster. And then they are present at the launch itself, at the launch site.

I think there is no allegation that the Chinese have an opportunity to dismantle the equipment, the satellite equipment.

It is also important that DOD monitors are required under the licenses to be present during technical discussions between the Chinese and the U.S. satellite manufacturer. Because, as you will also appreciate from your experience, sometimes the most important technology transfers take place not by dismantling the machine but by comments that are made and information that is passed.

Ms. SANCHEZ. So you are saying that the satellite is ready to go. You have a transportation plan. It is transported over. You have guards on it the entire time. You get it there when the discussions are made to ensure that there is actually good phasing between the launch and the equipment that we provide, that the entire time there are guards or monitors with respect to the equipment and with respect to any dialog that occurs with respect to the launch?

Secretary SLOCOMBE. Yes, and that is required under the licenses.

Ms. SANCHEZ. If there were significant technologies that could be taken from the process we just spoke about and there was technology that the Chinese would be interested in with respect to the military, what would be the most important crossover information they would be looking for?

Secretary SLOCOMBE. That would depend on where they thought they had the biggest problems in their systems, and I don't think I have any information on that.

Ms. SANCHEZ. Thank you very much for the information.

I yield back the remainder of my time, Mr. Chairman.

Chairman SPENCE. Mr. Rohrabacher.

Mr. ROHRABACHER. Thank you very much, Mr. Chairman.

Let me just—because of the last line of questioning, let me be very clear. The confusion on this issue stems from the fact that some people are talking about satellites and some people are talking about launchers.

Just for the record, I don't know anybody who is concerned that the Chinese have gleaned some information about how to make a satellite better. What people are concerned about is it appears that the Chinese Communist nuclear weapons delivery system, meaning missiles, or rockets, depending on how you want to define them, have been improved with American technology. The discussion of satellites, I believe, is being used to confuse the public on this issue as to the significance of what we are talking about today.

This is the third day of joint hearings and, unfortunately, our worst fears are being realized. It is evident that over the past half decade Communist Chinese missile capability has been improved with the assistance of American technology.

Witnesses from the first panel suggested, as differentiated from today, that under this administration there has been a collapse of standards and safeguards that protected the American people from deadly weapons technology transfers. And that was the first panel. And, as I say, we have heard a different interpretation from this panel. This panel of administration officials have a different interpretation.

But, of course, you are the individuals who have had the personal responsibility for seeing that our country is not put in jeopardy in its high-tech dealings with the Communist Chinese; and many people who have looked into this issue are convinced you didn't do your job because the Communist Chinese rockets have increased their capability under your watch.

We have been reassured, however, today, and by this panel, that if there has been improvements in the Communist Chinese missiles it is a result of practice makes perfect. Now, where did I come up with that phrase? Practice makes perfect is from the last hearing we had; and, as a result, because we have given them too much launch business, perhaps they have used that to perfect their launchers.

At any rate, according to—I guess it is Mr. Lodal, who is not with us today, but, again, reconfirmed by Mr. Slocombe today, although it has been admitted there is some overlap between these missiles—or rockets that carry satellites and missiles that carry warheads, that, frankly, they are very different pieces of technology, from what you are saying today.

Then we hear from Mr. John Holum that there is no worry about providing the Communist Chinese rockets with the ability to launch multiple payloads, and we hear that from Mr. Slocombe again today, because there is a difference, a dramatic difference, between launching warheads and payloads. This flies in the face of numerous briefings that I have been given as well as information that is readily available to members in this committee.

I am the chairman of the Space Subcommittee, and it is ridiculous to sit here and tell me that you can help a Chinese Communist rocket be developed that can have multiple payloads and that that does not significantly improve the Communist Chinese ability in the end to put MIRVs up and launch multiple warheads at the United States of America.

One of the witnesses, Mr. Tarbell, conveniently can't recollect any dissension by professional watchdogs under his jurisdiction. I don't know if maybe he has renewed his memory by going back and asking in between the hearings whether or not anybody had any disagreement of what was going on.

Well, I have a few specific questions that I hope, if it doesn't jog memories, at least will bring out some relevant facts.

Mr. ROHRABACHER. For this panel, do the Chinese Communist rockets now have ring laser gyroscopes? Come on, you are the experts.

Secretary SLOCOMBE. I do not pretend to be an expert on details of the Chinese systems.

Mr. ROHRABACHER. That is funny, a few minutes ago you were telling us that there is a big difference between the rockets and the missiles carrying the payloads and the warheads.

Secretary SLOCOMBE. For better or for worse, the answer to your question is I don't know the answer. I can readily find out.

Mr. ROHRABACHER. Excuse me, sir. You are not going to eat up my time. Does anyone know the answer to the question on the panel?

Secretary HOLUM. What was the question again?

Mr. ROHRABACHER. The question was, do the Communist Chinese rockets now have ring laser gyroscopes?

Secretary HOLUM. I don't know the answer to that.

Mr. ROHRABACHER. No one on this panel knows that. Thank you very much, sir. I am not going to let you eat up my time like I have seen you do at other hearings.

Do the Communist Chinese have exploding bolt technology?

Secretary SLOCOMBE. Yes.

Mr. ROHRABACHER. If they did, Mr. Slocombe, I am glad you know that answer, what type of exploding bolt technology do they have; when did they get it; did we help give them the exploding bolt technology? Were there licenses issued to give them exploding bolt technology?

Secretary SLOCOMBE. I will be happy to provide the answers to those questions for the record. Sitting here right now, I don't know the answer.

[The information referred to is classified and retained in the committee files.]

Chairman SPENCE. The time of the gentleman has expired.

Ms. Harman.

Ms. HARMAN. Thank you, Mr. Chairman. I arrived late at this hearing this morning because I attended the unveiling of Les Aspin's portrait at the Pentagon. He is now on that wall, as he is on this one. I was thinking about how he would approach this issue.

It seems to me he would want balance in our policy; and, Mr. Rohrabacher, my friend, he would want balance in the way we ask questions of the witnesses.

My congressional district in California used to be called the aerospace center of the universe. With defense downsizing, it is now still the satellite center of the universe. And as I think about this subject, I think about how important it is that we balance the objectives in our national interest. There are national security interests, there are foreign policy interests, and there are economic interests, and we need to fashion a policy which protects them all and sacrifices none. As I said before, I think by approaching this issue in a balanced and thoughtful manner, we can achieve a goal.

First, we need to protect our national security. It is imperative that we protect sensitive technology that cannot be obtained on the international market. We need to make sure that safeguards on our satellite exports are adequate to protect our national security interests.

I believe, as these witnesses have testified for hours before this committee, that the administration policy is adequate and appropriate to do this and that those who violate the administration policy should be appropriately punished. And I assume that if violations are found of the administration policy, they will be appropriately punished.

Second, we must consider foreign policy interests. This administration and I endorse a policy of engagement with China. We can be wary in our engagement, but rhetoric and Red-bashing is not going to improve our ability to work with China and make it a more responsible actor on the world stage.

Third, economic interests must be included in the equation. There are billions of dollars and thousands of jobs at stake. I believe we need to protect economic interests as well as national security interests. And it is important to note that the industrial base that produces commercial satellites is the same high-tech base we will need to call upon in the event of a future national security threat.

The balance of these interests is critical to any solution, and it is irresponsible to sacrifice any of them. We have argued in this committee about whether or not we gave the Chinese some valuable technology about their launches recently.

We had a chart inserted in the record last week, Mr. Chairman, a computer chart, that showed that the Chinese launch rate was increased or was totally successful after 1996. Well, I would like, Mr. Chairman, to insert that entire chart in the record, which shows that the Chinese launch industry, without our technology, had 16 successful launches in a row and no failures from I believe it is 1984 until 1990. So the Chinese have been able to launch successfully without U.S. technology as well. I just would ask permission to put that in the record.

Chairman SPENCE. Without objection.

[See Chart J in the appendix on page 379.]

Ms. HARMAN. Thank you, Mr. Chairman.

My question for these witnesses is just this. I believe strongly that balance is the key. I think that we should constantly reassess whether we have the best possible policy. I would like their comments on, one, the need for balance, and, two, on whether there are ways to improve the export licensing policy that we have so that we ensure in the future two things. One is that we do not have improper technology transfer, which obviously everyone on these committees abhors; but two, that we do the best job of balancing the interests, economic, political and national security, so that we have a robust capability in the United States, and I hope, Mr. Rohrabacher, in the future so that we also improve U.S. domestic launch capability and do not need these export licenses anymore.

Secretary HOLUM. Let me respond briefly to your first questions on the need for balance.

Clearly, all of these interests that we are weighing are important ones, but I think it is important as well to keep in mind that there are no other interests, commercial in particular, overrides or trumps international security interests. If there is a concern that we will be increasing the capabilities of a potential adversary, that is paramount. And commercial interests, I have never been part of a proceeding in which somebody said, well, no, let's have commercial interest prevail over that.

Ms. HARMAN. And I am not arguing for that, Mr. Holum. I am sure you know that.

Secretary REINSCH. If I could comment as well, Ms. Harman.

I certainly agree with your premises. These are comments that you have made before and comments that I have made before. And I think you have a very clear perception of the importance of this technology both from a security point of view and from an economic point of view. And I certainly echo Mr. Holum's comments.

When the Department of Commerce is involved in these debates, we put national security first, just like everybody else. I do think that it is appropriate for the interagency process to know what the economic consequences of its actions are going to be before it takes them. Those are not necessarily dispositive, as Mr. Holum indicated, but they ought to be something that is taken into account that people are aware of.

You asked for a comment about the licensing process, and I appreciate that. That is something that I give some thought to from time to time, since we are the agency that has to operate the process. I think that I would pick up on a question I think was raised by either Mr. Burr, I believe, or Mr. Ryun that talked a little bit about the licensing process. And I would certainly make the point that this is, in fact, the process of dialog. It is not simply a process of passing paper out and getting yes or no back from other agencies. It is a process of dialog. It is largely a process of dialog amongst technical people in each of the agencies who are engineers for the most part and are familiar with these specifications of whatever it is that is being exported and are familiar with national security requirements that we have.

I think probably what would make the system more useful in complex cases, and satellite cases certainly are complex cases, they tend to take a long time, is if we could do more to formalize the dialog early in the process, by which I mean get our engineers, Defense's engineers, State's technical people, and, if appropriate, the applicant together to talk over the specifications, the design issues, why the applicant is asking for a particular configuration or a particular thing that they want included.

What we have found over and over and over again over the years is that when you sit down and discuss these things, and you discuss them at a technical level, people begin to understand.

What, in fact, happens in these cases is that most of the discussion is about the conditions and the parameters of the export, it is not simply about yes or no. It is, you know, is there a way to get to yes. And sometimes there isn't, and we just simply say no. But if there is a way to get to yes, we discuss what that way is, and we discuss conditions and how these things can be configured in a way that doesn't raise any international security concerns.

That is best done, I think, through dialog. It is something that we are doing now, but I think it is something we can do more frequently and earlier in the process.

Chairman SPENCE. Time has expired.

Ms. HARMAN. Thank you, Mr. Chairman.

Chairman SPENCE. Mr. McKeon.

Mr. MCKEON. Thank you, Mr. Chairman.

I want to thank you for holding these hearings. I was happy to hear Secretary Holum and Secretary Reinsch say that while balance is important, the paramount importance in this committee and in our Nation and from our aspect is national security. That should be paramount. That should be No. 1.

In that regard, I would like to yield the balance of my time to Mr. Rohrabacher.

Mr. ROHRABACHER. Thank you very much, Mr. McKeon.

Mr. Slocombe, we said today that no one on the panel knows whether or not Communist Chinese rockets have laser ring gyroscopes. You have stated that they do have exploding bolt technology, but you don't know how they got it.

Just for my colleagues, exploding bolt technology is essential for MIRVing and essential for stage separation. So if a Communist Chinese rocket were to hit the United States, if it did not have stage separation technology that worked or MIRVing technology, it would be less likely to kill millions of Americans. So this is important technology.

Was there a license issued to permit the Communist Chinese to get their hands on exploding bolts?

Secretary SLOCOMBE. First of all, the Chinese have had stage separation on rockets for years and years before any of this satellite technology.

Mr. ROHRABACHER. What we are talking about is upgrading their abilities, making their rockets better. Better exploding bolts maybe perhaps would permit their rockets to have better stage separation. That would be upgrading their rockets. That is what this hearing is about.

Secretary SLOCOMBE. My understanding is that the—it is a Chinese launch vehicle. Therefore, the Chinese provide the equipment related to the launcher, including the stage separation system. But my understanding is the United States has provided explosive bolts to be used by the Chinese as a part of that separation system, but the bolts were provided as end items with only the technical data required to interface them with the existing payload separation system.

Mr. ROHRABACHER. So your testimony is that the Chinese were never in possession of exploding bolts?

Secretary SLOCOMBE. No, that is not my testimony. My testimony is that they were provided.

Mr. ROHRABACHER. And a license was granted to provide that the Chinese were able to get ahold of explosive bolts?

Secretary SLOCOMBE. I believe the license was not a separate license for the explosive bolts, but it is a license for the launch system.

Mr. ROHRABACHER. Was there ever a license denied because it would permit the Chinese to have exploding bolts?

Secretary SLOCOMBE. Not to my knowledge.

Mr. ROHRABACHER. Well, you better go back and check your records. There is a license that was issued for exploding bolts that gave them exploding bolts. There is an issue that was denied. And according to your own testimony earlier, some objections had to be worked out. Well, this is one of the things that had to be worked out. It was denied because some people were concerned about national security. It was reworded in a way, "Well, now they have got it." Maybe we can hear about this later.

Has any American know-how or technology been used to upgrade Chinese Communist MIRVing capabilities?

Secretary SLOCOMBE. We had an exchange on this subject, Mr. Rohrabacher, a few weeks ago. The Chinese have a very extensive and ambitious intelligence operation, which I assume targets U.S. technology in a lot of ways and around the world. I am not prepared to say, no, they have never been successful in stealing U.S. technology.

Mr. ROHRABACHER. Well, I am not talking about stealing American technology. I am talking about the system that you are testifying and have studied about to inform this committee.

Let me be more specific. Did the Long March 2C rocket have multiple payload capacity prior to its involvement with American companies?

Secretary SLOCOMBE. I don't know the answer to that. But I do know that if the multiple payload capacity was a Chinese-designed, Chinese-provided system, it was done to a requirement to be able to launch more than one satellite at a time off a single booster. But it was the Chinese who designed and built the system.

Mr. ROHRABACHER. So that the Long March 2C did not have multiple payload capacity prior to dealing with Americans, but after dealing with the Americans, it developed this multiple payload capacity. That just happened without any input from the American companies; is that what you're telling us today?

Secretary SLOCOMBE. My understanding is that, other than to state the requirement and provide information about what the satellites were that would have to be launched, that the Chinese—it is true that the Chinese didn't have a multiple satellite launching capability until they had a requirement to have it.

Mr. ROHRABACHER. How many consultations, meetings took place on the Long March 2C rocket between the American company and the Chinese, and were those monitored? We would like the records from those meetings and notes from those meetings. How many took place?

Secretary SLOCOMBE. I can provide that for the record.

[The information referred to is retained in committee files.]

Chairman SPENCE. The time of the gentleman has expired.

Mr. ROHRABACHER. Thank you very much, Mr. Chairman.

Chairman SPENCE. Mrs. Capps.

Mrs. CAPPS. Thank you, Mr. Chairman.

Mr. Spence, I do appreciate this joint hearing. I would like to focus, if we might, please, on the history of our efforts to stop the Chinese sales of missiles and missile technology over the years. I do not care who on the panel would like to respond, but if you could, please, just highlight briefly some of this history and particularly any successes that might have been registered.

Secretary HOLUM. Well, it has been a concern that we have had underway for a long period of time. The Chinese are, as you know, indispensable to any effort to control missile technology around the world. They have all the relevant technologies. Should they choose to do so, they could be a major problem.

In the late 1980's and earlier, going back a number of years, we have had concerns about, for example, actual sales of CSS–2 missiles, medium-range missiles to Saudi Arabia. There were discussions of M–9 missiles possibly going to Syria.

The Bush administration and the Clinton administration have both very actively engaged the Chinese on missile transactions. This included in 1991 the imposition of sanctions under the Missile Technology Control Regime category II sanctions, which were lifted in 1992. Two Chinese entities were sanctioned at that time.

As a condition of the waiver of those sanctions, the Chinese agreed to accept what were then the standards of Missile Technology Control Regime, the parameters. They are not a member of that regime. They agreed to accept its basic rules.

In 1993, the Clinton administration concluded that they were not living up to that obligation. They were sanctioned again. This related to MTCR category II transfers to Pakistan again. In that case, in October 1994, China agreed as a condition of lifting the waiver that they would no longer export MTCR-class missiles to any country.

Now, that goes beyond the MTCR rules. The MTCR contains a presumption of denial. This was a flat prohibition. All of our information leads us to believe that they are living up to that commitment. We have concerns with China about their interpretation of the breadth of that commitment as it relates to transfers of technology and smaller components of missiles, so we continue to discuss missile issues with them very aggressively. We are particularly concerned about the possibility of transfers both to Pakistan and to Iran.

In connection with Iran, we have also had concerns about transfers of sub-MTCR-class cruise missiles, C–801 and C–803 cruise missiles. We have raised those issues continuously with China. Last year, the Chinese provided to President Clinton, to other senior officials, Secretary of State Albright included, a commitment that they would not any longer transfer those missiles to Iran. And again, we have seen no indication that they are not living up to that commitment.

So I would characterize the overall record as one that is mixed, one that continues to require a great deal of attention, and it receives a great deal of attention, interaction with the Chinese, including, I am sure, the discussions President Clinton will have beginning Saturday. But we have made considerable progress.

Given the fact that China is indispensable to the solution to the missile proliferation problem, we think engagement is a crucial element of our ability to solve this problem.

Mrs. CAPPS. Thank you.

If I have another minute or two, could I ask a more specific thing about Mr. Sokolski's accusation that technology involved in satellite launches is directly applicable to ICBMs, such as kick-motor satellites, dispensing technology, and so forth. Do you have a specific response to this accusation?

Secretary SLOCOMBE. There obviously is a relationship between the two, and no one is denying that, but they are different kinds of activity. I talked about the payload dispersal system, which is one.

On the kick-motor technology, the answer there is that the kick-motors are subject to these very careful constraints to make sure that the Chinese do not have access to them, they can't take them apart, they can't learn how they function in the process of—in the process of their being delivered and put on the Chinese launch vehicle and launched into space or, in the case of a failure, crashed.

Chairman GILMAN. The gentlelady's time has expired.

Mr. Ballenger.

Mr. BALLENGER. Mr. Chairman, I would like to ask several questions of Secretary Reinsch, if I may.

First of all, did a leading U.S. satellite manufacturer suffer financial losses when the State Department imposed category II missile sanctions on China in August 1993?

Secretary REINSCH. As I recall at the time, Mr. Ballenger, there was a leading satellite manufacturer who made that statement publicly, yes.

Mr. BALLENGER. And their name was?

Secretary REINSCH. It was the Hughes Company that made a public statement.

Mr. BALLENGER. After the State Department imposed missile sanctions, did Commerce provide a legal interpretation of the impact of category II missile sanctions that permitted a U.S. satellite manufacturer who could not obtain a State Department export license to get an export license from Commerce for essentially the same satellite?

Secretary REINSCH. Well, it wouldn't be essentially the same satellite, no. And I think that is an important distinction, and I am glad that you asked the question that way. We provided our existing legal interpretation, which has been ours for some time, with respect to the status of satellites that fell under our jurisdiction. You will recall, as of that point in '93, that some satellites had been previously transferred to the Commerce Department. Those were satellites that did not meet or contain any——

Mr. BALLENGER. Is the difference rather substantial?

Secretary REINSCH. The technical capabilities of the satellites, yes. We are only at that point permitted to license satellites that did not contain or cross any of the nine technical parameters that the Bush administration created.

Mr. BALLENGER. Let me move on from there.

But that specific company again was Hughes, unless I'm mistaken.

Secretary REINSCH. The specific company that what?

Mr. BALLENGER. That was given the opportunity where they had been turned down by State, and even though it was not the same sort of satellites, that permission was granted.

Secretary REINSCH. Well, the legal interpretation that the Department of Commerce took it took with respect to the way we administered our rules and regulations. It wasn't taken with respect to an individual company, whether or not they fell under that category.

Mr. BALLENGER. Well, let me ask the question. Did Hughes fall under that category then?

Secretary REINSCH. Well, some of their satellites would. Some of their satellites wouldn't.

Mr. BALLENGER. Could you tell me who is the Chairman of the President's Export Council?

Secretary REINSCH. Michael Armstrong.

Mr. BALLENGER. And when was he appointed?

Secretary REINSCH. Well, I don't——

Mr. BALLENGER. How about February 1995?

Secretary REINSCH. Well, I thought it was somewhat earlier than that, frankly, but I'll agree with you if you say it was February 1995.

Mr. BALLENGER. What was his private sector role when he was appointed?

Secretary REINSCH. When he was appointed, he was the chairman and chief executive officer of Hughes.

Mr. BALLENGER. Right. Did you ever meet with this man to discuss missile sanctions, transfer of jurisdictional or commercial communications satellites, or Chinese launch failures investigations?

Secretary REINSCH. My contact with Mr. Armstrong, as I recall, was entirely in meetings with the President's Export Council, which I generally attended and participated in as part of my duties at the Commerce Department.

Mr. Armstrong, during his tenure—well, as Chairman, included on the agenda the issue of jurisdiction transfer. I don't recall that I had any discussions with him.

Mr. BALLENGER. What about representatives of his company itself?

Secretary REINSCH. Well, I have talked with representatives of his company from time to time. But the subjects again were what?

Mr. BALLENGER. Several things. Missile sanctions, transfer of jurisdiction of commercial communication satellites, and Chinese launch failures.

Secretary REINSCH. I've talked with representatives of his company from time to time certainly about the transfer of jurisdiction. That was an issue that was very much on our minds. With respect to the question of launch failures, I had a conversation with company personnel recently with respect to the 1995 failure Mr. Bereuter raised, which we discussed in some detail at the last hearing.

Mr. BALLENGER. When he took over Chairman of the President's Export Council, did he meet with Secretary of State Christopher and urge him to move the satellites off the U.S. munitions list to the Commerce Department?

Secretary HOLUM. I don't know if there was a meeting. He sent a letter to Secretary Christopher arguing that case.

Mr. BALLENGER. Right.

Secretary BARKER. If I could add something for the record. None of us were in our current position. Of course, none of us, at least in the State Department, were in our current positions then. But we understand that there was a meeting that Mr. Armstrong had with Secretary Christopher in which we understand that they talked about transfer of jurisdiction. And while I don't have the exact read-out from that meeting, I assumed he also expressed some concerns about the missile sanctions.

Mr. BALLENGER. Right. I've got a letter here dated April 10 that shows that.

Let me get back to Mr. Reinsch. Was this not the same company to which you provided your 1995 determination endorsing the company's participation in the launch failure investigation?

Secretary REINSCH. I don't think "endorsing" is an appropriate word. What I said last week—it is the same company. What I said last week was that we examined—in the case of the 1995 failure, we examined the results of their failure analysis, the material they proposed to provide initially to their insurance company and concluded that the information in it fell within the parameters of the licensing.

Mr. BALLENGER. My time is about out.

Is this the same company that is now subject to the Justice Department's criminal investigation for its role in the 1996 launch failure investigation?

Secretary REINSCH. I couldn't say, Mr. Ballenger, because I have not talked with the Justice Department. It has been publicly reported to that effect, but I don't have personal knowledge from the Justice Department what the subject of their investigation is.

Mr. BALLENGER. Thank you, Mr. Chairman.

Chairman GILMAN. The gentleman's time has expired. Thank you, Mr. Ballenger.

Mr. McHale.

Mr. MCHALE. Mr. Chairman, I was present in the hearing room a few minutes ago when Mr. Allen raised what I think was a profound line of questioning regarding the incorporation of our GPS technology into our Nation's missile targeting systems; and Mr. Slocombe quite correctly at that time mentioned that in an open session it would not be appropriate to respond.

The article that I read, the same article that was referenced by Mr. Allen, indicated the possibility of turning off that system in the event of hostilities. For a variety of practical reasons, turning it off is not really a viable alternative. The issue, Mr. Chairman, I believe, is selective access to that system in the event of combat situations.

I think that is a fundamentally important issue raising absolutely vital concerns of national security, and I would respectfully but strongly recommend to the Chairman that we in the very near future conduct a closed-door hearing to examine the national security implications of possible access by an adversary during a time of hostility consistent with our own need to use that system in the accomplishment of our assigned missions. I think Mr. Allen raised an extremely important issue and one that we need to promptly explore in an appropriate closed session.

Gentlemen, the only question I have relates to a possible conflict between Mr. Holum's testimony last week and Mr. Slocombe's testimony today. On Friday, following your testimony last week, Mr. Holum, an article appeared in the New York Times with the headline "Administration Officials Defend Satellite Export Policy." The article was by Eric Schmidt.

Let me read to you a portion of that article and ask for your comment as to whether or not it is accurate; and if it is accurate, then, Mr. Slocombe, I will ask you whether or not you agree with it. That article said, and I quote, the Government's Arms Control Agency warned last year that waiving a ban on satellite exports to China to entice Beijing to curb missile sales had accomplished "nothing." The warning by the Arms Control and Disarmament Agency, contained in a memorandum apparently sent to State Department officials, was made public at the House hearing. The memo undercut one of the White House's fundamental reasons for granting waivers to allow satellite exports to China despite the ban imposed after the 1989 massacre in Tiananmen Square. Quote, "History has proven that the only time we have gotten movement from the Chinese on missile proliferation has been in the face of a penalty being imposed," the memo said. "The carrots have gotten us nothing." But the agency's advisory is not included in the State Department's official recommendation to the White House earlier this year. President Clinton approved the waiver last February despite objections from the Justice Department that it would undermine an ongoing criminal inquiry into whether two satellite makers divulged sensitive information to Chinese rocket scientists.

And finally, and most importantly, "Holum, who heads the Arms Control Agency, told an usual joint hearing of two House committees, the National Security and the International Relations Committees, that the warning was not adopted because," and again I quote, "it would have been very disruptive," end of quote, "to broader American commercial interests."

Mr. Holum, is that an accurate summary of your testimony last week? And, Mr. Slocombe, if that is accurate, particularly in the closing quote, "disruptive to broader American commercial interests," do you agree with that as the rationale for the administration's position?

[The information referred to can be found in the appendix beginning on page 381.]

Secretary HOLUM. First—and I don't blame the reporters or the coverage for this difficulty. I gave a rather long answer, and that is a rather short summary of it, and it leaves out some, I think, very crucial elements.

My conclusion in late 1997 early '98 regarding this waiver was that it would not be advisable to specifically link denial of waivers to the nonproliferation efforts we have underway with the Chinese because I don't think the leverage will be effective in this case.

I am no shrinking violet when it comes to pursuing sanctions. We have had sanctions on the table a number of times with respect to China and missile activities. But in this case, I thought our leverage was very limited because of the fact that there is a shortage of launch capability in the world, satellite launch capability in the world. As a consequence, we would deny the Chinese essentially nothing. So the actual leverage from doing that would be limited.

Mr. MCHALE. Mr. Holum, I appreciate your response. But perhaps in the limited time that we have, if we could get to the core of the issue. Was the State Department's warning not sent to the White House because of commercial considerations?

Secretary HOLUM. No.

Mr. MCHALE. Mr. Slocombe.

Secretary SLOCOMBE. I have no way of knowing what—I've never seen any of these documents of base internal stage and active communications. So I'm not entirely sure what the question is. I agree with what John Holum says about the issue of leverage. It is a difficult question whether this kind of leverage works or not.

Mr. MCHALE. Mr. Slocombe, my time has expired. I guess what I'm really looking for, and I guess it goes to the fundamental issue in question of this hearing, I would like to have you passionately tell the members of this committee that this administration's policy with regard to technology transfer to China was shaped decisively by a fundamental adherence to American national security interests and was not substantially influenced by commercial considerations.

Secretary SLOCOMBE. That is, to the best of my knowledge, absolutely true.

Mr. MCHALE. Thank you, Mr. Chairman.

Chairman SPENCE. Mr. Hunter.

Mr. HUNTER. Thank you, Mr. Chairman.

Gentlemen, thank you for being with us today. I have passed out to you a Memorandum of Agreement on Satellite Technology Safeguards between the Government of the United States and the People's Republic of China. That has been mentioned several times in your testimony. That's basically our agreement with them, and it is signed at the back by a U.S. representative and by a Chinese representative, and it's the agreement that prohibits any transfer of technology from American companies or people to China except the so-called interface technology or the satellite mating up with the launcher.

Specifically, I wanted to direct your attention to the bottom line on page 8, "Conduct of U.S. persons while in the PRC." I want to read that to you and have you focus on that.

It says that "these persons shall not engage in business or commercial activities beyond or in conflict with the provisions of this agreement. U.S. persons shall not engage in activities that will harm launch safety or would lead to the transfer of Chinese launch vehicle and launch operations technology."

First, have you read the DTSA report? I would ask the panel that. Who has read the report? If you read the report, please raise your hand.

Secretary SLOCOMBE. You mean the report on Loral?

Mr. HUNTER. Exactly.

Those who have read the report, could you raise your hands.

Secretary SLOCOMBE. I have skimmed it briefly.

Mr. HUNTER. Do you agree with the report that U.S. security was damaged by the interaction following the 1996 crash?

Secretary SLOCOMBE. Mr. Hunter, with great respect, I feel obligated to abide by the instructions we have from the Justice Department not to discuss the details of that report because of the danger of interfering with their investigation.

Mr. HUNTER. I don't want you to discuss the details. I'm just asking you if you agree with their conclusion that American security was harmed?

Secretary SLOCOMBE. I have not formed a judgment on that question quite deliberately because, as Mr. Rohrabacher objects to, but I am not a missile expert.

Mr. HUNTER. Mr. Holum, you read the report. Do you agree that American security was harmed?

Secretary HOLUM. I would give the same answer. This is something that will be evaluated in the Justice Department review.

Mr. HUNTER. Well, I understand the Justice, but you're a defense expert of sorts, and you've read the report. And if you take the facts as true in the report, do you agree with the report's conclusion that American security has been harmed?

Secretary HOLUM. I don't have any comment on that.

Mr. HUNTER. Mr. Barker.

Secretary BARKER. I agree with the previous statement.

Mr. HUNTER. Thank you.

Now let me just ask you gentlemen, this agreement that I passed out to you basically sets forth that no American citizen shall give any aid or technology to Chinese officials or workers that assist them in launch capability.

Now we had three launch failures, counting the 1996 failure, that are fairly well-known. We had one in 1992, a Chinese LM–2–E launch. We had one in 1995. It was a Chinese LM–2–E launch. Both of those exploded. And it was written extensively that following those explosions, those failures, there was extensive engagement by Hughes in a 7-month investigation, what they deemed, and I am quoting Hughes, 7 months of vigorous and cooperative investigations.

Now, these were cases, as was the 1996 case, where launchers failed. Now our policy, according to this agreement, says you cannot help them with launchers. The only thing American companies can do is work on the satellite package itself and that satellite interface with the launcher, but you can't help them with launchers.

I want to ask you gentlemen, and start with Mr. Tarbell and work to the left, what did you do to ensure that this U.S. agreement of satellite technology safeguards was being carried out; that is, that there was not technology being transferred in those postlaunch joint analyses between Hughes, in some cases Loral, and China? What did you do?

Mr. TARBELL. Let me start with each in turn and tell you what I know about it.

There was a launch failure of the Optus B2 in December 1992. That launch was monitored by DOD pursuant to a State Department license. And subsequently a launch failure analysis was conducted by the company with the Chinese, and that analysis was also monitored by DOD to ensure that no technology was transferred.

Mr. HUNTER. OK. 1995.

Mr. TARBELL. In 1992, sir.

Mr. HUNTER. No, I am saying let's go to '95.

Mr. TARBELL. Apstar 2 in January '95. That launch was not monitored by DOD, and we do not know the extent of any launch failure analysis that was conducted postlaunch there with insurance companies.

Mr. HUNTER. Let me ask you this question. You have, somebody up there has, a responsibility to the people of the United States of seeing to it that we are not inadvertently or advertently transferring launch technology in those postfailure analyses.

Now, it was fairly well-publicized in Wall Street Journal and Defense Daily that these, quote, 7-month joint cooperative investigations into the failure were taking place between U.S. industry and China.

Were you aware of this engagement by U.S. companies following these failures? Obviously, you were with the first failure, right, because the failure analysis was monitored by an American employee?

Mr. TARBELL. Yes, by one of the DOD monitors.

Mr. HUNTER. OK. The second failure.

Secretary REINSCH. Maybe I can comment on that, Mr. Hunter, since that was one of ours.

Mr. HUNTER. I want to see what Mr. Tarbell did first, and then I will let you have a shot, Mr. Reinsch.

Are you aware of that failure?

Mr. TARBELL. In 1995, this matter was not brought to my attention by the individuals on my staff who are responsible for these matters.

Mr. HUNTER. OK. Should we have had a monitor in this postlaunch investigation?

Mr. TARBELL. Yes, sir, I believe we should have.

Mr. HUNTER. And we failed to do that?

Mr. TARBELL. There was no monitor there.

Mr. HUNTER. Are we asking Loral or Hughes—or Hughes in this case what transferred, what they gave to the Chinese?

Mr. TARBELL. This, I understand, is a matter of some investigation.

Mr. HUNTER. OK. But that is not a prohibited investigation where nobody can talk about it, right? You are saying there is some investigation?

Mr. TARBELL. There is a subsequent investigation, and that investigation also involves the Justice Department, yes, sir.

Mr. HUNTER. But is there any DOD investigation to see if American security has been hurt by that postfailure crash analysis?

Mr. TARBELL. Not currently. But we only became aware in the last couple weeks that there was an actual launch failure analysis that was completed and the context in which it was provided.

Mr. HUNTER. Let me just read this.

A February 1995 article in the Wall Street Journal referred to a Hughes official who said, a thorough investigation of the explosion would take months; and another July 26, 1995, article regarding the failed January 1995 launch indicated that the joint statement released by Hughes and China Great Wall that they would, quote, work together to eliminate the two mentioned possible causes of failure.

We have an international paper quoting Hughes officials saying, we are going to do a postlaunch investigation, and we are going to figure out what went wrong with the launch. And you are not aware of that until a couple days ago?

Mr. TARBELL. Sir, the matter that I was not aware of was that there was a launch failure analysis that was conducted by Hughes that was provided to the Chinese and insurance companies.

It is the case that, in certain circumstances, independent analyses can be conducted by the satellite manufacturer and independent analysis conducted by the Chinese, and then that information is provided to the insurance companies. Where there would not be any interaction between the Chinese——

Mr. HUNTER. My question is, when you have the companies saying that they are going to work together to eliminate the two possible causes of launch failure, that does not send up a red flag that the agreement is about to be violated? They are not supposed to do that, according to our agreement.

Mr. TARBELL. You are stating the facts as I know them. And it is true that, for some reason, the staff did not raise this or were not aware of it and did not bring it to our attention.

Mr. HUNTER. Mr. Slocombe, are you aware of any actions that were taken to prevent important technology from going to China pursuant to a 1995 launch failure we're talking about?

Secretary SLOCOMBE. Beyond what Mr. Tarbell said, no.

Mr. HUNTER. Mr. Holum.

Secretary HOLUM. No.

Mr. HUNTER. Mr. Barker.

Secretary BARKER. No, not beyond what has been said.

Mr. HUNTER. Mr. Reinsch.

Secretary REINSCH. Yes, Mr. Hunter, I can comment on that, if I may. This was the one that we discussed with Mr. Bereuter at the previous hearing.

Following the failure, the company conducted a launch failure analysis, which they represented to us as something that they had done independently of the Chinese. They presented us with that, our licensing officials with their analysis. We determined that the analysis did not contain any information that exceeded the terms of the license that had been issued, and we authorized them to provide that analysis to the Chinese. I cannot address the statements they made to the press.

Mr. HUNTER. Did DOD or CIA or State review that analysis?

Secretary REINSCH. No, sir.

Mr. HUNTER. Why not?

Secretary REINSCH. We believe that we are competent, that our engineers and licensing officials are competent to make judgments about the parameters of our own licenses.

Mr. HUNTER. But can you make judgments with respect to DOD technology, security?

Secretary REINSCH. We are not licensed to provide that kind of technology. The only thing we looked at was whether the study exceeded the bounds of our license, and we determined that it did not.

Mr. HUNTER. Let's follow this a little bit further. Was there any personal interaction, were there any meetings, or was this simply a matter of moving paper from one company?

Secretary REINSCH. Interaction between whom?

Mr. HUNTER. Was there interaction between the Hughes team of engineers and China?

Secretary REINSCH. I believe they represented to us that they had done the launch failure analysis independently of the Chinese. That is probably a question that is best put to them, though.

Mr. HUNTER. Well, did you ever inquire of them when they made the public statement that they would work together with the Chinese to eliminate the two mentioned possible causes of launch failure? Did you inquire of Hughes as to whether or not they were, in fact, meeting with the Chinese?

Secretary REINSCH. We didn't inquire at that time, no, sir.

Mr. HUNTER. Could you find out if they met with them?

Secretary REINSCH. I am happy to ask them and tell you what they say. You may want to ask them directly as well.

Mr. HUNTER. Let me say this, Mr. Reinsch. You folks signed this memorandum saying you are going to safeguard and American technology will be safeguarded. Now, I do not know any other American citizens that can ensure that the technology is safeguarded except the folks seated at the table.

Now, there was no monitoring of this postcrash failure in 1995. I want to know if there was a postcrash failure analysis where there were meetings with American technical people and the Chinese. That is kind of an important aspect of this.

Let me ask you one last question with respect to your meetings. Mr. Reinsch, you talk in your statement about the industry group that met. And if you could go to your statement. I am quoting:

"With respect to the transfer of jurisdiction, Secretary of State Christopher, in a June 8 statement, in the Los Angeles Times makes clear his 1995 decision was to task State and the Industry and Advisory Group to continue the process begun by President Bush."

Now, the Industry Advisory Group, I take it that that includes a number of people in the industry who are, in fact, leaders of these companies that are involved in the trade is the subject matter?

Secretary REINSCH. Yes, Mr. Hunter. Most of the various agencies have, as matters of both law and policy, various advisory groups that help them on a variety of things, and at the Commerce Department we usually refer to them as technical advisory committees. This is a State Department group, and I refer to them as its composition.

Mr. HUNTER. Mr. Bernard Schwartz, he is a CEO of Loral, right?

Secretary REINSCH. Yes, he is.

Mr. HUNTER. Was he a member of that advisory group?

Secretary BARKER. I'm sorry, we don't know that. We'll take that for the record. There were certainly representatives from the aerospace industry that were involved in that advisory group. He may have been one of them.

Mr. HUNTER. Do you—any of you folks know Mr. Bernard Schwartz? Have you been in meetings with him?

Secretary REINSCH. I never met him.
Secretary BARKER. I have never met him.
Mr. TARBELL. I met him in an earlier capacity, not in my current capacity.
Mr. HUNTER. What was that capacity?
Mr. TARBELL. I worked as the Director of International Economic and Energy Affairs in the Department and did defense trade matters for the Department and met him in the course of meetings of the Defense Policy Advisory Committee on Trade.
Mr. HUNTER. Mr. Slocombe.
Secretary SLOCOMBE. I had been in meetings of DPAC with 50 or 60 people in the room. It's conceivable he was in the room, but I have no conscious recollection of ever having seen him.
Secretary HOLUM. I don't have any recollection.
Secretary BARKER. I have never met him.
Secretary REINSCH. I have never met him.
Mr. HUNTER. Mr. Lewis, presumably the same thing?
Mr. LEWIS. No.
Mr. HUNTER. Are you aware of any meetings involving the waiver of Chinasat–8, of the sanctions on the Chinasat–8 launch that involved Mr. Schwartz? In other words, were there any meetings with him concerning whether or not they were going to get this waiver for Chinasat–8?
Secretary REINSCH. I am not aware of any, no, Mr. Hunter.
Secretary BARKER. No.
Secretary HOLUM. No.
Secretary SLOCOMBE. Not as far as I know. You mean meetings with anybody?
Mr. HUNTER. Any meetings that Mr. Bernard Schwartz would have with any members of the administration concerning the waiver of Chinasat–8?
Secretary SLOCOMBE. I don't know of any.
Mr. HUNTER. Could you try to get that for the record?
Secretary BARKER. Mr. Chairman, Loral, I believe, is a member of the Defense Trade Advisory Committee, basically an advisory committee for the State Department. I am not aware offhand that he attended any of those meetings, but certainly his company representatives would have.
Mr. HUNTER. OK. Thank you, Mr. Chairman.
Chairman GILMAN. Ms. Danner.
Ms. DANNER. Thank you, Mr. Chairman.

I remember some of the comments made more than the particular individual who made the comments, so that I would expect you to weigh in if you are the individual who said, with regard to waivers, we consider innocent until proven guilty.

That bothers me a lot. This is not, you know, the American judicial system we are talking about. We are talking about national security. I regret that there is no one here from the Department of Justice so that I could pose this question to them. But if indeed there is an investigation by the Justice Department and the waiver is granted, what then is the real incentive for the Justice Department to go forward if the waiver has already been granted? So why is it that we have a policy of innocent until proven guilty when it is dealing with our national defense?

Secretary HOLUM. I was the person who used that phrase, and I was defining our approach to deciding whether a future waiver, a waiver involving a launch yet to be conducted, should be held up if there is an investigation into a past event, a past circumstance under a previous license.

Our practice, and I think it is a sound practice, is to not deny this company's right to do business on the basis of there being an investigation. However, if there is an indictment based on past behavior, if it moves that far along, and that's why I suppose innocent until proven guilty is highly qualified in that circumstance, because they haven't been convicted of anything when there is an indictment, but at that point we will deny licenses because there is sufficient probable cause to justify an indictment. That for us is sufficient in most cases to conclude that licenses should be denied on a moving-forward basis.

Ms. DANNER. Is it your knowledge that this has been the approach of past administrations? I know the administration is very eager to say this happened under Reagan and Bush. But we had this kind of transfer in technology in previous administrations without having a green light from the Justice Department, or let me rephrase that, having a red light from the Justice Department.

Secretary HOLUM. Well, we have looked into precedence in connection with this specific case to see what past practices were. I don't think there were any past practices having specifically to do with satellite launches in this kind of waiver. But the practice has existed for many years in connection with our export licensing rules under the International Traffic and Arms Regulations that we have not—previous administrations have not denied licensing, suspended licenses, or denied licenses on the basis of an investigation short of an indictment.

Ms. DANNER. Considering the concern that has been raised with this issue, do you not think that it might be prudent to revisit the subject matter and see if we can go forward? You are constantly referring to what has transpired in the past. That does not necessarily mean that that is right for our future. Should we not revisit this issue and see if what we are doing is the prudent method and manner?

Secretary HOLUM. I think any policy is always subject to review and should be. No approach should be cast in stone.

On the other hand, it is important to keep in mind, and I think it was part of our—I am aware it was part of our consideration in this case, that if an investigation should lead to an indictment, to further legal proceedings, that we have the option of revoking outstanding licenses, preventing any further activity. So you can come down hard at that stage.

Ms. DANNER. But in this instance, in my vernacular, coming from the Midwest, that would be the proverbial locking the barn after the horse is out.

Secretary HOLUM. That is not the case, because these launches usually occur many months after the license is granted. It takes a great deal of preparation. They do not go forward with the activity until they have a license in hand. But then they have many months of preparation before the actual launch can occur. So you can interrupt the proceeding before it is completed.

Ms. DANNER. Thank you very much.

My last question will be for Mr. Tarbell. I believe you are the individual that said, and I will have to look at the transcript to ascertain that—I got the impression from what you were saying that many decisions were made before you sent them on to a higher level, I believe was the phrase you used.

I am wondering—and then you used the word "bureaucrats"—if we indeed have a situation here where the bureaucrats are making decisions regarding policy instead of the policymakers?

Mr. TARBELL. That is my job, Congresswoman, to discern whether a policy issue is raised by a particular license matter.

As I explained at the last hearing, we review 21,000 licenses per year in my agency that are referred by the State Department and the Commerce Department. Not all those licenses come to my attention. Not all those licenses raise policy matters. But when they do raise policy matters, then I do take it for higher levels for review.

Ms. DANNER. So when Justice raised the issue, you took it forward to a higher level for review?

Mr. TARBELL. In the circumstance of the waiver, as Under Secretary Slocombe explained, Defense Department does not review these waivers for determination of whether or not they meet a national interest. We review them to make sure that they have accurately described the conditions that are in the licenses that are associated with the waivers, the safeguards that are attached to them.

Chairman GILMAN. The gentlelady's time has expired.

Mr. Weldon.

Mr. WELDON. Thank you, Mr. Chairman.

I want to follow up on the testimony of last week. I tried to press our panelists to discuss whether or not there was encryption technology contained in the 1996 failure of the Long March. We got the answer that the 1995 failure did not have an encryption capability on board.

But even though I had had committee staff meet with you before your testimony, in fact a week before, and committee staff told you-all that I would be asking that question about whether or not there was an encrypted capability in the 1996 launch, when I asked that question last week, all of you said you had no answer even though you had been told by the staff previous that you were going to be asked that question and be prepared to answer it. No one, in fact, would answer it. I asked you before the end of the day to answer that question for me. I don't know whether you ever did or not, but I've gotten the answer, and I will discuss that in a moment.

But, Mr. Reinsch, in response to my question, you made the statement that—in wire service reports of the transcript, they incorrectly identified your testimony as coming from Mr. Lodal, and I would ask the wire service to correct that—it was you who responded, "I have had some experience in encryption. The answer, in my judgment, is no. It wouldn't have had any effect on our national security," which is a pretty emphatic no.

Well, since last week we have been in touch with the NSA, and I have a document I would like to insert in the record at this point in time, Mr. Chairman, from the NSA relative to the lost encryption board and also the launch failure comment from DOD, or State Department——

Chairman GILMAN. Without objection.

[The information referred to can be found in the appendix beginning on page 381.]

Mr. WELDON. In fact, the statement from DOD is pretty emphatic. What I could not get from you-all last week is the fact that the command processor box as a whole were recovered in the 1996 incident, but not the circuit cards, which contained the encryption that were in those boxes.

It would have been very easy for you to give me that answer last week. I cannot believe that none of you knew that answer last week. And yet, in one phone call, I was able to get that answer. So my question is, I wonder why that tack was taken last week for this committee, because this is a critical element. If we can't trust the Chinese, when we have a failure and they hold our people back for 5 hours, as was stated in this hearing last week, they went in and evidently recovered the processor box; but when we got the processor box opened, the circuit board which contained the encryption was missing.

Now, Mr. Reinsch on the record said that, in his opinion, that would not jeopardize our national security, but that is not what the NSA and DOD has said. And I have entered both those documents in the record. And even if we were to buy Mr. Reinsch's position that it would not jeopardize our national security, I would ask the question why the NSA has subsequently changed the space encryption algorithms since that failure? If it was not a problem, why would our National Security Agency then have to change the encryption algorithms; supposedly, I guess, because of the security breach there?

These are very troubling questions because this gets at the heart of our question that we have to answer. Was the security of this country jeopardized; was it compromised in any way?

And we do not know the answer to that and the select committee is going to investigate that, but I want to raise this because I just don't think you were candid last week. It is hard for me to believe that none of you knew the answer to that question when I asked it last week. Even though I had been assured by Mr. Tarbell that we would get that by the end of the day, to my knowledge we never did get that answer. We had to finally go get it for ourselves over the weekend.

Mr. WELDON. But it is in the record now. Does anyone have a comment?

Mr. Slocombe, you weren't here last week, so I would like to have the others answer it who were here.

Mr. Reinsch, do you disagree with my characterization now and do you think that perhaps NSA and the DOD are overreacting on the encryption piece?

Secretary REINSCH. We have to get the right piece of paper, Mr. Weldon.

I also had a discussion with the NSA on the same subject because I wanted to make sure that what I said was accurate. I think what I can best do is—I haven't seen the statement that you have put in the record, but I can read you what they have provided to us.

Mr. WELDON. Let me ask you, since that crash, they have changed their algorithms.

Secretary REINSCH. Let me read the two sentences that I have here, if that would be helpful. And if, in fact, they are the same, then perhaps we have made some progress.

There is some chance that a third party could reverse engineer such devices to gain some knowledge about the engineering principles for making such chips. However, any loss of the chips and associated encryption algorithms would have had only minimal impact on national security, because the older satellite in the Intelsat case, which is the one you were alluding to, used older algorithms for which the keys are unique.

Mr. WELDON. I am aware of that. But my point is, if there was no concern, why would the NSA change the algorithms?

Secretary REINSCH. I think that that is something they do as a matter of routine, periodically, but I think you can discuss it with them.

Mr. WELDON. The statement I put in the record says it is a result of the crash.

Secretary REINSCH. That they changed the algorithms?

Mr. WELDON. Yes.

Secretary HOLUM. I am venturing into an area I don't understand very well, but isn't that a demonstration of a minimal impact on national security? Because we have the ability to change the algorithm so that the chip is not the utility.

Mr. WELDON. My point was, as Mr. Reinsch testified last week, there was no damage to our security. And if there was no damage, why would NSA think it serious enough to change the algorithms?

Secretary REINSCH. Perhaps I could comment, if I may.

I said what I said for two reasons. One, because the encryption in question, as I understand it—if I am wrong, let me know. As I understand it, the encryption in question is not data encryption but encryption that influences the direction and movement of the satellite.

Mr. WELDON. But the process of that encryption I think is what is in question, of putting that encryption into a chip and not just the ultimate use of that encryption.

Secretary REINSCH. But I don't understand, what would that give them—assuming they could recover that and take that chip, what would that give them?

Mr. WELDON. Well, that is going to be the subject and focus of the Select Committee's action. In fact—what additional capabilities did, in fact, China get as a result of this?

And what troubles me more even than the issue itself, if we have this reliable relationship with the Chinese and we have a very sophisticated capability that no one wants to give them—because I don't think for a moment any of you wanted to give them any of our encryption capabilities. I mean, I have the highest respect for you in that regard, but when we realize that the box containing that is recovered but that device is missing, doesn't that tell us something? I mean, it sure tells me something.

And for all of what are supposedly good intentions for improving commerce, we better call the Chinese on this issue if, in fact, we recovered the device. But that chip, which they obviously must know what it is, is missing. That, to me, is a very serious concern because it indicates intentions.

And maybe what they really are about doing, not this relationship of, you know, old mother and apple pie and everything is OK and we just held you back for 5 hours from the site because we didn't want to see you at risk. I mean, maybe I am a little pessimistic, but I don't buy that argument when I have been informed that we recovered the command processor boxes as a whole but not the circuit cards which contain the encryption.

If any of you have any comments on that, I would be very much interested in hearing them.

Secretary SLOCOMBE. If we fully trusted the Chinese, we wouldn't have all of these elaborate arrangements with monitors and people sitting in meetings and so on.

Mr. WELDON. I know my time is up, and Mr. Gilman is going to pursue a line of questioning along that line, Mr. Slocombe, where I would agree with him that some of the monitoring after some of these has not been what you would profess it to be, which raises additional concerns, and I have a number of other questions.

Chairman GILMAN. The gentleman's time has expired. We want to commend the gentleman for being on the task force that will be able to pursue this issue.

Mr. Burton will wind up the first round of questions.

Mr. Burton.

Mr. BURTON. Thank you, Mr. Chairman.

You know, Mr. Slocombe, you said, in response to a question from Mr. Weldon, that we don't trust them. That is why we have to be careful about these things. And you talked about the theft of information, I think, when Mr. Rohrabacher was talking.

You know, I don't understand why we are giving these technology transfers when you can't trust them and they are stealing this stuff from us. I mean, if they are doing that, why in the world are we even doing business with them in this area? I don't understand that.

Anyhow, let me move on to a couple questions I have.

I asked Mr. Milhollin last week, do you believe that the technology that we lost puts us at risk on the next battlefield? And he said, I think that the Gulf War taught the whole world, including us, an important lesson, and that was that the person or the side who can process information faster is going to win. If you can see over the horizon and your opponent cannot, if you have smart weapons and your opponent doesn't, if you can pick him up on radar and he can't pick you up, you have real-time battlefield information for remote sensors and he doesn't have, all of which can be processed or must be processed by computers, then you can see what you said is absolutely right, that the next battlefield, in fact, is already here. The Gulf War was such a battlefield. It was an information technology battlefield.

Mr. Reinsch, I would like to ask you this question. Do you recall in an interview with staff investigators from the Government Reform and Oversight Committee, my committee, in late September,

1997, do you recall the subject of technology transfer in the case of Uchi America to the People's Republic of China in 1994?

Secretary REINSCH. I recall that meeting, Mr. Burton. I recall that conversation, and that one has been laid to rest a long time ago. It is sort of cobwebby. But go ahead.

Mr. BURTON. OK. Do you recall that the Department of Commerce found that Uchi America was found to be outside the bounds of regulations by the Bureau of Export Administration and was fined $200,000 for breaking the law?

Secretary REINSCH. That is correct.

Mr. BURTON. And do you recall that after that, after they were fined $200,000 for breaking the law, they went ahead and sent two sophisticated 5 XS C&C horizontal machining centers to the People's Republic of China after the penalty was imposed? And can you explain why, since they were penalized, that the transaction took place, went forward and there was no prosecution in the case? Why did that technology go onto China?

Secretary REINSCH. We have been over this with your staff, Mr. Burton.

Mr. BURTON. Well, let's go over it again.

Secretary REINSCH. Well, I am happy to.

The answer, as I recall, and we will have to get you additional details, if you would like, is that the fine was because of their attempt to export these items without securing the proper license and without going through the proper procedures.

Subsequent to the fine—hold on. Subsequent to the fine and the action that we took, because they were violating the statute, the interagency process—they submitted an export license application for the same items. The interagency process reviewed that export license application and decided to grant it.

Mr. BURTON. So you fined them for breaking the law, but then you went ahead and granted the export license?

Secretary REINSCH. We fined them for breaking the law. That is correct. And they did break the law. They then subsequently acted pursuant to the law, and we reviewed their application pursuant to the law on its merits.

Mr. BURTON. And that involves seamless aircraft parts and possible missile applications from that technology, right?

Secretary REINSCH. I am not sure that that is correct. I would have to go back and get the record.

Mr. BURTON. My staff tells me it is.

Do you recall the RSA–SDI issue?

Secretary REINSCH. The which? I don't recall having a discussion about that, no.

Mr. BURTON. RSA encryption and SDI, what is the name of the company?

Secretary REINSCH. This is a new one on me, Mr. Burton. I don't recall that in the——

Mr. BURTON. Rivest, Shamir, Adleman, Deal and what's the S?

Secretary REINSCH. I know what the RSA algorithm is. I don't recall any discussion with your staff on that subject.

Mr. BURTON. Well, would you explain that issue for us?

Secretary REINSCH. I don't know what the issue is. It is a reference to the algorithm that is used in encryption.

Mr. BURTON. It is encryption technology that has gone to the People's Republic of China, and it was previously banned, as I understand it. You are not familiar with that?

Secretary REINSCH. This is not an issue that we discussed with your staff, and I am not quite sure what you are asking me.

Mr. BURTON. I didn't say that you discussed this issue with my staff. I am asking you this question now.

Secretary REINSCH. The RSA algorithm is a public—you can download it off the web, Mr. Burton. It is not a classified or closely held algorithm. It is proprietary, but it is available on the Internet.

Mr. BURTON. It was previously banned, though, is that correct? But now it——

Secretary REINSCH. I don't think so, no. I have never heard that before. I am happy to look into it.

Mr. BURTON. Well, I wish you would; and I wish you would give us an answer on that.

Going back to the Uchi America, you don't think there was any real concern about that technology transfer after the $200,000 fine was levied?

Secretary REINSCH. Well, what I can tell you is that the agencies reviewed the export license application and determined it was an appropriate one to approve.

I would be happy to go back and review the application and let you know my conclusions.

Mr. BURTON. I would like to have your answer on that.

Thank you, Mr. Chairman.

Chairman GILMAN. The gentleman's time has expired. Thank you, Mr. Burton.

We will now start our second round of questions.

Mr. Skelton.

Mr. SKELTON. I have a question that relates to licensing. I think, Mr. Slocombe, you might be the best one to answer this; but, if not, I will refer to the others.

Relating to the licensing system, Chairman Spence, in his opening remarks last week, said Department of Commerce procedures effectively downgrade national security considerations and elevate commercial and trade factors.

Now, Mr. Hamilton, in his opening remarks, who is a member of the other committee and as the ranking member of the other committee was here the other day, last Thursday, said that Gordon Oehler, former nonproliferation chief at the CIA, confirmed the present export licensing system because the CIA now sees all the Commerce licenses; and when it was under the State jurisdiction, he did not see them. Would you comment as to which statement is closer to the truth, in truth and fact, and explain why?

Secretary SLOCOMBE. I saw that report of Mr. Oehler's testimony. Maybe Mr. Tarbell knows whether it is true. I assume it is true.

Secretary REINSCH. I can respond. It is true that the Commerce Department shares all its licenses with the nonproliferation center at the CIA. I can tell you that.

Mr. SKELTON. The question is, is it tighter now than it was before?

Secretary SLOCOMBE. I think the current system is an improvement over the old system.

The very first question at this hearing was from Mr. Hefley, and he asked, is the Defense Department view, as a national security perspective, properly reflected now that the jurisdiction is in Commerce?

And I think the present system, in our experience, fully ensures that we know about the licenses when they come in. We have the opportunity to raise issues when they need to be raised, if they can't be worked out at a lower level, which is the way all of these licensing systems have to work. There are thousands of licenses involved. If they can't be worked out—not for satellites, but what DTSA has to do. There are thousands of licenses—what can't be worked out at lower levels as a good appeals system to get it up for decision where policy issues are raised or where there are real differences.

I am fully satisfied that, procedurally and substantively, the national security perspective is well heard in this system and that important improvements were made in 1996 when the changes were made. If you like, I can go back over what those improvements are, but it is in the record already.

Mr. SKELTON. Thank you.

Chairman GILMAN. Thank you, Mr. Skelton.

Mr. Spence.

Chairman SPENCE. Thank you, Chairman Gilman.

Let me pursue, just a minute, a question that was raised by Mr. Weldon relative to the crash and the lost encryption board. Just that one part of it. Is this technology, the lost encryption board that we are talking about, one that would require a license for export to China?

Secretary REINSCH. Yes.

Chairman SPENCE. That being the case, have the Chinese now acquired this controlled technology illegally?

Secretary REINSCH. Well, we don't actually know what happened to it, but that is certainly a possibility.

Chairman SPENCE. That is the central question I think we have been involved in here today.

Mr. Slocombe, I think you might have answered the other day, a while ago, at least, a question. We have been talking about export controls and all the rest of these things, and we get a disconnect sometimes when so many different answers are given to different questions and we miss the central point, sometimes, I think that we are trying to get to.

Mr. Slocombe, you said, I think, if I heard you correctly a while ago, that if there was a missile technology transfer which might be harmful to national security we can't talk about that report because it is being investigated by Justice. If that were the case, this transfer was illegal and not authorized by the export controls, is that a fair summary of what your position was?

Secretary SLOCOMBE. Yes.

Chairman SPENCE. Thank you.

Chairman GILMAN. Thank you.

We are going to go to give the panelists an opportunity to take a break. We will take a 5-minute recess.

Secretary SLOCOMBE. Thank you.

[Recess.]

Chairman GILMAN. The hearing will come to order. Witnesses please take their seats. Members please take their seats.

Mr. Bateman.

Mr. BATEMAN. Thank you, Mr. Chairman.

I am going to raise a particular question today, during this opportunity, to ask for a response on the record. I am more interested in that than I am in an ad hoc, off-the-top-of-your-head kind of response.

I also want to take a moment to express further my sense of aggravation that we are dealing here with what is said to be a policy in keeping with our national security interest and much of which is bottomed upon the necessity of getting satellites into orbit, and yet I have not heard a single commentary from any witness about anything that we have done or are doing to enhance America's own domestic commercial satellite capabilities.

I am much more interested in the jobs for Americans, the boost to our economy, than I am hundreds of millions of dollars of additional business for the Chinese. That doesn't make sense to me as a national policy, that we go and leave it unaddressed.

The question I will ask at this point for the record is an account, or an article I should say, in the Wall Street Journal today.

Secretary SLOCOMBE. Mr. Bateman, it is a little hard to hear.

Mr. BATEMAN. I am sorry. I will get closer to the microphone.

Secretary SLOCOMBE. Thank you.

Mr. BATEMAN. There is an article in the Wall Street Journal today under the byline of Brian Duffy which asserts, I don't know whether it is accurate, that the Pentagon officials didn't monitor the launches of U.S. satellites on three Chinese rockets because of bureaucratic confusion. The article goes on to indicate that we don't know which launches were unmonitored.

I find this very, very difficult to accept. And I also would like—in addition to an explanation as to whether this article is factually correct, I would like further explanation on this matter of our satellites and the things that are licensed for export to the Chinese, if they are guarded or monitored.

Mr. Slocombe, I believe I am correct that you said they aren't monitored or guarded under these transportation plans by DOD personnel. Are any government personnel involved as monitors or guards? And, if not, who are they, and are they the very people who represent or work for the companies that wish to export not only this particular technology under this particular license but who may have pending billions of dollars in contractual work that they are seeking from the Chinese Government?

I would like that answered for the record in some detail.

Secretary SLOCOMBE. I will provide more detail, but let me give you a general answer to both parts of your question.

First, as to who does the monitoring, my understanding is that the monitors are U.S.—mostly military personnel, sometimes civilian employees, who volunteer for this duty. So the actual monitors are U.S. DOD employees.

Mr. BATEMAN. Not to argue the point—that isn't my purpose today—but it is your understanding——

Secretary SLOCOMBE. I want to check and confirm that, but that has been what has been represented to me.

Now with respect to who is the guard, so to speak, during transportation, it is my understanding that, in general, the satellites are shipped on U.S. carriers. And when it is a U.S. carrier during the time the satellite is on the airplane, contract guards are used. But that if, as is sometimes the case, a non-U.S. carrier is used, then one of the official monitors, the government officials would do the monitoring.

But I would get you a more complete answer as to just how the mechanics of this monitoring works.

[The information referred to is retained in the committee files.]

Secretary SLOCOMBE. With respect to what I understand to be three cases, which I think are what is the subject of the Journal article that you mentioned, there was a period of time from about 1993 to 1996 when the Commerce Department had jurisdiction to launch—I am sorry, had jurisdiction to license launches of satellites, as Mr. Reinsch testified a little while ago, that did not have any of nine identified technical characteristics that the Commerce Department could license those without a State—without a—they required a Commerce license, not a State license. Those licenses did not require DOD monitors and, therefore, there weren't DOD monitors.

In the normal course, there was an accompanying license which was a State license having to do with the launch arrangements, and that did require DOD monitor; and when the monitor was required, they were there. In these three instances—and, to the best of my knowledge, it is three instances which we can specify as to what they were—the company took the position that it didn't need a license for the launch information because it was only what is called form, fit and function information, and it came within the scope of the Commerce license. That is the reason that there were no DOD monitors on those three launches.

One of the ways in which I think the system has been improved since that period with the new arrangements is that there are always DOD monitors provided for and paid for by the launching—by the satellite—I guess the satellite manufacturer in both Commerce and State licenses. So that this gap, if you will, in the monitoring requirement has been fixed.

We will get you more detail on the particular instances, but that is the general answer. It is not that there were supposed to be monitors and nobody showed up to monitor. There was a position, which was not an unreasonable one, that no monitors were required; and, therefore, there weren't any monitors. But we will get you more detail on the particular story.

Mr. BATEMAN. I would like that because what you are telling me would seem to indicate that the Wall Street Journal reporter is not accurate in what he is reporting.

[The article referred to can be found in the appendix on page 342.]

Chairman GILMAN. The gentleman's time has expired. Thank you, Mr. Bateman.

Secretary HOLUM. Could I also provide a brief response on the question of assistance to the domestic launch industry? Because I know you have been concerned about that since the other hearing. That is something we are interested in.

I think Under Secretary Reinsch has made the point here and in other hearings that we have approximately two-thirds of the satellites to be launched and, domestically, about 40 percent of the launch capacity. There is a domestic launch industry, but that is the gap we need to fill up with foreign launches, and that number of launches is expected to grow over a period of time.

One of the ways we have worked to protect and advance the interests of the U.S. launch industry is through the agreements we have negotiated with the Chinese which provide quotas on the number of launches that they can conduct and also provide that they can't undercut by a dramatic amount the prices charged by others. So there is a limitation on their ability to undercut domestic pricing. The idea is to level the playing field for domestic launchers.

The Department of Defense and the Department of Transportation are working with industry on the evolved expendable launch vehicle program to develop the next generation launch vehicle, which is something we are interested in that would help support the domestic launch industry. And we have met, from the Department of State, with a number of entities who are working on developing new launch capabilities, including in cooperation with other countries.

Chairman GILMAN. The gentleman's time has expired. Thank you, Mr. Bateman.

Mr. Taylor.

Mr. TAYLOR. Thank you, Mr. Chairman.

I am going to read a portion of the testimony that the staff put together talking about, on February 12, 1998, the President signed another waiver of Tiananmen sanctions for the launch of a Loral satellite. This approval was reportedly given over the objections of the Justice Department which expressed concerns that the ongoing criminal investigation into the 1996 transfer could be compromised by approving another Chinese launch for Loral while the company was under investigation.

The administration had argued that, under a 1993 Department of State memorandum, a license can be denied only if an indictment has been issued or if there is reasonable cause to believe that U.S. export control laws have been violated. Despite sensitivity surrounding the waiver, the President's National Security Adviser recommended approval in a memo to the President arguing that the advantages of this project outweigh the risk.

On April the 23rd, Chairmen Spence and Gilman sent letters to the different agencies asking for the materials that led up to this. Surprisingly, on April 29, the Deputy Under Secretary of Defense, John Hamre, responded that the Justice Department has asked that we not release the DOD report at this time because it has concluded that doing so, even in a redacted form, would compromise an ongoing criminal investigation.

Gentlemen, this sounds like the Keystone Cops. One week it's OK to transfer this technology and grant another waiver to the Chinese. The next week, no, we can't even tell you why we did that because there is an ongoing criminal investigation. It's one or the other. It's right or it's wrong. Now, that it has happened, that is water under the bridge. There is an ongoing criminal investigation.

And I really hope if someone sold this country out for any reason that they fry for it.

My question to you is, what is being done so that this doesn't happen again? What is being done to either encourage commercial satellite launching in this country, using existing missiles that we have in our inventory? What is being done to prevent this from happening again in what is proposed—a joint Russian-United States space launch proposal being done at the Equator on an offshore oil platform? What is being done so that this doesn't become the norm, rather than something we find appalling? And what recommendations are you going to make to this committee or should we come up with our own?

Secretary HOLUM. Well, I won't go back through what I just said about our efforts to encourage the domestic industry.

One of the efforts that—one of the areas that we are exploring is U.S. entities in cooperation with other countries, using an offshore platform to launch space-launched vehicles.

Mr. TAYLOR. If I may, why always in cooperation with other countries? I am as deeply concerned as anyone in this room that it doesn't really matter if you tell this person or that person sensitive information it suddenly becomes in a position where anyone can get a hold of it.

The fact Loral faxed a 200-page memorandum to, first, an insurance company, apparently without either the Department of Defense or the Department of State looking at it and, in turn, faxed it to the Chinese Long March missile people, that troubles me, because I am not so sure of just how patriotic that insurance company is. They might have turned around and sold the information. I think, in most instances, somebody was asleep at the switch. My question to you is, what is being done so that this doesn't become the norm?

Secretary HOLUM. One of the things that is crucial to do—and I think this is not a function of satellite licenses or waivers. It is a function of enforcing the law. If someone is determined to share U.S. national security secrets with China, if they have access to information of that kind, they will do it, unless——

Mr. TAYLOR. If I may, please.

Secretary HOLUM. I would like to finish my answer.

Mr. TAYLOR. They are not sneaking a note to somebody. They are faxing this over the telephone.

Secretary HOLUM. Mr. Chairman, is it permissible to answer?

Chairman GILMAN. Allow Mr. Taylor to make his inquiry, and then you may respond.

Mr. TAYLOR. Sir, my point is, this was not done secretively. It was not done through espionage. It was done over a fax machine. Isn't someone monitoring the conversations between Loral and the Chinese? Isn't someone monitoring the conversations between Loral and the insurers to find out whether or not sensitive information is being passed along? And whose job is it and what is going to be done so this doesn't happen again?

Secretary HOLUM. It is all of our jobs. And what I was getting to is you can't link a violation of law to any specific kind of commercial activity. It can occur anywhere. And the answer to it is to enforce the law, which is what we are doing.

As soon as this instance came to the attention of the appropriate authorities, it was referred promptly to the Department of Justice, and that is what we need to do. We need to make clear on a continuous basis that this kind of activity, in any context, whether it is in satellite launches or on the part of anyone who has access to this information, that they are not—that it is not going to be permitted and it will be prosecuted vigorously.

Mr. TAYLOR. Was anyone supposed to be monitoring conversations between Loral, Hughes and the Chinese or did you only decide after the fact they were going to start doing that? Again, this wasn't done secretively. It was done over a fax machine.

Chairman GILMAN. The gentleman's time has expired.

Go ahead.

Mr. TAYLOR. Mr. Chairman, I would request that if they can't get me the answer today, I would sure like to have it in writing.

Chairman GILMAN. Can someone give a response to the inquiry? Mr. Slocombe?

Secretary SLOCOMBE. Well, the licenses require that all technical meetings between the U.S. satellite manufacturer, the person who has the license, and the Chinese be monitored by DOD personnel. So the general answer is that is a requirement. And if it was violated in a particular case, it should be prosecuted vigorously, as you said at the beginning of your inquiry.

Chairman GILMAN. Thank you.

The gentleman's time has expired.

I am going to take my 5 minutes of my time.

Secretary Reinsch, on Page 4 of your prepared statement you address the issue of an analysis conducted in the 1995 launch failure. You stated that you determined U.S. participation in the 1995 launch investigation was consistent with the terms of the Commerce export license. Is it your view, therefore, that a Commerce license permits U.S. persons to participate in launch investigations?

Secretary REINSCH. We were informed in that case that the launch failure analysis that we reviewed had been conducted independently by the company.

Chairman GILMAN. And that was under the license? Is that appropriate under the license?

Secretary REINSCH. Yes, at that time.

Now, following the President's decision in 1996 to transfer jurisdiction of all these things to the Commerce Department, we have added a condition to our licenses at that point that specifies that, in the event of a launch failure, the license that we have provided doesn't extend to beyond that point and that the company needs to go to the State Department for a separate license if it is to participate in a launch failure analysis.

Chairman GILMAN. Well, at the time of this investigation, did you coordinate your determination with DOD and with State?

Secretary REINSCH. That was Mr. Hunter's question earlier; and the answer is, no, sir, we did not.

Chairman GILMAN. When did you tell DOD and State that you would authorize the launch investigation?

Secretary REINSCH. We didn't authorize the investigation. I think the question might be, when did we tell them what was going on?

And I will have to get back to you with a precise date, if I can, Mr. Chairman.

[The information referred to is retained in committee files.]

Chairman GILMAN. Well, who was in charge of the investigation? Who started the investigation? Who initiated the investigation?

Secretary REINSCH. The company conducted a launch failure investigation. The U.S. Government hasn't conducted—well, the agencies sitting here haven't conducted an investigation of that particular case. As far as I know, no allegations have been presented regarding that.

Chairman GILMAN. Well, did any of the agencies authorize the company to do the investigation after the launch failure?

Secretary REINSCH. I would say that, within the terms of the license that we granted, further authorization wasn't needed. They didn't exceed the terms—in terms of the technology that we were looking at, they didn't exceed the terms of the license that they had been given previously.

Chairman GILMAN. How do you make that determination if there was no monitoring of this?

Secretary REINSCH. We examined the report that they provided, the launch failure analysis, and reviewed it before they were allowed to give it to anybody and determined that its content did not contain information that exceeded the bounds of the license that we had provided.

Chairman GILMAN. Let me understand that. The investigation was pursued by the company?

Secretary REINSCH. The company pursued a launch failure analysis.

Chairman GILMAN. Then the information was given back to you before it was released to anyone?

Secretary REINSCH. That's correct.

Chairman GILMAN. And you authorized its release?

Secretary REINSCH. Because it did not go beyond the limits of our license. That is correct.

Chairman GILMAN. Who was it that authorized the release of the information?

Secretary REINSCH. Our licensing officer.

Chairman GILMAN. And who would that be?

Secretary REINSCH. Gene Christiansen.

Chairman GILMAN. Licensing officer of which department?

Secretary REINSCH. Department of Commerce.

Chairman GILMAN. So Christiansen was authorized formally to release it to the Chinese?

Secretary REINSCH. Mr. Christiansen made the judgment the information in the report did not exceed the terms of the license that had been previously awarded, so he told the company they could provide the document to their insurance companies and subsequently to the Chinese.

Chairman GILMAN. And who cleared that with national security?

Secretary REINSCH. What do you mean by national security?

Chairman GILMAN. Was there any clearance?

Secretary REINSCH. It was our license. This was a judgment we made on our own.

Chairman GILMAN. You made it on your own. Was any national security interest reviewed with regard to the release of that information?

Secretary REINSCH. The license itself defined the terms of what could be transferred. The license was approved by the other agencies in question. The license specified what technology could be transferred. The information in this report did not exceed the terms of what the agencies had approved when the license was granted.

Mr. HUNTER. Would the gentleman yield?

Chairman GILMAN. Yes, I would be pleased to yield to the gentleman.

Mr. HUNTER. Thank you, Mr. Chairman.

You talk about a post-failure report that was transferred. You got a chance to look at it before it was transferred to the Chinese?

Secretary REINSCH. We examined it and determined that it did not exceed the terms of the license and then told them that they could transfer it to their insurance companies.

Mr. HUNTER. Did any meetings take place between the Chinese and the company officials?

Secretary REINSCH. We will have to check. I don't have the answer to that question right now.

Mr. HUNTER. If you don't have the answer, if you don't know whether or not they had face-to-face meetings and talked to each other, how can you tell us you are certain that no military technology or prohibited technology was transferred?

Secretary REINSCH. We can tell you that, based on what we know, based on the license analysis, based on what was discussed in the analysis and what they covered in that analysis, which related entirely to the satellite and the ferrying around the satellite, we have no reason to believe that any technology was transferred, other than that that was permitted by——

Mr. HUNTER. But you don't know if any meetings took place. It's apparent now that lots of meetings took place between people.

Secretary REINSCH. I don't at this moment, Mr. Hunter, know of any meetings taking place. I will have to find out whether the Department knew if meetings took place.

Chairman GILMAN. Regaining my time, Mr. Hunter.

Mr. HUNTER. Thank you, Mr. Chairman.

Chairman GILMAN. Mr. Slocombe, do you think that your agencies should have been consulted before the release of that information was made available to the Chinese?

Secretary SLOCOMBE. As I understand it, under the procedures that were in effect at that time, that was not necessary.

Chairman GILMAN. I am asking you now, looking back at that investigation, would it have been far preferable to submit that information to your agencies?

Secretary SLOCOMBE. Without reference to that particular case, I think the system works much better when all the agencies are involved and it is a transparent process among the agencies. That is why I believe the post-1996 licensing system is a good system. Commerce administers it, but it is an open process, as far as we are concerned.

Chairman GILMAN. Mr. Holum, what is your opinion?

Secretary SLOCOMBE. But I don't want to be misquoted. I am not in any way suggesting that there is anything wrong with what happened in this case. I prefer a system in which there is a routine system——

Chairman GILMAN. It would have been far preferable to have both State and Defense involved in this. Is that what you are saying to us?

Secretary SLOCOMBE. I prefer the present system in which there is interagency involvement, but I accept that at the time this incident took place, it was not required.

Chairman GILMAN. Mr. Holum.

Secretary HOLUM. I think it really depends on what is done—we may get tangled up a bit in semantics in terms of what the inquiry is called. If the inquiry goes to the launch vehicle in analyzing why the launch vehicle didn't work, then that requires—any post-launch failure analysis would require a State Department license.

It is conceivable to me, not having reviewed the facts of this case, that there could be a post-launch failure analysis that would focus only on the satellite, that would not implicate the launch vehicle. Under those circumstances, I can understand why it wouldn't be brought under the ITARs and included.

Chairman GILMAN. But if it did include the launch vehicle, the missile technology, would you then change your opinion?

Secretary HOLUM. If it involves a launch vehicle, then it requires a State Department license, and we should be involved.

Secretary REINSCH. In those facts, I would change my opinion, too. The rules are very clear. If it involved launch vehicle technology, we would not have been able to make the decision that we did, and we would have referred it to the State Department.

Chairman GILMAN. Didn't it involve here the launch vehicle technology?

Secretary REINSCH. No, sir.

Chairman GILMAN. What made that determination—who made that determination?

Secretary REINSCH. The licensing official that I named earlier reviewed the document in question and determined it did not contain technology that went beyond the terms of our license. We are not in a position, either then or now, to authorize the transfer of launch vehicle technology. That is Reserved for the Department of State. Had the company, at the time of its license or subsequently, proposed a transfer of that kind of technology, we would have referred them to the Department of State.

Chairman GILMAN. So your opinion then, or Mr. Christiansen's opinion, was based upon the report given to you by the company, is that correct? And since there was no monitoring——

Secretary REINSCH. He examined the report, and that was what he authorized them to transmit.

Chairman GILMAN. What company manufactured the satellite that was destroyed in the failed launch?

Secretary REINSCH. In the 1995 failure?

Chairman GILMAN. Yes.

Secretary REINSCH. The Hughes Company.

Chairman GILMAN. And what was the name of the satellite that was scheduled to be launched after the failed February of 1996 launch that Justice is now investigating as a criminal matter?

Secretary REINSCH. The name of the satellite, the one that exploded?

Chairman GILMAN. We are talking about following the explosion.

Secretary REINSCH. The next one?

Chairman GILMAN. The one scheduled to be launched after the failed February 1996 launch that the Justice Department is now investigating. Was that Apstar 2?

Secretary SLOCOMBE. The paper in front of me says it was Intelsat 708. 702 is the January 1995.

Chairman GILMAN. Our information is it was a huge Apstar 2, is that correct?

Secretary REINSCH. That is the one that blew up in 1995, I believe, Mr. Chairman.

Chairman GILMAN. And then the next one was——

Secretary REINSCH. The one that blew up in 1996 was Intelsat 7—whatever the rest of the numbers are.

Secretary SLOCOMBE. 708, I think.

Chairman GILMAN. 708 is the next one, not Apstar 2, is that correct? What was scheduled following the failed 1996 launch?

Secretary SLOCOMBE. I am not sure I understand what you mean, what was scheduled following it. You mean, what was the next one that was——

Chairman GILMAN. That is right. That is what I asked. I asked, what was the name of the satellite that was scheduled to be launched after the failed February 1996 launch that the Justice Department is now investigating as a criminal matter? And it was my impression that was Apstar 2.

Secretary SLOCOMBE. According to this sheet, it is Apstar 1A, which was a huge satellite which was scheduled to be launched and was launched—I don't know when it was scheduled to be launched. According to this, it was launched on the third of July, 1996. Apstar 2 is the one where the launch failed, that was according to this, on the 25th of January 1995.

Chairman GILMAN. Have you or anyone else at Commerce had any contact with the company, Mr. Reinsch, with the company or the other U.S. company that is the subject of the Justice investigation, considering their participation in that launch failure or the Justice investigation into their participation?

Secretary REINSCH. If I understand the question correctly——

Chairman GILMAN. Have you been in contact with anyone in the company?

Secretary REINSCH. I think I indicated in response to Mr. Hunter's question that I have had a conversation with the Hughes Company regarding the 1995 explosion.

Chairman GILMAN. Who was that that you were in contact with?

Secretary REINSCH. Loretta Dunn, I believe.

Chairman GILMAN. When did that take place?

Secretary REINSCH. In that particular conversation, last week, the week before.

Chairman GILMAN. And what was that in reference to?

Secretary REINSCH. Pardon me?

Chairman GILMAN. What was that conversation in reference to?

Secretary REINSCH. As I indicated before, I would have to gather my thoughts and recollection, Mr. Chairman. We had a discussion about the general question of the whole history of this.

Chairman GILMAN. Was that in preparation for your testimony today?

Secretary REINSCH. It was in preparation for my testimony last week.

Chairman GILMAN. If you didn't inform DOD or State at the time of this investigation, did you talk to anyone at the White House about all of this?

Secretary REINSCH. About which investigation now are we talking about?

Chairman GILMAN. The failure of the 1996 launch—'95 launch, I am sorry.

Secretary REINSCH. I haven't—no, sir, I don't think I have had conversations, except in the context of preparing for this testimony.

Chairman GILMAN. And you did, in preparation for this testimony, speak with someone in the White House?

Secretary REINSCH. With respect to 1995? I reported to them the same information I provided to the committee in the last several questions.

Chairman GILMAN. And who did you confer with in the White House?

Secretary REINSCH. Primarily, Maureen Tucker, one of the National Security Council representatives.

Chairman GILMAN. When was that conversation?

Secretary REINSCH. There would have been several of them over the past couple weeks.

Chairman GILMAN. How recent was your last conversation with Ms. Tucker?

Secretary REINSCH. On any subject?

Chairman GILMAN. On the subject we are discussing today.

Secretary REINSCH. I imagine yesterday, or perhaps Friday.

Chairman GILMAN. Specifically, with the 1995 launch investigation?

Secretary REINSCH. No, we haven't discussed that. I think what happened in that situation is clear, and we haven't had a discussion about that for a while.

Chairman GILMAN. Have you made the Justice Department aware of your role in the 1995 investigation?

Secretary REINSCH. The Justice Department has been in contact with us on that matter and has requested the documents and we are complying.

Chairman GILMAN. And who in Justice have you been conferring with?

Secretary REINSCH. We have received written communications from the Justice Department.

Chairman GILMAN. Have you conferred with anyone personally in Justice in preparation for this hearing?

Secretary REINSCH. I had one conversation with them about what the appropriate response would be with respect to matters that are under their investigation. Because, like my colleagues, I don't want to deviate from their instructions.

Chairman GILMAN. And who did you confer with?
Secretary REINSCH. Craig Misko.
Chairman GILMAN. And what is his title?
Secretary REINSCH. I don't know.
Chairman GILMAN. And how recent was that conversation?
Secretary REINSCH. That was prior to testimony last week.
Chairman GILMAN. And did you provide the committee—our committee and the Justice Department the documents that consumed your role in the 1995 failure investigation?
Secretary REINSCH. We are in the process of complying with the Justice Department's request. Its exact status as of this minute, I am not sure, but we are going to provide those documents. To the extent that your committee requested them, we are happy to provide them as well. I am not sure they were in your initial request, but if they are in your current request, then we will provide them.
Chairman GILMAN. To this date, have you provided any documents to our committee with regard to this investigation?
Secretary REINSCH. We have provided documents to your committee, Mr. Chairman. I would have to review them on whether they related directly to this investigation. As I recall—just a second. As I recall, your committee's initial request to us for documents did not cover this case. If your pending request covers this case, then we are happy to provide the documents.
Chairman GILMAN. Well let me make it clear. We would like you to provide documents with regard to this pending case——
Secretary REINSCH. Fine.
Chairman GILMAN [continuing]. And provide it to our committee.
Secretary REINSCH. I am happy to do so.
Chairman GILMAN. Thank you. And I hope you could expedite that request.
So let me summarize then. You quietly authorized a United States company to share information with China concerning a launch failure investigation in 1995 and did so without consultation with any other U.S. Government agency, including the Justice Department, whose criminal prosecution you may have undercut. Is that a correct statement?
Secretary REINSCH. No, I don't think so, Mr. Chairman.
Chairman GILMAN. Tell me what is wrong with that statement.
Secretary REINSCH. I think the simplest—I think where you went astray, primarily, Mr. Chairman, is with respect to the last sentence. I have not been aware of any Justice Department examination of the 1995 launch until some point last month. The authorization that you are referring to and the launch failure analysis conducted by the company that you refer to were conducted 3 years previously, in 1995. I don't believe—I am quite confident there was no Justice Department investigation of this incident at that time.
Chairman GILMAN. What about the investigation of 1996? Are you aware of that?
Secretary REINSCH. We have not been involved in the investigation of 1996.
Chairman GILMAN. Were you aware of the 1996 investigation by Justice?
Secretary REINSCH. I became aware of it in the media. Our licensing officers were aware in 1996—I understand, were aware

that the company had—in this case, the Loral Company—had come in and made a self-disclosure, but my primary awareness of the investigation of the Justice Department came through the media earlier this year.

Chairman GILMAN. The Loral Company is the subject of a criminal investigation by Justice for a launch failure investigation 12 months later because of potential harm to U.S. national security, is that correct?

Secretary REINSCH. That is my understanding from the media, Mr. Chairman.

Chairman GILMAN. Then one last question.

Mr. Reinsch, is it still your view that administration policy and your agency's central role in the policy is a smooth policy and one that does not need any change?

Secretary REINSCH. Yes, it is, Mr. Chairman. I think the President's decision in 1996 strengthened the procedure significantly across the board and particularly in this case. I think these procedures have operated well.

There is one allegation in the midst of all this which concerns the 1996—one allegation of national security being compromised which concerns the 1996 case. We shall see, as a part of the Justice Department's examination of other cases, whether there are any other allegations. None have been made with respect to the 1995 case, and there is only one that has been made with respect to the 1996 case. The procedures that the President has put in place occurred subsequent to both of those events, and I think they have solved the problems you are alluding to.

Chairman GILMAN. Thank you, Mr. Reinsch. I know I have exceeded my time.

Mr. Snyder.

Mr. SNYDER. Thank you, Mr. Chairman.

I just have a few questions of clarification about some of the things that have come up today. Maybe you can help me, Mr. Holum. Mr. Hefley earlier mentioned President Bush had three to four waivers and President Clinton had approximately 11 waivers. My understanding is that President Bush's waivers were in batches, and the number of launches was nine under the Bush administration. Was the policy of President Clinton to do a waiver per satellite or per launch, rather than batches? Was that a deliberate thing or is that just something that worked out? Is there any significance to that?

Secretary HOLUM. You are correct in the numbers. There were nine launches under the waivers issued by President Bush, 11, I believe, under President Clinton. I don't know if there was—I suspect it was the way they came in. And when they were grouped together, they would grant the waivers at one time.

Can you amplify?

Secretary BARKER. Maybe I can amplify that.

When we go back and look at the record, the Bush administration waivers happened to be that most of them were bundled. All of them were for more than one satellite. Most of the Clinton administration waivers have been for one satellite or one satellite program only. In 1996, there was one waiver which covered three satellites.

Although we can't speak to how exactly these work, generally, when the licenses are completed and a waiver is necessary, it is sent to the White House. If a number of satellites come in at the same time, it is efficient to batch them together. I would assume in the Bush administration, they decided at some point it was efficient to batch, I think, maybe six of them together at one particular waiver for administrative convenience, but there is no particular significance to it.

Mr. SNYDER. I have this great front row seat here. Unfortunately, in the exchange of Mr. Rohrabacher, it was hard to hear both of you talking at the same time. But from what he was saying—I am trying to pick up on what he was saying—the concern is with the missile technology and not the satellite technology. But I think, obviously, in Mr. Weldon's questioning, some of us have concern about the encryption device and the explosion at the village, and there is concern in both areas. Is it fair to say when you all say national security concerns you are referring both to satellite technology and launch technology?

Secretary SLOCOMBE. Anything that could come out of the transactions.

Mr. SNYDER. And in your formal process, Mr. Holum, speaking for the group, you look at both those areas?

Secretary HOLUM. That is right.

The one area of distinction between Commerce and State licenses has been that the State Department, under our procedure, conducted, pursuant to a see-through, where we controlled the license that was embedded or included as separate ITAR items. That hasn't been the case under Commerce.

Mr. SNYDER. I want to go back to this chart we were talking about last week that ends up with 100 percent launch reliability since 1996. And I was aware of Ms. Harman's statements before I got here today, because our chart only begins in December 1988, the one that was provided to us last week. And, in fact, if you go back 4 years earlier, you end up apparently with 16 successful launches in a row. I think it would be helpful to the committee if you all or somebody in the administration were to analyze this charge, to point out is there any statistical significance to it or not.

Mr. Holum, do you have any comments from last week about that?

Secretary HOLUM. Well, I think it's important to have a longer view. The particular missile, the DF–5 that the Long March space launch vehicle is related to or derived from goes back to 1981, so it inevitably or certainly has a long history of launches, and it seems fair to me to include the entire life of the related booster to determine where the reliability gains came in.

Even in the chart that was handed out last week, one of my colleagues at State pointed out last night the vast bulk of the launches between May 1990 and March 1997 or into 1998—or up through March 1997 were Chinese launches. They were gaining experience with their own satellites.

The few blue dots on that chart are outweighed, heavily, by China's own launches. That leaves aside, of course, whether, if the U.S. weren't launching on those satellites, somebody else might not have.

Mr. SNYDER. Mr. Tarbell, you made mention of your 21,000 a year licenses that you do. How many different companies is that?

Mr. TARBELL. How many different companies?

Mr. SNYDER. Are we talking thousands of companies?

Mr. TARBELL. Yes. It is hundreds, at least.

Mr. SNYDER. It seems like, I mean, I think there are legitimate concerns about the whole process of technology, as it relates to defense issues, but I don't think this is going to get any better, is it? I mean, as the technology that changes, whether satellite technology or launch capability technology or Internet technology, as the commercial technology becomes increasingly similar and not that far behind the defense technology, your job is going to get tougher and tougher as time goes by in terms of sorting that out. What you think would be a national security threat this year, 2 years from now, it may be something you can pick up downtown in a Wal-Mart. Do you have any comments about how you see your job in a changing world?

Mr. TARBELL. I think you stated it pretty accurately, that is, that technology is a moving target, and it is something that you can't particularly wrap your arms around completely.

Our objective in most of the areas we review is to keep, in most of the commercial technologies, the dual-use technologies, keep our potential adversaries and keep some of the countries that we control technology to a couple steps behind us to maintain U.S. military technological advantage.

There are circumstances where we seek to prevent any transfer of technology, that is in the case of weapons of mass destruction, for example, but even in that area our ability to control that unilaterally in the U.S. is something that is very, very difficult and requires the cooperation of a lot of countries around the world.

Mr. SNYDER. Thank you, Mr. Chairman.

Chairman SPENCE. Mr. Rohrabacher.

Mr. ROHRABACHER. Thank you very much.

As the chairman of the Space Subcommittee, let me just note for Mr. Holum that anything that upgrades the capability of Chinese rockets is an enemy of the development of American rockets here, especially when we are trying to help develop new launch capabilities here in the civilian sector. So anything we are doing over there to help them perfect their rockets hurts us and hurts the development of our own domestic launch industry, dramatically hurts us.

And one thing that I feel has been a betrayal of this, other than national security, it's been a betrayal of our own aerospace workers and such, when we are talking about specific technologies that have been transferred. And, again, we are not talking about expertise that has been developed by practice makes perfect. I don't think anybody buys that.

One of the reasons why have I been personally upset is the fact that during this administration—I mean, frankly, in other areas we have seen this administration has been an expert at obfuscation and evasion and stonewalling. And that is why I am sorry if earlier I wanted to get to my questions and I only had 5 minutes, and if someone can give me a direct answer rather than eating up time, I had to make sure that the questions were answered.

So before I get into these last questions, let me say, after hearing all the testimony today, it is disconcerting to me that any public official who has responsibilities such as yourselves has the nonchalant attitude toward MIRVing and toward upgrading of rockets, claiming that, well, actually the rocket carrying the satellite is so different than the rocket carrying the warhead. It is disconcerting seeing officials with such responsibility have such a nonchalant attitude toward the transfer of this type of technology.

Secretary SLOCOMBE. My attitude to this is not nonchalant at all, sir.

Mr. ROHRABACHER. It seems to me over and over again you tried to just pass it by as saying, well, after all, there is these significant differences between the rockets carrying the payload and the rockets carrying the warheads.

Well, that is not what we have learned from other experts. And it helps somebody developing their MIRVing system and saying, well, it can help with multiple payloads, but do not worry, it will not help with multiple warheads. That is not what other experts are telling us.

For example, the Long March 2C rocket had 14 launchers, and then it went into serious consultation with American companies, and those launchers for the American companies, unlike the other launchers, had multiple payload capacity. Isn't that improving a rocket capability, and doesn't that have some implications for our national security?

Secretary SLOCOMBE. Well, that is, of course, the issue, and I think I've given my answer, and I don't want to waste your time by repeating it.

Mr. ROHRABACHER. Well, I would just say that your answers will have to be judged by the American people whether they are acceptable or not. I think the American people understand that when you have a rocket that is not capable of launching multipayloads, and then it is capable, that that has some national security implications, not to mention the implications for these aerospace workers who are trying to develop domestic launch capabilities.

Has American know-how or technology been used to improve Chinese rocket control and guidance systems?

Secretary SLOCOMBE. I think my answer to that is the same as before. You mean American technology transfers?

Mr. ROHRABACHER. Has American technology know-how been used to improve Chinese Communist rocket control and guidance systems?

Secretary SLOCOMBE. We have no reason to believe that that has happened.

Mr. ROHRABACHER. Is that the answer from everybody here? Isn't this the subject of a criminal investigation?

Secretary SLOCOMBE. I am sorry, unlicensed. I should have made clear that we are all—all of my answers are subject to the qualification of the particular incident of a potentially unlicensed and improper transfer are under investigation.

Mr. ROHRABACHER. All of your answers to my questions about the different technologies that may have been transferred, now there is a caveat thrown in?

Secretary SLOCOMBE. I have said many times in my answers, and so have my colleagues, that there is a criminal investigation going on of a potential unauthorized and illegal perversion of technology.

Mr. ROHRABACHER. That is not what you just said, something about license and a caveat.

Secretary SLOCOMBE. I am satisfied that, insofar as the license procedures have been followed, and the only place where there is an allegation that they have not been is in this one instance, that there has been no opportunity for the Chinese to use the technology to improve their guidance system.

Mr. ROHRABACHER. I would submit to you that whether it is guidance or control, or whether it is exploding bolts that are for MIRVing and for stage separation, or whether we are talking about laser gyroscopes, which we have not determined, I mean, these things affect America's security, and all this gobbledygook about license complications, we are looking for straight answers.

Secretary SLOCOMBE. You have gotten straight answers, sir.

Mr. ROHRABACHER. What we have gotten here is coached answers. It is evident from what has been said that you guys have been coached on this, but you did not do your homework enough to answer to me whether or not Chinese rockets have laser gyroscopes or not.

Secretary SLOCOMBE. That is exactly the kind of question that if we have advance notice you are interested in the subject, we can certainly find out. And we will find out now that we know you are interested.

Mr. ROHRABACHER. Well, maybe that was the subject of this hearing was technological help for the Communist Chinese rocket system.

Chairman SPENCE. Mrs. Fowler.

Mrs. FOWLER. Thank you, Mr. Chairman.

My questions are for Mr. Tarbell.

Chairman SPENCE. Before we proceed, let me state that we are trying to get through here. We appreciate the indulgence of all of our witnesses. We are on a tight schedule. We are trying to complete all the people we have here today. We had to schedule this meeting again because we could not get through it in a day. We will try to get through it by 1:30. Just so the remaining Members can take that into consideration.

Mrs. FOWLER. Could you restart the clock for me, Mr. Chairman? You know I talk slower, like you do.

Mr. Tarbell, last week I asked you whether the U.S. Government licensed the sale of any ground station technology to any Chinese entity that would support the entity's ability to encrypt its transmissions to and from a satellite. And your answer was you thought the question would best be answered in a closed setting. It just seems to me that any Chinese entity that received such technology, they already know they received it.

It is my understanding that on March 21, 1996, just 2 weeks after the United States sent two carrier battle groups to the Taiwan Straits following the PLA missile test that threatened Taiwan, DTSA recommended approval of a license to provide telefinding earth stations to the China Electronics System Engineering Corporation. Now these are ground station terminals for use in transmitting and receiving signals from satellites that are hard-wired to receive an encryption device. Can you tell us why DTSA signed out such a recommendation? I would like that answer at the end of my questions.

According to DOD directive 2040.2 regarding international transfers of technology, goods, services and munitions, and I quote, "In all technology transfer cases referred for review, the DOD component shall conduct policy reviews, technical evaluations, operational and military mission impact assessments, and intelligence assessments of proposed transfers."

The directive also directs the Defense Intelligence Agency to "conduct and provide intelligence reviews on technology goods, services, and munitions transfer cases."

Can you tell us in the case of the licensing of telefinding earth stations to China Electronic Systems Engineering Corporation, was DIA consulted for an intelligence assessment? I would like a yes or no answer. Were they consulted?

Mr. TARBELL. Congresswoman, I am not familiar with that particular case; and I will have to do some research for you and get back to you on that particular case and give you an exact answer on exactly who recommended what and how it was recommended category.

Mrs. FOWLER. OK, that's fine. Apparently they were not. But I would like for you to do some research.

[The information referred to is retained in the committee files.]

Mrs. FOWLER. I have a chart here. This is an unclassified chart that was published by the DIA in October 1995.

On this chart, right over here, it lists the China Electronics Systems Engineering Corporation as a company that is wholly owned by the PLA general staff department. Now it seems hard to believe that, if asked, the DIA would have supported the sale of encryption technologies to a PLA-owned company 2 weeks after the U.S. and China almost came to blows over Taiwan.

So this administration not only approved the sale of U.S.-built satellites by which the PLA has performed its command and control functions, a situation that the White House has admitted in a June 13 statement is occurring, but it also approved the sale of ground station equipment which has facilitated the encryption of these PLA's command and control communications via these U.S.-built satellites. So when you combine this with the high-performance supercomputers that we have already sold to Chinese entities, the PLA has obtained encryption capabilities they would not have otherwise been able to develop for many years. Would anyone like to comment on that?

And I would like an answer to my first question why you signed out such a recommendation. My first question, Mr. Tarbell, was why you signed out such a recommendation to provide these telefinding earth stations to a PLA-owned company? You have no memory of that either?

Mr. TARBELL. Well, because I did not personally review that case. And I'd have to go and ask people who did sign out that recommendation why they did that, and I'll get you an answer to that.

Mrs. FOWLER. I would certainly like an answer for the record, because this goes to the heart of the matter and what the problem we have right now with China, because somebody has been asleep at the switch. Whether it has been your department or where, procedures have not been followed, permission has been given, and they now have capabilities they would not have been able to develop for many, many years because the process that is supposed to work so well has not been working, obviously. So I would appreciate an answer.

[The information referred to is retained in committee files.]

Secretary HOLUM. Could I give you a general without referring to specific cases, but referring to the matter of encryption in ground stations generally, and in part to amplify what I answered last week, applying general guidance to a specific case where it didn't fit?

Before the transfer of jurisdiction to Commerce of encryption, which happened in December 1996, encryption devices generally were on the munitions list. State did license ground stations with encryption items for China during that period leading up to December 1996. And we need to follow up in more detail with your staff on the basis for doing that. It's not something that we can pursue in open session.

Mrs. FOWLER. I agree. My understanding is that during that process DTSA did have a role in the approval process that existed. And according to their charter, they are to be looking out for national security category. So if you've got a PLA-owned company, I have some questions about that. Thank you.

Chairman SPENCE. Mr. Gejdenson.

Mr. GEJDENSON. Thank you.

Let me ask if when these approvals occurred, is there a process by which the Congress of the United States is notified each and every time a waiver is issued?

Secretary HOLUM. That's correct, each waiver was notified to the Congress.

Mr. GEJDENSON. And of these 11 waivers, how many did Congress object to?

Secretary HOLUM. None.

Mr. GEJDENSON. So in 11 instances the administration asked for waivers, and the duly authorized committees were notified in each case?

Secretary HOLUM. That's correct.

Mr. GEJDENSON. And in none of those instances did any committee of the Congress or individual of the Congress object to the waiver?

Secretary HOLUM. Not to my knowledge.

Mr. GEJDENSON. Does anybody have other information?

Secretary SLOCOMBE. I have no different information from—I have no information different from what Mr. Holum has just said.

Mr. GEJDENSON. Let me ask you another question. The technology in question, are we the exclusive holders of that technology? There are—are there no other nations that have the technology that the Chinese could get by launching our satellites?

Secretary SLOCOMBE. There are certainly other countries that have comparable technology. The United States has a substantial technological advantage in this and in other fields in general, but there are plenty of other countries that build satellites.

Secretary REINSCH. Well, there are other satellite manufacturing companies largely in Europe, and there are also other foreign launchers, people who build launch systems; the French, the Russians primarily.

Mr. GEJDENSON. So then we find ourselves in this situation. We are back in the battle that we have been in for the last at least 20 years. We have an inadequate international control system in a number of areas, conventional weapons, chemical, biological and technology areas. The country then ends up making decisions as to which technologies it makes sense for us to sell, and at least until the political climate changed in this country, neither the Congress nor the administration and all the technical people within the administration expressed any concern or any serious concern about these 11 approvals. Is that correct?

Secretary SLOCOMBE. The underlying licensing all went through this process, yes.

Mr. GEJDENSON. And if we could go back to the process for a moment, again, one of the things that always troubled me is that in the past one agency could almost stop every sale, and that we now have a process where the executive is pressed to make a decision in reality, isn't it, that if you folks disagree amongst yourselves—Mr. Reinsch, you might be most familiar with this ongoing struggle since we have been on the same sides and different sides on this issue—that at the end of the day, what we now have is the way that any rational corporation would operate.

No corporation in the world would let one of their divisions veto an entire corporate policy, and that the, you know, cabinet-level or subcorporate-level people would have their input. If there was a debate, it gets bumped up to the chief executive, and a decision is made. But we do not refuse to take action just because there is a debate within the government or within the company. Is that correct?

Secretary REINSCH. That's correct, Mr. Gejdenson. What the President did in 1995 was exactly what you described, create a system in which everybody was at the table, and decisions were forced up the line until one was made.

Prior to that time, your committee, as you will recall, somewhat uncharitably but correctly characterized the process as having a large black hole into which licenses periodically went and never returned. We believe we've largely corrected that problem.

Mr. GEJDENSON. But in this case we would not even have had to have gone to that process because there was unanimity on these licenses.

Secretary REINSCH. That's correct.

Mr. GEJDENSON. So we have a process here where the present administration approved two more satellites than the previous administration, but as I understand it, only 9 of those 11 have launched. So the Bush administration approved 9 satellites. The Clinton administration approved 11; 9 launched. There was no debate between the perspectives of the Defense and Commerce people or State people. And what we really have is a question of whether or not Loral, which to commend its general counsel, when it recognized they may have done something wrong, came to the Government; is that correct?

Secretary HOLUM. That's correct.

Mr. GEJDENSON. So the way the Government found out that Loral may have erred in its actions was the corporation came to the Government; is that correct?

Secretary HOLUM. That's something we can't go into much detail on.

Chairman SPENCE. The time of the gentleman has expired.

Mr. Bartlett.

Mr. BARTLETT. Thank you very much. You gentlemen have been very patient. Thank you very much.

Secretary SLOCOMBE. Not as patient as you, sir.

Mr. BARTLETT. Thank you.

We learned through I think it was a New York Times article in the last several weeks that the Chinese have for some period of time—unknown to me had 13 of their intercontinental ballistic missiles targeting our cities. When did we first know that?

Secretary SLOCOMBE. The Chinese have had I think the number is 18 ICBMs deployed for years and years.

Mr. BARTLETT. But when did he we learn that 13 of those 18 are targeted on our cities?

Secretary SLOCOMBE. I don't know when that exact determination was made. I've always been under the understanding that if all 18 are not targeted today, they could be retargeted. But I will find out. I will give you an answer for the record on when the determination that we think 13 or so targeted was made.

Secretary REINSCH. My understanding Mr. Bartlett is it has been since the early 1980's.

Secretary SLOCOMBE. The missiles have been deployed certainly since the early 1980's.

Mr. BARTLETT. You cannot tell us when the administration first knew that 13 of these missiles were of a certainty targeted on our cities?

Secretary SLOCOMBE. It is difficult to discuss this issue in the open. Subject to a few technical questions, there's nothing that keeps the Chinese from targeting or retargeting any or all of their 18 ICBMs, which can reach the United States to wherever within the range of the missiles they want to target them.

I will get you an answer on the question of why we think 13 is the right number and when we thought it.

[The information is classified and is retained in committee files.]

Mr. BARTLETT. Thank you. My question has to do with the propriety of continuing to contract with the Chinese for launching our satellites, when at a minimum they benefit militarily from each of these launches since practice makes perfect, and every launch they make they now have more experience. And so, if that is the only benefit to them, it is a benefit to them.

I have a concern that we are now continuing to help the Chinese; if only by providing them more opportunities for launches, we are continuing to help the Chinese improve their capabilities to target our cities.

Could you tell us how many pending licenses or waiver requests are now under review?

Secretary BARKER. We're aware of at least one Commerce Department license, which we're reviewing now. And I'm sure there are other ones which are in the preliminary stages of review. As you can probably appreciate, a number of companies come in to ask for licenses for individual components. We will be reviewing those right now, too.

Mr. BARTLETT. What would the Chinese need to do before we would refuse to license additional commercial satellite launches? If targeting 13 of our cities, one of them presumably would be Washington D.C., we are now about 40 minutes away from the detonation of a weapon over our city if they chose to do so. What would they have to do before we decided it was not within our national security interest to continue licensing these launches? If having 13 of our cities targeted is not enough, what would be enough?

Secretary SLOCOMBE. The Chinese have had ICBMs for years and years. We don't for that reason—it would not any way help or reduce the danger of the United States if we stopped licensing satellites. They would still have the ICBMs. I can—I can think of—I can think of a variety of things. If the agreement with the Chinese broke down, if they stopped allowing monitoring, that kind of thing, it would certainly make us stop issuing the licensing.

Bear in mind, which I believe, and this, I think, is the technical judgment of the people in the government whose responsibility it is to look at these issues, that the premise is wrong; that is, that the satellite launch arrangements are such that they prevent the Chinese from gaining a military advantage from doing these launches. If we came to the conclusion that the—in the Department of Defense that either a particular occasion or the practice in general involved a military danger to the United States, we would oppose the license, and we would take it to the President.

Mr. BARTLETT. If they are able to develop MIRV capability, how many weapons could they put on each of their Long March missiles?

Secretary SLOCOMBE. It would depend on the characteristics of the system.

Mr. BARTLETT. Roughly; half a dozen?

Secretary SLOCOMBE. I am not in a position to make a judgment that that. It depends on the characteristics of the RVs and the characteristics of the dispensing system.

Mr. BARTLETT. I would like to note that developing the ability to launch multiple satellites I would think would have a bearing on their ability to develop MIRV.

Mr. GIBBONS. Thank you to my colleague Mr. Bartlett. Let me say that the answer to the question that the Department of Defense couldn't answer is that the Chinese have 18 ICBMs, 17 of which are targeted at U.S. cities. That is the answer to the question.

Secretary SLOCOMBE. I know they have 18, and I understand that there is some technical question as to whether one or more——

Mr. GIBBONS. The answer is 17.

Secretary SLOCOMBE. That's why I was interested in the 13. I always assumed the number was 18.

Mr. GIBBONS. That's the point I'm trying to make here is that there is information out there that needs to be brought out as well.

Let me ask a quick question. This should be a yes or no answer. Do any of you gentleman here have any information that technology transferred by the United States has improved the capability of the SR, short-ranged ballistic missile capability of China?

Secretary SLOCOMBE. I have no information to that effect.

Secretary HOLUM. Nor do I.

Secretary REINSCH. Nor do I.

Secretary BARKER. No information.

Mr. GIBBONS. Are any of the members of this panel attorneys?

Secretary SLOCOMBE. Yes.

Mr. GIBBONS. How many? Would you raise your hand.

To you gentlemen, had you known that granting the Chinasat-8 waiver and license would jeopardize the investigation of the 1996 accident currently pending before the Department of Justice, would you make a recommendation to grant China—the Chinasat-8 license?

Secretary SLOCOMBE. I would want to know why the Justice Department made that judgment. I don't honestly know why the Justice Department made that judgment.

Mr. GIBBONS. Do you have any information that they consulted the Department or any of your Departments with regard to that investigation?

Secretary SLOCOMBE. I think you would have to ask the Justice Department. My understanding is that they consulted with the White House, which is where the decision was made.

Mr. GIBBONS. Do you know whether or not the White House weighed or balanced the investigation over the advantages of issuing the license?

Secretary SLOCOMBE. I have no direct knowledge.

Mr. GIBBONS. Mr. Reinsch, do you?

Secretary SLOCOMBE. I can make inference, but I have no direct knowledge.

Secretary REINSCH. I am in the same situation as Mr. Slocombe. I believe one of the Members, I'm not sure which it was, it may have been Mr. Gilman, read an excerpt that indicated that was precisely what they did.

Mr. GIBBONS. They weighted the balance of risking the investigation for issuing the license, Chinasat-8.

Secretary REINSCH. I'm sorry, I may have misunderstood the question. I think I better—Mr. Gilman was alluding to something else. I don't have any information on that, Mr. Gibbons.

Mr. GIBBONS. Let me say for the record, Mr. Chairman, I'd like to submit to this committee a document dated February 13, 1998, on the Presidential national interest waiver for the China satellite—Chinasat-8 satellite issuance, and as well as a memorandum to the President from Phil Caplan dictating that there was serious concerns about the issuance of that license that went ahead anyway with regard to the jeopardy, and I will submit these for the record, Mr. Chairman.

Chairman SPENCE. Without objection.

[The information referred to is retained in the committee files.]

Mr. GIBBONS. Thank you.

Let me ask if any of you gentlemen are aware of technologies in the computer field that have been transferred that may impact or have been used by China in their satellite or missile technology program?

Secretary SLOCOMBE. Mr. Gibbons, the answer—I would prefer to answer that by saying I have not made a personal inquiry into the computer side, which is a big and complicated field. I just can't give you an offhand answer. It's not that I have in my head something that I know that was awful that happened, but I would want to review the whole story before I gave you an answer.

Mr. GIBBONS. The problem I've got here is that there is more than just satellite technology that's been transferred under some of these licenses by the Commerce Department to China, and I'm very concerned about it.

Mr. Reinsch, what does CSPP stand for?

Secretary REINSCH. CSPP, as far as it may stand for several things. What I know it stands for is the Computer Systems Policy Project.

Mr. GIBBONS. Very good. Have you had a role in CSPP?

Secretary REINSCH. Not in the any sense that I could imagine. I talk to them from time to time. I don't belong to them. It is a private sector group.

Mr. GIBBONS. Do they meet at the White House?

Secretary REINSCH. They have. It's their habit for their practice for their CEOs, as I recall, to come to town approximately twice a year.

Mr. GIBBONS. Have you met with them on occasion as you said?

Secretary REINSCH. Yes.

Mr. GIBBONS. Have you discussed anything classified with them?

Secretary REINSCH. No, not that I know of.

Mr. GIBBONS. Mr. Chairman, may I make a request in my last question here as my time runs out, I would like Mr. Reinsch or anyone at the White House be able to provide us a copy of each CEO organization or individual that has been a member or is a member of CSPP or participant in one of those meetings and what the subject of their discussions may have been at those meetings.

Chairman SPENCE. Time of the gentleman has expired. If you could comply with that request, we would appreciate it.

[The information referred to can be found in the appendix beginning on page 381.]

Chairman SPENCE. The time has arrived and passed. We have provisions for a third round. Our good friend and neighbor, Mr. Skelton, wants to make some closing remarks.

Mr. SKELTON. Thank you so much, Mr. Chairman. I do have a curiosity question that should have been asked earlier. And as I understand it, China, Russia, Ukraine and France are the other countries that launch beside us; is that correct?

Secretary SLOCOMBE. Those countries do. At least I have a list which indicates there are some other countries that have at least a limited launch capability.

Mr. SKELTON. That we've been providing satellites for?

Secretary REINSCH. Those are the other commercial launches. I think the Ukraine is not there yet, but they will be shortly.

Mr. SKELTON. Out of curiosity, what if we passed a law and the President signed it saying there should be no more waivers. It will all be American launches? Come on, 25 words or less.

Secretary REINSCH. There would be a shortfall in our ability to launch if we could only launch on American rockets right now.

Secretary SLOCOMBE. If there was a good national security reason to do that, we should pay the penalty. But there would be a penalty for the telecommunications industry and for the satellite industry.

Mr. SKELTON. In other words, we have a lot more satellites?

Secretary SLOCOMBE. We at present have a larger share of the world satellite market than we have of the launch market.

Mr. SKELTON. Thank you very much.

Let me, Mr. Chairman, very, very briefly summarize where we are after these 3 days of hearings on our 10-year-old satellite national policy. The first issue concerns the impact of our satellite export policy on our national security. Critics have charged that sensitive technology could have been passed to the Chinese military as a result of the policy of satellite sales and cooperative launches with China. The administration witnesses explained in great detail the safeguards currently in place to prevent the transfer of such technology to China. In my opinion, that case has not been made.

The second issue concerns the transfer of satellite licensing jurisdiction from the State Department to the Commerce Department. Much was made that the transfer to Commerce in 1996 was originally opposed by the State Department. We then learned in an oped piece published June 7 in the Los Angeles Times by former Secretary of State Warren Christopher that the matter was fully resolved to his satisfaction when he was Secretary of State, prior to the 1996. Excuse me. In fact, I am convinced that the process is stronger today than it was prior to 1996, as we've discussed a few moments ago.

A third issue concerns the specific case of Loral and possible release of export-controlled information in the review it conducted subsequent to the February 15, 1996 crash of the Chinese space launch vehicle which attempted to carry one of its satellites into space. We do know as a result of a voluntary disclosure by Loral that it participated as part of an independent review committee (IRC) and transmitted the results of the review to the Chinese authorities. It did so without obtaining the approval of the State Department in two counts, either to participate in the activities of the IRC or to transmit the results of the review. The key issue here is to determine what damage may have committed to United States national security.

The fourth issue concerns the February waiver for the launch of another Loral satellite from China. The waiver was granted despite concerns expressed by the Justice Department about the potential impact of the waiver and its ability to persuade a jury to convict Loral in the event of prosecution. The administration witnesses address those. First the waiver was handled in a normal manner and the State Department, Defense Department, ACDA, and the other consulted agencies recommending approval of the export license and waiver.

In his testimony Mr. Holum said that the waiver is consistent with national security and foreign policy. So over these 3 days after hearings we have been provided much needed information on these issues, and I thank the witnesses here today as well as those that preceded. This whole issue will now become the principal focus of the Select Committee on China, and I hope that we can now allow that committee to conduct its work in a fashion that will do credit to us here in the Congress. We should have completion of its work before changes are enacted or before we rush to judgment.

So with that, Mr. Chairman, I thank you for allowing me to summarize, and I thank the witnesses for their patience as well as for their information.

Chairman SPENCE. We thank you for that nonbiased summary. With that, we will conclude these hearings. We do appreciate your indulgence. I know you have been waiting for lunch, and some of us have, too. Thank you very much for your contributions.

[Whereupon, at 1:37 p.m., the committee was adjourned.]

APPENDIX

JUNE 17, 18, 23, 1998

STATEMENT OF REP. SILVESTRE REYES
HEARING ON CHINA AND SATELLITE EXPORTS
June 18, 1998

As I stated yesterday, I am concerned that our review of these issues may not be motivated entirely by national security concerns.

Today this House working to establish a bi-partisan committee to look into these issues in detail over a long period of time. These Committees, however, may be reacting too fast and may not be considering all sides of the issue.

Yesterday both Jim Ryun and I asked the witnesses for suggestions on how to fix the problem - if there is one. No one seemed to have any proposals. We in Congress cannot be the experts on these issues. I thought that the witnesses before us yesterday were the experts. But - apparently - they were not.

Mostly, they seemed simply to want to escalate the rhetoric, declaring a problem, but offering no solutions. This is why I am concerned that the activities of these Committees and of others in the House may be partisan in nature.

As Members of Congress, we have a duty to protect the national security of this nation. We should ensure that in the process we put partisanship aside and review the facts rather than the rhetoric. Our relationship with China is too important for Members of Congress to be making drastic statements until we have calmly and carefully reviewed the facts.

China's 1.2 billion people represent nearly one-fourth of the planet's total population. Their population will increase by the size of America's current population every 20 years. China represents our sixth largest export market, and with its staggering population will very likely soon be our largest foreign market. In addition, China strategically dominates East Asia, and is situated on one of the world's largest land masses with tremendous untapped natural resources. Our country simply must engage the People's Republic of China. Furthermore, we must do so face to face.

During a recent visit to China, I observed first hand the Chinese people, culture, economy, and government. I observed enormous disparities across China from impoverished rural areas to booming metropolitan cities. Further, I observed that with China we are dealing with a country whose history and culture spans not simply centuries, but millenia. As Dr. Freese noted yesterday in her testimony, traditions and practices are unique and vastly different from those of the West.

Business opportunities for American companies are numerous and increasing. In fact between, 1991 and 1996, U.S. exports to China have increased by 90.5%. In addition, China has announced that they expect to spend $750 billion over 3 years in infrastructure development. This would involve commercial areas where the United States maintains a global competitive advantage in Energy, Transportation, and Telecommunications.

No one will dispute that there are a number of disagreements between our nation's interests and those policies and actions undertaken by the Chinese government. Our nation must and should be concerned about these issues including human rights, religious freedom, standards of labor, transfers of technology, nuclear proliferation, and developments in commerce and trade.
If we are not engaged with China and actively working to resolve our differences, our nation will be harmed economically from loss of financial opportunities and defensively from the lack of ability to affect missile proliferation.

I support looking at national security issues like the satellite transfers that are the subject of the hearings yesterday and today. However, I believe that we need to ensure that we are truly seeking the facts and that if a problem if discovered, that we are truly seeking positive remedies.

Statement by Congresswoman Lois Capps

Good morning. I would like to thank the Chairman and Ranking Member for convening this important and timely hearing.

It is imperative that Congress conduct a comprehensive examination of the issues before us today because of their national security implications as well as the effect they could have on our international trade relationships.

It is also important that we use this opportunity to discern fact from fiction, distortion from reality. This hearing has the potential to become highly politicized, so I call on my colleagues to approach these issues in a spirit of fairness and bi-partisanship. There is too much at stake to turn this important proceeding into a partisan political side-show.

Thank you Mr. Chairman.

Congressional Research Service • Library of Congress • Washington, D.C. 20540

A Joint Hearing of the Committees on
International Relations and National Security
U.S. House of Representatives

U.S. Policy Regarding Satellite Exports to China

Testimony by Shirley A. Kan
Analyst in Foreign Affairs
Foreign Affairs and National Defense Division

June 17, 1998, 9:30 am
Rayburn House Office Building, Room 2118

Good morning. I am Shirley Kan of the Foreign Affairs and National Defense Division of CRS. I am here to present a statement that will provide background on foreign policy and national security issues concerning Chinese launches of U.S. satellites and Chinese missiles. I request that a CRS Report on the background and chronology leading to the current controversy, a CRS Report on China's ballistic and cruise missiles, and a CRS Issue Brief on commercial space launches be included for the record.

China Great Wall Industry Corporation

Important for our understanding is the Chinese company at the heart of various concerns. China Great Wall Industry Corporation is the only Chinese state-owned corporation that offers launch services to foreign customers. It is also a subsidiary of China Aerospace Corporation, which heads the Chinese defense-industrial aerospace complex and oversees both space and missile research and development efforts. China Great Wall uses the Long March series of rockets to launch satellites. The Long March boosters are also produced as China's DF-4 and DF-5A intercontinental ballistic missiles (ICBMs). U.S. sanctions have been imposed twice on China Great Wall for missile proliferation in Pakistan.

China gains several benefits from launching satellites for foreign customers: foreign capital, technical expertise, and international prestige. Some satellites or parts exported to China for launch have been for Chinese use. However, China has also been embarrassed by a number of costly failed satellite launches. In 1992, the engine for a Chinese rocket failed to ignite while attempting to launch a satellite, and another rocket exploded and destroyed a second satellite. Both satellites were built by Hughes Electronics Corporation. In 1995, a Long March rocket exploded and destroyed another satellite built by Hughes. In 1996, yet another Chinese rocket exploded and destroyed the Intelsat satellite built by Loral. Aside from the dramatic explosions, other problems have prevented Chinese rockets from successfully launching satellites into the correct orbits. More recently, since 1997, China had success in all satellite launches. Initially, China's response to the failures was to blame foreign satellite companies. China later shifted to a willingness to work with U.S. companies to determine the causes of the explosions and other failures.

Multiple Issues

Loral's Case. There are several issues surrounding the recent controversy on the use of Chinese Long March rockets to launch U.S. satellites. The immediate legal issue which prompted the controversy concerns whether Loral Space and Communications Ltd. provided expertise to China that it can use to improve its ballistic missile force. The *New York Times,* in April 1998, cited a classified Pentagon report that reportedly concluded in May 1997 that American expertise was transferred to China that significantly enhanced the reliability of its ballistic missiles and that U.S. national security was harmed. Since last September, the Justice Department has been conducting a criminal investigation into whether Loral violated export control laws. Loral acknowledges that, after the explosion of a Chinese rocket that destroyed a Loral-built satellite in February 1996, it formed an independent committee to review the Chinese findings about the cause of that explosion. The company admits that the committee provided a report to the Chinese in May 1996 without first getting approval from the State Department. Loral says that it promptly informed the Departments of Defense and State of having provided the report and is cooperating with the Justice Department and Congress. Loral says that the report did not transfer sensitive information or technology.

Policies, Sanctions, and Export Controls. A second issue concerns how the Clinton Administration has shaped and implemented U.S. policies on China, on satellite exports to China, and on missile nonproliferation. Specifically, the Administration reviews satellite exports to China in making decisions on two sets of sanctions (one for the Tiananmen crackdown and one for missile proliferation) in conjunction with the export control process.

The Administration transferred control over the licensing of satellites from the State Department to the Commerce Department in November 1996.

Some, including the President, have argued that engagement with China is a better policy than isolation of China. Others say that the issues are how the policy of engagement is carried out and whether satellite exports need to be included in that policy. In part to promote ties with China, the Reagan Administration, in September 1988, first notified Congress of its decision to approve licenses for exports of satellites for launch from China. That Administration acknowledged the need to protect legitimate U.S. national security interests and the U.S. commercial launch industry, while promoting relations and trade with China. Over the years, some in Congress have supported the use of Chinese space launch vehicles to launch U.S.-origin satellites. President Bush first issued waivers of post-Tiananmen sanctions. President Clinton continues the long-standing policy of engagement with China, including satellite launches to expand multifaceted relations. The Clinton White House argues that satellite exports to China have benefits for commercial competitiveness, nonproliferation goals, spread of democracy, and the policy of engagement.

Let me clarify that, for satellite exports to China, considerations of Presidential waivers for post-Tiananmen sanctions and missile proliferation activities have been added to the export control process. U.S. sanctions imposed in 1990 for the Tiananmen crackdown have meant that Presidential waivers allowed by the law were needed before licenses could be approved for export of satellites or parts to China. As shown in the chart on those waivers, Presidents Bush and Clinton have issued 13 waivers for 20 satellite projects, by reporting to Congress that it is in the national interest to do so.

Those waivers have been issued fairly routinely, except for the consideration of the Justice Department's concerns about the latest waiver or other sanctions for missile proliferation. After China transferred technology for the M-11 short-range ballistic missile to Pakistan, the Bush Administration, from June 1991 to March 1992, denied exports to China of high-speed computers, satellites for launch, and missile technology or equipment. The Clinton Administration determined that China again transferred M-11 technology to Pakistan and imposed sanctions from August 1993 to November 1994 denying exports of missile equipment and technology to China, affecting exports of satellites for launch by China Great Wall. During those two periods of sanctions, one Presidential waiver was issued (in July 1994) for the export of satellites for launch by China Great Wall. Some debate has arisen as to whether the Administration is required to impose sanctions again, since reports say that China has continued to transfer missile equipment and technology to Pakistan.

Commerce vs. National Security. A third and broader issue concerns the underlying debate over the national interest: weighing the benefits of trade and commercial competitiveness versus the risks to national security. Some are calling for a stop to the export of satellites to China. Others say that such exports benefit American companies, consumers, and the country.

The White House says that licenses or waivers for satellite exports to China do not authorize the transfer of sensitive missile technology. Administration officials state that export licensing procedures and strict security measures (including escorts of satellites by Pentagon officials) preclude any assistance to Chinese missile programs. The policy is intended to allow American exporters to compete while protecting U.S. security interests. Some also consider the benefits of the Pentagon having access to the secretive Chinese aerospace and missile complex. According to the March 23, 1998 *Washington Times*, a secret National Security Council memo discussed the use of satellite launches as a "carrot" for China to curb missile proliferation. Some supporters also argue that satellite broadcasts promote the spread of democratic values. The Chinese argue that commercial space launches are part of what they call defense conversion, defense industries doing civilian business.

Critics say that allowing China to launch U.S.-origin satellites indirectly subsidizes and assists China's missile research and development efforts. The technology and know-how used in launching satellites are similar to those used in ballistic missiles. Secondly, some say that although an individual piece of technology or technical assistance may not be significant, assistance given over time has a cumulative effect of conveying valuable expertise. Third, U.S.-China cooperation has been characterized by a tension between a certain unreliability of Chinese rockets and the interests of American companies to have successful launches. Working with American and European engineers, Chinese engineers may have learned important quality controls and other lessons. Some also say that allowing the Chinese to launch satellites rewards or ignores their missile proliferation and any U.S. expertise gleaned through satellite launches may then be transferred by China to other countries. Others say that while satellites may promote a more open Chinese society, they may also be used by the Chinese military. Some experts point out that China's defense industries have not converted, but diversified into both military and civilian research.

As for potential benefits to China's ballistic missile program, some observers point out that launching satellites has not provided breakthroughs and China has deployed ICBMs since 1980, before launching U.S. satellites. As seen in my second chart on Chinese ballistic missiles, the Chinese have a diversified nuclear-armed missile force in what they call the Second Artillery, consisting of intercontinental, medium-range, submarine-launched, and short-range ballistic missiles. Other experts say that China has experienced problems with its missile force and reportedly is improving the reliability and accuracy of its missiles. The Second Artillery is expected to deploy, after the turn of the century, new ICBMs, possibly with MIRV (multiple independently targeted reentry vehicle) capability, and a new submarine-launched ballistic missile. MIRV technology is similar to that used by China to launch multiple satellites. I will end my presentation of background here. Thank you.

U.S. Satellite Technology Transfers To China: What's At Issue

by

Henry Sokolski
Executive Director
The Nonproliferation Policy Education Center

Testimony Presented Before

A Joint Hearing of

The House International Relations Committee and

The House National Security Committee

17 July 1998
Washington, DC

The Nonproliferation Policy Education Center
1718 M Street, N.W., Suite 244, Washington, D.C. 20036
phone: 202-466-4406 / fax: 202-659-5429, E-mail: npec@ix.netcom.com

The Nonproliferation Policy Education Center
1718 M Street, N.W., Suite 244, Washington, D.C. 20036 phone: 202-466-4406 / fax: 202-659-5429, E-mail: npec@ix.netcom.com

U.S. Satellite Technology Transfers to China: What's At Issue?

Q. In allowing U.S. satellite technology transfers to China, isn't' the Clinton Administration simply doing what the Bush Adminstration did before it?
A. No, there are differences. Although Clinton and Bush both waived Tainamen Square sanctions to allow U.S. satellite technology transfers to China, the Clinton administration went much further in loosening controls over them by moving virtually all commercial satellite technology from State to Commerce Department controls. This shift has eliminated systematic government monitoring of prelaunch conversations between U.S. contractors and Chinese space firms and, according to the General Accounting Office, marginalized the previously important licensing input of the Defense Department.

Q. But isn't the most important point that Reagan and Bush started this commerce?
A. It's hard to see how. Either the transfer of critically important military technology is an inevitable part of launching U.S.-made satellites or it's not. If it is, then persisting in the business (with 14 launches) is just as blameworthy as starting it (with 4). Indeed, it could even be worse if you began such commerce in the vain hope that you could control its diversion but persisted in transferring satellites knowing that it couldn't. On the other hand, if you believe U.S. controls can prevent the Chinese military from benefiting from such trade, then, not who started such commerce, but who has allowed what to be transferred are the critical issues.

Q. OK. But if China could hit Los Angeles with a nuclear warhead before it launched its first U.S.-made satellite,what more of military significance could it gain for its missile forces from U.S. satellite makers?
A: Plenty. China has repeatedly objected to U.S. development of missile defenses that could neutralize China's missile forces. In this regard, China's mastering of how to place single and multiple satellites in exact locations in space (with precision satellite kick motors and multiple satellite dispensers) is directly applicable to its current efforts to develop missiles with post boost vehicles (PBVs) and multiple independently targetable reentry vehicle systems (MIRVs) designed to overwhelm or evade such defenses. Its latest solid-rocket missile projects -- the DF-31, DF-41, and J1-2 -- in fact, are all believed to be perfecting such systems. Also, China is keenly interested in making their theater missiles accurate enough to destroy targets (e.g., in Taiwan) without having to use chemical, nuclear or biological weapons. Acquiring such accuracy is intrinsic with placing sophisticated satellites precisely on station -- something China has had a 100% success rate with since the Loral and Hughes failures of 1996.

Q. Putting missile improvements aside, though, isn't it clear that keeping U.S.-made satellites from prying Chinese eyes has effectively kept the Chinese military from exploiting such commerce for its own satellite needs?
A. Not really. Increasingly, the Peoples Liberation Army (PLA) is using U.S.-made satellites to transmit military encrypted message traffic. China has especially relied upon U.S.-made satellites over which it has operational control -- e.g., the APTStar and Chinastar satellite series. China is also trying to develop advanced communications satellites of its own with Western help (a project Hughes wanted permission to work on before the recent Loral-Hughes flap). This, in the final analysis, however, is pointless unless China develops rocket systems that are reliable, gentle, and precise enough to place such satellites safely on station. U.S. satellite launch revenues to China (approximately $500 million since 1990), launch experience with advanced U.S.-made satellites (nearly half of China's last 30 launches), and transfer of associated technology have all served this purpose.

Q. But doesn't Chinese use of U.S.-made satellites make it easier for our intelligence agencies to intercept whatever encrypted messages the Chinese military might transmit?

A. Not enough to justify U.S. satellite transfers. Indeed, supplying the PLA with satellite communications is hardly the preferred way for the U.S. to intercept China's message traffic since it enables China to use a variety of antenna to transmit an increasing volume of encrypted messages at a lower cost. To the extent Chinese encryption may be strong and the satellite antenna used are narrow in beam width, our intelligence agencies' task of intercepting Chinese communications will become *more*, not less difficult with increased PLA use of our satellites.

Q. At the end of the day, though, don't U.S. satellite makers have to use Chinese launchers to stay competitive since Chinese launchers are so much cheaper than alternative launch services?
A. Hardly. Even by the U.S. satellite industry's own calculations, Chinese launch services save them only 10 to 15 percent over alternative providers and this advantage is expected to evaporate over the next 36 to 48 months when newer American., Russian, and Ukrainian alternative launch services come on line. What makes U.S. satellite makers competitive, moreover, is not lower cost launchers, but higher quality satellites: These frequently cost one or two times more than the launch and produce even more in revenues over their lifetime.

Q. But if the U.S. forgoes satellite business with China,won't this merely open the Chinese satellite market to America's competitors?
A. None that we know of. Over 75 percent of the world's commercial satellites, in fact, are either made in the U.S. or require U.S. permission to be launched because they contain key U.S. components. Virtually all the remaining commercial satellites come from member nations of the European Space Agency (ESA), who are inclined to launch their satellites on ESA launchers. As for making satellites in China, this market is already saturated: Dailmer-Benz is already working with the Chinese military to develop an advanced communications satellite but so far the system (after two launchings) has yet to work.

Q. If there is a technology transfer problem, isn't putting all commercial satellites back under State controls the preferred fix?
A. Not necessarily. If State controls are not enforced with proper monitoring or if compliance concerns are not taken up by senior political officials and the contractors challenged to address possible control infractions (two complaints some government monitors had from 1993 through 1995), shifting back to State controls will do little good. Also, Secretary Daley could make Commerce controls identical to State's now if he chose to. Finally, if the customer (e.g., the PLA) is intent on diverting technology, even the best controls cannot prevent it. In this case, barring satellite technology transfers or, at least, transfers of the most advanced satellite types may be appropriate.

Q. What, then, should Congress do?
A. Keep things from getting worse until it gets all the answers. The House and Senate should insist that no further satellite transfers be allowed until its investigations and deliberations over this matter are complete. There simply is no substitute for determining what, if anything, of military significance was transferred to China and whether or not such transfers could have been prevented by tighter controls. If nothing was transferred, then, all's well. If something was transferred, then, Congress must determine what it was, what else of value might be at risk, and whether or not these risks can be eliminated by tighter export controls or if barring future transfers is required.

Q. How should Congress go about doing this?
A. Stop pointing fingers or proposing easy solutions and get to work. It would be useful to depose all of the current and previous officials tasked to monitor this commerce along with their private corporate counter parts. It may be necessary to grant the later immunity. In no case, however, should Congress be reassuring the Executive Branch or private industry now that politically easy solutions -- such as merely moving commercial satellites back to State controls -- are the agreed way out. Indeed, too much partisan political capital has already been expended on this controversy for Congress to do anything less than a substantive oversight effort unprejudiced by preconceived solutions. Certainly, if Congress fails to do this, the credibility of both parties -- Democratic and Republican -- will suffer along with pubic regard for Congress and, more important, for the rule of law and our nation's security.

WISCONSIN PROJECT ON NUCLEAR ARMS CONTROL

Testimony of Gary Milhollin

Professor, University of Wisconsin Law School
and
Director, Wisconsin Project on Nuclear Arms Control

Before the Committees on International Relations and National Security
United States House of Representatives
June 17, 1998

1701 K STREET, NW SUITE 805
WASHINGTON, DC 20006
PHONE 202-223-8299 FAX 202-223-8298

I am pleased to appear today before these distinguished Committees. I have been asked to discuss U.S. policy on satellite exports to China. I will direct my remarks to the effect of that policy on Chinese transfers of missile technology to countries that are trying to make weapons of mass destruction.

We have heard a lot about satellites lately, especially American-made satellites going to China. We have also heard about campaign contributions, about waivers of export prohibitions, and about the transfer of control over satellites from one government agency to another.

What we have not heard, and what I will talk about today, are decisions by our government that have given Chinese companies a green light to sell missile technology to countries like Iran and Pakistan. The Administration has made three crucial decisions in this regard.

First, it has decided to transfer control over satellite exports from the State Department to the Commerce Department, an action that effectively pulls the teeth from any future U.S. sanctions against Chinese companies guilty of missile proliferation.

Second, it has decided to suspend, without any legal basis, the implementation of U.S. statutes that require sanctions to be imposed against Chinese companies for past sales of missile technology to Iran and Pakistan.

Third, it has decided to invite China to join the Missile Technology Control Regime, an invitation that, if accepted, would immunize Chinese firms from any future application of U.S. sanctions laws for missile proliferation.

When we look at the cumulative effect of these decisions, we see something both surprising and alarming. Our government has enabled Chinese companies to proliferate missile technology with little fear of punishment. More specifically, Chinese companies have been able to sell Iran and Pakistan components for nuclear capable missiles without worrying about losing U.S. satellite launch contracts.

Satellites

Our sanctions laws, as written by Congress, are based on a simple idea. A foreign company cannot import American missile technology with one hand and proliferate missile technology with the other. If a Chinese company decides to sell Pakistan or Iran a nuclear-capable missile or the means to make one, that company has to forget about importing any missile-related technology from the United States. U.S.-made satellites were originally part of this equation, because they contain missile-related American components.

That simple idea has now been abandoned by the executive branch. When, for example, the Administration transferred licensing authority over satellites from the State Department to the Commerce Department, satellites were effectively removed from the list of U.S. exports subject to missile sanctions.

Let me explain why this is so. If a Chinese company sells whole missiles to a country like Iran or Pakistan, the company would be guilty of what is known as a "Category One" violation. This is because whole missiles are listed on Category One of the Annex to the Missile Technology Control Regime, an agreement among countries that are trying to curb missile proliferation by controlling their exports. For selling a Category One item, the sanctions bar the guilty company for at least two years from importing any item controlled by the Export Administration Act. Category One sanctions would thus bar satellites even if controlled by the Commerce Department, which administers that Act.

China did export whole missiles to Pakistan in the fall of 1992. Since then, however, China has changed its tactics. It now exports missiles piecemeal, as components. An example would be a piece of guidance equipment. These items are on Category Two of the MTCR Annex. A "Category Two" violation bars only the export of U.S. "missile equipment or technology " under the Export Administration Act and under Section 73 of the Arms Export Control Act.

But what is "missile equipment or technology"? According to the licensing practice of the State Department, a missile-related item *retains its identity* as a missile item even if it is embedded in a commercial satellite. Thus, if a Chinese company were sanctioned, the export of satellites would be blocked by the State Department because satellites have missile-related items embedded in them. In the view of the Commerce Department, however, a missile-related item *loses its identity* as a missile item if it is incorporated into a commercial satellite. Thus, the export of satellites would not be blocked by the Commerce Department even though the satellites contained items that would be considered missile-related if not embedded. These embedded items are such things as radiation-hardened computer chips, gyroscopes, and accelerometers.

The Administration first transferred export controls on satellites from State to Commerce in 1996, and in its implementing regulations, it also moved a number of missile-related items associated with satellites. Then in 1998, the Administration moved the rest of the missile-related items associated with satellite launches, even though the items are not embedded in satellites. These items included ground support equipment, test equipment, replacement parts, and non-embedded kick motors.

The result is that now, satellites are insulated from missile sanctions because control over virtually everything associated with launching them has been transferred to the pro-export Commerce Department, where sanctions will not be applied.

It is important to realize that the same Chinese companies that launch U.S. satellites also sell missiles to places like Pakistan. Who are these companies? China Great Wall Industries, China Aerospace International Holdings Ltd. (CASIL, of Hong Kong) and their parent, China Aerospace Industry Corporation. These companies launch satellites on China's Long March rockets. The United States has sanctioned both China Great Wall and China Aerospace Corporation in the past for supplying missile technology to Pakistan.

It is also important to realize that a satellite launch contract is one of the most lucrative things a Chinese aerospace company can get from the United States. Thus, by removing satellites from the threat of sanctions, the Administration has surrendered one of the most important levers America has to stop Chinese missile proliferation. Chinese companies are free to sell missile technology to Iran or Pakistan without risking their most lucrative source of revenue.

Failure to apply sanctions laws

China's exports remain the most serious proliferation threat in the world. Since 1980, China has supplied billions of dollars worth of nuclear weapon, chemical weapon and missile technology to South Asia, South Africa, South America and the Middle East. It has done so despite U.S. protests, and despite repeated promises to stop. The exports are still going on, and while they do, they make it impossible for the United States and its allies to halt the spread of weapons of mass destruction.

I have attached tables to my testimony that list China's exports of nuclear, chemical and missile technology since 1980. The tables reveal that China has consistently undermined U.S. nonproliferation efforts for nearly two decades and is still doing so today.

In the early 1990s, Chinese companies were caught selling Pakistan M-11 missile components. The M-11 is an accurate, solid-fuel missile that can carry a nuclear warhead about 300 kilometers. In June 1991, the Bush administration sanctioned the two offending Chinese sellers. The sanctions were supposed to last for at least two years, but they were waived less than a year later, in March 1992, when China promised to abide by the guidelines of the Missile Technology Control Regime.

But by December 1992, in violation of its promise, China had shipped 34 M-11 missiles to Pakistan. Waiving the sanctions was a mistake.

In August 1993, the Clinton administration applied sanctions for two years, after determining that China had violated the U.S. missile sanctions law a second time. Then in October 1994, the United States lifted the sanctions early again, when China pledged once more to stop its missile sales and comply with the MTCR.

Since late 1994, the stream of missile exports has continued. U.S. officials say that China's missile exports have continued up until the present moment, unabated. These exports include the sale of missile-related guidance and control equipment to Iran as well as Pakistan.

In fact, our officials have learned that they were duped in 1992 and 1994. China was not promising what we thought it was. Our officials now realize that China interprets its promises in 1992 and 1994 so narrowly as to make them practically meaningless. It is clear that China has not complied with the MTCR in the past, that it is not complying now, and that it probably never will comply unless something happens to change China's attitude on this question.

In its latest venture, China is helping to build a plant to produce M-11 missiles in Pakistan. U.S. officials say that activity at the plant is "very high." If the Chinese continue to help at their present rate, the plant could be ready for missile production within a year.

By the autumn of 1996, the intelligence community had completed an air-tight finding of fact on China's missile transfers to Pakistan. There was clear proof that the transfers had happened. All the factual analysis necessary to apply sanctions had been finished. A similar finding on China's missile exports to Iran had also been made.

And roughly one year earlier, an important legal analysis had been completed. The legal analysis established that sanctions could be applied where a foreign person "conspires to or attempts to engage in" the export of any MTCR equipment or technology. Thus, sanctions could be applied *without* a finding that hardware or technology had actually been exported. A conspiracy or even an attempt to transfer such items would be enough. One did not need a photograph of a missile with "made in China" written on the side.

The findings of fact and the legal analysis showed clearly that China should be sanctioned. Both the findings and the analysis had been circulated to the relevant agencies by the autumn of 1996. Both Pakistan and Iran were covered. The process, however, was short-circuited at that point.

The next step would have been for the National Security Council to call a meeting at which each agency could submit for the record its views on whether sanctions should be imposed. The NSC would then forward these views to the Department of State, which would prepare a decision memorandum for the Under Secretary, who has the legal authority to impose sanctions.

But none of these steps were ever taken. The State Department simply chose not to complete the administrative process. Thus, the sanctions law is not being implemented as Congress intended and, in fact, is being circumvented. It is obvious that the law can never take effect unless the administrative process is completed, so the failure to complete it is manifestly illegal.

Last week, in testimony before the Senate Committee on Foreign Relations, Mr. Gordon Oehler, who headed the CIA's non-proliferation center when the findings were made, confirmed that the evidence of the missile transfers was clear. Mr. Oehler said that there was no doubt in his mind that the transfers happened, and that the rest of the intelligence agencies agreed.

The State Department has now admitted this fact by implication. The State Department is no longer saying that there is "not enough evidence" to apply sanctions to China. It is now saying that it has "not yet made a determination" to apply sanctions, which is quite different. In effect, the State Department is saying that it has not applied sanctions because it has not chosen to complete the administrative process. I recommend that Congress take steps to see that the law is enforced.

Now that Pakistan has demonstrated its nuclear weapon capability, and announced that it will mount nuclear warheads on missiles, this matter has become urgent. The Chinese-supplied M-11s will actually carry nuclear weapons. President Clinton has said that the world should try to prevent India and Pakistan from putting warheads on missiles, but his Administration refuses to apply a U.S. law designed to prevent Pakistan from acquiring missiles in the first place.

I would also like to mention the Iran-Iraq Nonproliferation Act of 1992. Under that Act, a foreign company is sanctioned if it exports "destabilizing numbers and types of advanced conventional weapons" to either Iran or Iraq. The sanctions bar any American product that needs an export license, including satellites.

In 1995, the China Precision Machinery Import Export Corporation sold Iran advanced C-802 anti-ship cruise missiles. Estimates of the number sold range from 60 to 100. The State Department admitted that the shipment occurred but concluded in 1997 that it was not of "a destabilizing number and type." Thus, it declined to apply sanctions, despite the fact that Admiral John Redd, our naval commander in the Gulf, took the unusual step of complaining publicly about the sale. The missiles threaten our ships and sailors in Gulf and also commercial shipping.

If sanctions had been applied, the export of U.S. satellites would have been cut off. Under the Iran-Iraq Nonproliferation Act, the sanctions reach the parent organization of the guilty party. The parent of China Precision Machinery is the China Aerospace Corporation, which launches U.S.-origin satellites. This raises the question whether the State Department, when ruling that the missiles were not destabilizing, was simply trying to protect U.S. satellite makers.

China in the MTCR?

This past March, the Administration invited China to join the Missile Technology Control Regime. In a memorandum dated March 12, White House staff member Gary Samore stated the reasons for making the offer. Sanctions figured prominently among them. If China joined, the memo stated, China could expect "substantial protection from future U.S. missile sanctions."

Mr. Samore could have said "complete protection." Under Section 73 of the Arms Export Control Act, sanctions would not apply to a Chinese company if China joined the MTCR even if the company transferred complete missiles to Iran or Pakistan. Sanctions would be avoided if the sale were legal under Chinese law, or if China took action against the company, or if China found the company to be innocent. In effect, the Administration offered China a complete shield against U.S. sanctions law.

The Administration was also offering China a second benefit. As things stand now, if a U.S. exporter gets an order from a known missile maker in China, the exporter cannot make the sale without notifying the government and getting an export license. This is required by Section 6 of the Export Administration Act. If China were to join the MTCR, however, no license would be required for such a sale. U.S. firms could deliberately outfit Chinese missile manufacturing sites without telling anyone.

It is difficult to see how such an offer is prudent. China has repeatedly failed to comply with MTCR guidelines since promising to do so in 1992 and 1994. There is no real evidence that China has changed its ways. Thus, the offer seems to be yet another effort to insulate Chinese aerospace companies from U.S. sanctions laws so that satellite launches can continue.

Conclusion

When we look at the history of U.S. sanctions policy, we see a willingness to sanction China for missile proliferation in both the Bush Administration, which applied sanctions in 1991, and in the Clinton Administration, which applied sanctions in 1993. But in 1995, the Clinton Administration policy changes. China supplied the C-802s to Iran in 1995; China supplied other missile components to Iran and Pakistan in 1995 and has continued to supply up to the present time. And in 1996, the Administration refused to act on explicit findings by the intelligence community that the transfers occurred. Also in 1996, the Administration started transferring control over satellites exports from the State to the Commerce Department, insulating them from the application of missile sanction laws in the future. For some reason, the Administration decided in either 1995 or 1996 that missile sanctions would no longer be part of U.S. policy in dealing with China.

Why did U.S. policy change? I don't know the answer to that question, but I urge the Committees to look into it.

We do know the result of the policy change. It gave China a green light to proliferate. Our government's policy on sanctions has enabled Chinese satellite launch companies to sell missiles and missile components to Iran and Pakistan without fear of punishment. Thus, it may be that we are asking the wrong question about how our satellite export policy affects missile proliferation.

Whether or not our satellite exports caused U.S. missile technology to go to China, they have made it easier for Chinese missile technology to go to Pakistan.

India, of course, has watched this happen. India watched China help Pakistan make not only missiles but the nuclear warheads to go on them. India also watched the United States invent every excuse possible not to do anything about it. America asked the Indians to show restraint in nuclear testing, but America was unwilling to put restraints on its own satellite companies by sanctioning China. The Indians no doubt concluded that Uncle Sam was against the spread of the bomb unless it might cost him something. It should not surprise us if our non-proliferation policy lacks credibility.

CHINA'S DANGEROUS EXPORTS

IRAN

1980-1984	Sold production capability for the Oghab short range rocket
1985-1989	Trains Iranian nuclear technicians in China
	Sells Silkworm anti-ship missiles
	Supplies a miniature reactor, a subcritical facility, and tributylphosphate useful in plutonium extraction
1990-1998	Supplies a calutron and a copper-vapor laser that could be used for uranium enrichment research
	Contracts to sell 25-30 MW research reactor
	Helps prospect for uranium in Iran
	Contracts to sell nuclear reactor and isotope separator
	Supplies a nuclear fusion research facility and scientists and engineers to help install it
	Delivers poison gas ingredients
	Delivers missile guidance systems and computerized machine tools
	Sells virtually complete chemical weapon factories including precursor chemicals and glass-lined vessels
	Hosts Iranian nuclear specialists to study a uranium hexafluoride plant for export to Iran
	Delivers components for missile guidance and ingredients for rocket propellant
	Supplies 400 tons of poison gas ingredients
	Agrees to sell gyroscopes, accelerometers and equipment to build and test missile guidance components
	Supplies two tons of calcium hypochlorate, used in chemical decontamination
	Continues to supply chemical precursors
	Supplies C-801 and C-802 ship-based anti-ship cruise missiles
	Supplies through Hong Kong high-grade seamless steel pipes used to make chemical weapons or explosives
	Contracts to supply hundreds of tons of anhydrous hydrogen fluoride, a key ingredient needed to process natural uranium to nuclear weapon grade

the RISK REPORT — CHINA'S DANGEROUS EXPORTS

PAKISTAN

1980-1984	Supplies nuclear bomb design and its fuel
	Helps build Hatf missiles
	Helps with gas centrifuges to enrich uranium
1985-1989	Agrees to sell tritium gas to boost the yield of fission bombs
	Ships equipment for M-11 nuclear-capable missiles
	Starts building a 300 MW nuclear reactor at Chashma in spite of de facto international supply embargo
1990-1997	Secretly delivers more M-11 missile components
	Trains Pakistani nuclear technicians in China
	Continues to deliver components for M-11 missiles
	Supplies more than 30 M-11 missiles now in crates at Sargodha Air Force Base near Lahore
	Helps build a secret 50-70 MW plutonium production reactor at Khusab, and a nearby fuel fabrication or reprocessing plant
	Supplies blueprints and equipment for a missile factory near Rawalpindi, now under construction
	Supplies ring magnets used in gas centrifuges to enrich uranium
	Supplies heavy water to Kanupp nuclear reactor
	Sells a high-tech furnace and diagnostic equipment with military applications
	Ships rocket fuel seized en route in Hong Kong
	Agrees to build Chashma-2, a second 300 MW nuclear reactor

China's Dangerous Exports

	ALGERIA	ARGENTINA	BRAZIL	INDIA	IRAQ	ISRAEL	SYRIA	SAUDI ARABIA	SOUTH AFRICA
1980-1984	Secretly agrees to supply a nuclear research reactor	Sells over 60 tons of heavy water Sells uranium concentrate and low-enriched uranium hexafluoride		Sells at least 130-150 tons of heavy water	Nuclear bomb design supplied to Pakistan makes its way to Iraq				Supplies 60 tons of enriched uranium, undercutting U.S. pressure on South Africa
1985-1989	Trains Algerian scientists and technicians; starts building reactor	Sells enriched uranium	Sells enriched uranium Agrees to provide liquid-fuel and guidance technology for missiles in exchange for solid-fuel technology		Helps make magnets for centrifuges to enrich uranium Sells "Silkworm" anti-ship missiles		Contracts to sell M-9 nuclear-capable missiles	Sells CSS-2 medium-range, nuclear-capable missiles	
1990-1997	Completes reactor and supplies heavy water and uranium fuel Agrees to aid plutonium research			Agrees to supply Tarapur reactors with enriched uranium	Agrees to sell rocket fuel ingredients Agrees to sell lithium hydride useful for nerve gas, missiles and nuclear weapons	Agrees to fund development of a cruise missile with a range of 400 kilometers	Sells ingredients for missile fuel Agrees to supply Syria's first nuclear reactor and train nuclear technicians		

Joan Johnson-Freese
Professor of International Security Studies
Air War College, Maxwell AFB, AL 36112
Joint Committee Hearing of the Committee on International Relations
and Committee on National Security
17 June 1998

The views stated represent those of the author, and are not the opinions of either the United States Air Force or the Department of Defense.

Having recently spent two years researching and writing a book analyzing the motivations and intentions of the Chinese space program, I would urge great restraint in approving legislation banning the export of U.S.-made satellites to China. The title of my book, *The Chinese Space Program: A Mystery Within A Maze,* reflects not only a Byzantine Chinese bureaucracy, but also that my research on China was the first time in all my research experience where I felt I had encountered deliberate subterfuge. Yet my analysis of the situation in China also led me to decisively conclude that U.S. policies toward China since 1986 have significantly strengthened our strategic position there, and subsequently U.S. national security interests, and ought to be continued and expanded.

In recent weeks, it appears that several domestic and foreign policy issues related to China have been inappropriately intermingled and confused. It is my fear that subsequently, actions concerning China may be undertaken that will be markedly counterproductive to the long-term national security interests of the United States. Therefore I offer you my analysis as testimony today toward achieving everyone's goal: the protection of our nation's national security.

The first task is to sort out the issues that have been raised. Domestically, there are two. A federal grand jury is currently seeking to determine if any U.S. export control laws may have been violated by Hughes and Loral in connection with the transfer of technology after the Intelsat 708 launch accident in February 1996. This is an important issue and one which is under investigation by the appropriate legal authorities. There is also the issue of whether campaign contributions influenced decision making regarding the transfer of sensitive technology to the Chinese by the Clinton Administration. Campaign fundraising and contributions and their impact on the American political system is an always-important issue, both here and in a broader sense as well. However, I believe that even in a worst case scenario of undue influence by corporations dealing with China, our national security was not jeopardized. At worst, contributions from either Loral or the Chinese would relate to potential commercial gain, not to military advantage.

The primary issue to be addressed concerning China can be simply stated as has the transfer of commercial space technology by the Reagan, Bush or Clinton Administrations jeopardized the national security of the United States with a resultant need to change our policy of engagement with China? I believe the answer is unequivocally, no.

Recently, sensational accusations have been made that through nefarious acts of the Clinton Administration since 1993, the United States has sold militarily sensitive technology to the Chinese. Perhaps the most disturbing allegation is that significant improvements in the Chinese ICBM capability are directly tied to the recent U.S. technology transfer. These allegations are based on half-facts that have been distorted and sensationalized. The Chinese do have ICBM's, the Chinese do launch satellites for the United States. However, the Chinese began their ICBM program in 1956, simultaneously

55-784 99 - 8

tested their DF-2A rocket and an atomic bomb in 1966, and first launched their DF-5 ICBM in 1979. American cities have been targeted since the 1980's, and as we all know, with nuclear weapons pinpoint accuracy is not a requirement in a countervalue nuclear strategy. As a vital part of U.S. national security interests, three presidential administrations have made it a priority to ensure through a policy of peaceful engagement that the Chinese do not use this arsenal. To abdicate that priority now could bear grave consequences.

The transfer of technology to China has two aspects, technical and political. The technologies cited as militarily vital, however, (including clean rooms, payload farings, coupling load analysis information, and precise navigational equipment) that are alleged to have compromised U.S. national security, are common to commercial satellites generally and have been disseminated around the world for several decades. At the urging of the United States, Intelsat was formed in 1965 as an international consortium to own and operate a global satellite communications system. As part of its organizational mandate, the satellites it owns and operates are internationally developed and launched. Since Intelsat IV, launched in 1971, both U.S. and non-U.S. components have been and are being used in the construction of Intelsat state-of-the-art satellites.

Beyond Intelsat, companies from Japan, Germany, Canada, Russia, France and a multitude of other nations are actively engaged in competition to gain a foothold in the lucrative market that China's 1.3 billion people offers. Comparatively, U.S companies are the most constrained among the vast array of competitors. A German firm exports momentum wheels to China for use on its DFH-3 communications satellite. A Canadian company has formed a joint venture in China to build onboard satellite electronic and ground facilities. Russia and France sell high-resolution imagery. All this means that the technology and know-how for communication satellites that can be used by the Chinese military are not a solely-owned U.S. commodities to give away or even control.

On the other hand, it would be completely inconsistent and insupportable to have a policy that allows U.S. satellites to be launched abroad, from any country, without allowing the technical exchange necessary to maximize the chances of success. Indeed the U.S. government will be using Russian launch vehicles to lift many technologically significant items in support of the International Space Station program, and would be remiss not to do everything possible to assure a successful launch. That does not mean, however, that a national security breech is inevitable or even a concern. It would also be unrealistic and economically insufferable not to allow foreign launch of U.S. satellites because of both lower costs and demand which cannot be met domestically. The U.S. has insufficient launch assets to allow economic competitiveness in the communications satellite field without using foreign launchers. Today, coping with dual-use technology is a fact of life. Let us not throw out the baby with the bathwater.

It is also both important and interesting to note that in the United States we are currently spending considerable time, energy and public and private funds trying to accomplish technology transfer at many different levels, military to civilian, public to private sector, and finding it far easier in theory than practice. Why then we expect that it would somehow be easier for the Chinese, with their entirely different cultural approach to everything from systems engineering to maintenance, and their debilitating propensity toward compartmentalization and bureaucratization, is rather astounding.

Further, most of the technology in question is commercial, not military, and though dual use, one cannot extrapolate too far. Just as nuclear fuel from a light-water reactor is not weapons grade, not all satellite technology is military technology. Just as the United States can launch space station components on Russian launchers without inherently compromising national security, so too can we launch commercial satellites on Chinese launchers without inherently compromising national security. In any case, regulations already ensure that a Pentagon official oversees all U.S. satellites launched by the Chinese 24-hours per day, until such time as they are actually on the launch pad. Consequently, there can be no possibility for the Chinese to obtain technological information by covert means.

Clearly the Chinese are adept at copying. They also have as long a history of cross utilization of space technology as the Russians and the United States. Indeed the location of the Xichang launch site is not entirely serendipitous. That the latitude coordinates of Xichang are similar to those of Kennedy Space Center (KSC) is not likely coincidental. After the Sino-Soviet split, the Chinese still had considerable Soviet space hardware to work from, and the biggest loss was in utilization knowledge. For that they had to rely on published data, which came primarily from the United States. Launching from the same latitude as KSC allowed the Chinese to emulate the expectations for such technical points as the proper rocket attitude and altitude. Should we then retract the First Amendment on the grounds that the Chinese acquired this knowledge from published information? I think not.

Politically, the scandal is perpetuated primarily through the argument that satellite waivers imposed on China after Tiananmen Square were improperly granted by the

Clinton Administration. First, I would suggest that linking commercial policy, security policy and social policy has always been problematic for the United States. It is harder to change the culture of a country than it is either its political or its economic system. Expecting countries to abandon cultural norms, in the case of China norms going back thousands of years, overnight and in response to commercial threats, is rarely effective. Second, leverage to encourage countries to eventually recognize the benefits of change can become severely diminished when bound by sanctions. The waivers that were granted weighed and balanced the goal being sought by the United States, a long-term, stable relationship with China where the U.S. could be influential. Through this engagement, we have the opportunity to demonstrate the value and efficiencies of our democratic system.

From a Chinese strategic perspective, the government must maintain the ability to deal with domestic problems by providing returns (currently economic more than political) that contribute to real economic development for the people. Without this, the stability of the government becomes tenuous. Government stability and economic development therefore are entangled into a synergistic relationship, the hybrid combination of which becomes the number one priority for all decision making in China. Within that context, external decision making is relatively easy to predict. At the strategic level, China can be said to be vigorously pursuing the same development and modernization campaign that Japan undertook so successfully earlier.

It is the thesis of my book that careful analysis is warranted within the U.S. policy community toward identifying ways in which the United States might urge China toward a more acceptable position on the international spectrum, along the lines followed by Japan. Specifically regarding space, during the Cold War period, China emphasized the military and arms sales potential of space and space hardware. Now, in the post Cold War period, the United States ought to focus on the development of constructive policies aimed at encouraging China to emphasize space as an element of peaceful domestic development. The question then is, can this be done without compromising U.S. national security. The answer is yes, and we are on the right road in encouraging the commercial use of space and space technologies.

From a U.S. perspective, "engagement" has been a consistent theme of both the National Security Strategy and the Defense Strategy in recent years. History shows engagement to be a far more successful policy in securing national security goals than disengagement, and hence its embrace in the past, and recognition at all levels of government that it will be even more critical to the United States in the future. To ignore this core strategic premise now would be both unwise and dangerous.

Commercial space has been an area in which the United States has actively engaged China. That engagement has benefits and costs in several areas, including proliferation, economics, and politics. The first two inherently involve technology transfer. Whether the known costs, based on the actions of three administrations, have suddenly today exceeded the benefits is certainly an issue that is properly the subject of Congressional review. I would strongly suggest, however, that the overall engagement policy should be continued as it has many advantages.

The open accident review process initiated after their launcher failed while carrying the Intelsat 708 satellite was a major step forward in leading the Chinese away from the past opaqueness of their program. They engaged in such an open review reluctantly, and only because they feared the repercussions of the space insurance industry and resultant loss of commercial launch business that clearly would have occurred had they not. Although the new process of openness is an evolving one, the premise should not be discouraged. The shroud of secrecy that has previously surrounded the Chinese space program is being lifted, no matter how slowly and reluctantly, primarily because of western companies doing business in China. Indeed the private sector has allowed access to and provided far more information to U.S. intelligence than has ever been available before. The Chinese resist these inroads culturally and bureaucratically, but inroads are nonetheless steadily being made.

Equally important are the non-proliferation benefits of engagement. If the Chinese commercial launch market declines because of new regulations imposed by the United States, they will undoubtedly turn to supplementing their coffers through increased arms and missile sales. If they are not allowed to launch U.S. satellites they will increasingly turn to other countries. The United States government then will subsequently lose the ability to influence Chinese actions simply by being a non-player.

Clearly, the market potential in China is so enormous that most international corporations are competing relentlessly in order to gain a foothold. Since U.S. companies work within more stringent controls than do most other countries, the pressure to compete is considerable. That makes it more imperative for government engagement, not disengagement.. If control measures in areas concerning space technology transfer have not been adhered to and appropriately applied, then that is a matter for the courts. Yet, the policy itself remains sound.

The Chinese desire for foreign technology wherever and whenever they can get it makes it imperative to persuade the Chinese that it will benefit them in the long-run to favor commercial aspects of space, and cost them to do otherwise. That approach has already shown its effectiveness, especially in terms of adherence to the Missile Technology Control Regime (MTCR). The M-11 missile transfers from China to Pakistan in 1992 are less likely in 1998 because of U.S. policies. It is more productive to stay focused on China in the present and future, rather than the past. The United States has embraced a new era and relationship with Russia, but we seem far more reluctant to do so with China. If Congress is seeking to identify cause and effect patterns related to U.S. actions in China, this is one to strongly consider. If the U.S. retreats from its policy of engagement with China, then we can expect proliferation problems to increase, not decrease.

The Chinese culturally understand the premise of reciprocity, and are willing to negotiate to preserve their own precarious domestic balance. It is up to the United States to make sure that the Chinese are then clear as to what we consider our priorities in particular instances of negotiation. Non-proliferation, human rights, market access, democratization, Taiwan, political influence in Asia, military modernization and other issues are all important, but cannot each be our number one priority simultaneously, and, as much as we might want them to be, they are not always *quid pro quo* issues.

National security is increasingly a multifaceted concept, involving military, political and economic aspects. The key is to balance these individual aspects, and to realize that focusing on one while neglecting the others is disadvantageous and contradictory to our National Security Strategy. As international economic competitiveness is increasingly recognized as a vital element of national security, as acknowledged in recent National Security Strategies, involving the Commerce

Department in export reviews with the Departments of State and Defense seems pragmatic rather than nefarious. I would suggest that since 1986 U.S. policy concerning satellite transfers to China has worked well toward achieving a more stable and predictable peace with China and Asia generally, allowing further negotiation in the areas stated above from a position of increased leverage. Banning the export of U.S.-made satellites to China would likely result in the Chinese being pushed to a position diametrically opposite from a stable and predictable peace, if for no other reason than the domestic problems in China which could result or be exacerbated.

Defense Secretary Cohen's 1998 Annual Report to the President and Congress describes a U.S. defense strategy which includes a list of three elements requisite to shape the international security environment in ways favorable to US interests between 1998-2015. "Shaping" is described as efforts to promote and protect U.S. national interests through efforts to build coalitions, promote stability, reduce conflicts and threats, and deter aggression and coercion. This concept applies not only in the macro scale of relationships between nations, but also and more relevant in this case, in the internal challenges facing a large nation. Reduction of internal economic pressures helps nations become much more effective strategic trading partners.

The three policy recommendations that I make in my book aim to persuade Beijing that in order to sustain a long-term relationship with the outside world requisite for economic development and hence internal stability, it is in its best interests to make certain changes. Suggestions on how this might be done fall into three inter-related, general categories.

- American aid to China that would help the country implement a regulatory and legal system that includes a binding commercial code.
- A U.S. demand for more reciprocity in public and private sector relations.
- Increased space cooperation, including controlled transfer of selected U.S. satellite technology that helps China solve domestic problems without compromising U.S. national security.

These recommendations are intended to build confidence in both China and the United States.

Lack of a viable legal system to govern business is what separates China from other Asian countries, like Japan. China remains a country where business dealings are ruled by complex, interpersonal relationships based on Confucian tradition. It is what allows *Guan Xi*, the web of necessary personal connections, to dominate business practices and slow the move from opaqueness to transparency, to the detriment of international commerce and ultimately understanding their key strategic objectives. Removing this opaqueness is critical, as opaqueness fosters technology diversion. Moving away from *Guan Xi* to a binding commercial code might also have spillover effects into areas of democratization and human rights.

In general, a policy of greater reciprocity would allow Beijing to pursue its own internal goals while at the same time tolerating U.S. pursuance of its own priorities. This can be particularly important regarding MTCR adherence in the future. It would also, however, require the United States to make some difficult political choices.

International cooperation recognizes that you cannot influence another person without leverage, and seeks to build such leverage. Satellites for domestic use are particularly high on Beijing's list of priorities, and also a matter of considerable internal dispute. Beijing's concerns are several: bureaucratic competition between the China Aerospace Corporation (CASC) and the Ministry of Post and Telecommunication (MPT), domestic need, the ever present concerns about preserving domestic stability, international economic competitiveness, and a debate over foreign component use and satellite design.

Beijing is eager to link China's urban and rural areas, recognizing that political discontent in the past has come from rural areas feeling disconnected. Yet less than 5% of China's population has phone service. Plus, although 1.3 billion people is a large market, providing them jobs is a major domestic concern. Therefore, whether to build more CASC supported DFH-3 satellites, which requires major components from abroad (not necessarily the U.S.), or go with totally new, foreign-designed more efficient satellites, supported by MPT, is a current discussion. The U.S. can influence this decision, but only as a player.

Better maps, commercially available satellite imagery, global position information, weather forecasting hardware and data, all can arguably be said to increase Chinese military capabilities. The United States, however, is not the sole or even the primary purveyor of these or other technologies or services available in the world market. Turbofan engines for jet trainer aircraft can be applied to cruise missile technology. Yet turbofan engines are available from a variety of industrial and third-world countries, including Russia, France, and the United Kingdom. How far do we want to reach into

American industrial competitiveness and jobs, in a futile attempt to put the genie back into the bottle? Controlled licensing of technology offers the United States more leverage than a free-wheeling technology grab by the Chinese from countries and companies overly anxious to establish their place in the 1.3 billion person Chinese market. In the post Cold War world, leverage is more important in posturing for national security than false illusions about controlling technologies available in the open market.

In conclusion, I would strongly urge you not to retreat from the engagement policy with China, including U.S. satellite launches, which has enhanced our strategic position there, in Asia, and in the world generally. It is only through engagement that we can transition from the Cold War era into an era of cooperation. That is the goal from which our focus must not stray.

Joan Johnson-Freese

Dr. Johnson-Freese has been a Professor of International Security Studies at the Air War College since 1993. Prior to that time she was Director of the Center for Space Policy & Law at the University of Central Florida and a member of the Political Science Department faculty there. She has taught Space Policy at undergraduate, graduate and PME institutions, and has served as the Co-Chair of the Policy & Law Department of the International Space University SSP since 1994. Professional experience includes: Office of Technology Assessment, US Congress, Advisory Panel for US Space Launch Capabilities Study 1994-1995; Guest lecturer on Space Policy, Space Tactics School, 1996; Army War College, 1994, 1995, 1998; Air Force Academy 1994; SPACECAST 2020, Advisory Board, 1993-94; and National Academy of Sciences, Space Science Board, Committee on International Programs, 1994-current. Recent publications include the following:

THE CHINESE SPACE PROGRAM: A MYSTERY WITHIN A MAZE, forthcoming, (Melbourne, FL: Orbit Books, Krieger Publishing, 1998).

SPACE, THE DORMANT FRONTIER: CHANGING THE SPACE PARADIGM FOR THE 21ST CENTURY, Joan Johnson-Freese and Roger Handberg, (Westport, CN, Praeger Publishers, 1997).

THE PRESTIGE TRAP: A COMPARATIVE STUDY OF THE U.S., EUROPEAN AND JAPANESE SPACE PROGRAMS, Joan Johnson-Freese and Roger Handberg, (Dubuque, Iowa, Kendall-Hunt Publishing, 1994).

OVER THE PACIFIC: JAPANESE SPACE POLICY INTO THE 21ST CENTURY, (Dubuque, Iowa, Kendall-Hunt Publishing, 1993).

CHANGING PATTERNS OF INTERNATIONAL COOPERATION IN SPACE, (Melbourne, FL: Orbit Books, Kreiger Publishing, 1990).

Joan Johnson-Freese & Roger Handberg, "The Return of the American Military to Crewed Spaceflight: Hypersonic & Other Visions," SPACE POLICY, May 1998.

Joan Johnson-Freese & Roger Handberg, "Searching for Policy Coherence: The DOD Space Architect as an Experiment," JOINT FORCES QUARTERLY, Summer 1997.

Joan Johnson-Freese, Roger Handberg & George M. Moore, "Space Convergence and Conversion: The Issue of Organizational Culture," Vol. 13, No. 3 (1997), DEFENSE ANALYSIS.

Joan Johnson-Freese and Roger Handberg, "The Return of the American Military to Crewed Spaceflight: Hypersonic and Other Visions," forthcoming, SPACE POLICY.

Joan Johnson-Freese & Roger Handberg, "Making Sausages & Space Policy: The Winning X-33 Bid," SPACEFLIGHT, September 1996.

Joan Johnson-Freese & Roger Handberg, "The Myth of the National Space Transportation Policy: *Deja Vu* All Over Again," DEFENSE ANALYSIS, August 1996.

"Expanding Remote Sensing Utilization: Issues and Initiatives," EARTH OBSERVATION MAGAZINE, April 1995.

Joan Johnson-Freese & George M. Moore, "A Disturbing Trend in Space Science: Process Over Product," SPACEFLIGHT, 1995 .

George Moore, Victor Budura, & Joan Johnson-Freese, "Joint Space Doctrine: Catapulting Into the Future," JOINT FORCES QUARTERLY, Summer 1994.

United States General Accounting Office

GAO

Testimony

Before the Committees on International Relations and on National Security, House of Representatives

For release expected at 9:30 a.m., EDT Wednesday June 17, 1998

EXPORT CONTROLS

Issues Related to the Export of Communications Satellites

Statement for the Record by Katherine V. Schinasi, Associate Director, Defense Acquisitions Issues, National Security and International Affairs Division

Mr. Chairmen and Members of the Committees:

I am pleased to submit for the record my testimony on the evolution of export controls on commercial communications satellites. The allegation that a major U.S. satellite manufacturer provided China with sensitive technologies that may have applicability to its missile programs has highlighted how the United States controls the export of such technology and how this policy has changed in recent years.

My testimony today is based largely on our January 1997 report, prepared at the request of the Chairman, House National Security Committee, on the military sensitivity of commercial communications satellites and the implications of the 1996 change in export licensing jurisdiction.[1] I will discuss (1) key elements in the export control systems of the Departments of Commerce and State, (2) how export controls for commercial satellites have evolved over the years, (3) the concerns and issues debated over the transfer of commercial communications satellites to the export licensing jurisdiction of the Department of Commerce, and (4) the safeguards that may be applied to commercial satellite exports. Lastly, I will share some observations on the current export control system.

SUMMARY

The U.S. export control system–comprised of both the Commerce and State systems–is about managing risk. Exports to some countries involve less risk than to other countries and exports of some items involve less risk than others. The planning of a satellite launch with technical discussions and exchanges of information taking place over several months, involves risk no matter which agency is the licensing authority. Recently, events

---

[1]Export Controls: Change in Export Licensing Jurisdiction for Two Sensitive Dual-Use Items (GAO/NSIAD-97-24, Jan. 14, 1997).

have focused concern on the appropriateness of Commerce jurisdiction over communication satellites. This is a difficult judgement. By design, Commerce's system gives greater weight to economic and commercial concerns, implicitly accepting greater security risks. And by design, State's system gives primacy to national security and foreign policy concerns, lessening–but not eliminating–the risk of damage to U.S. national security interests.

BACKGROUND

The U.S. export control system for items with military applications is divided into two regimes. State licenses munitions items, which are designed, developed, configured, adapted, or modified for military applications, and Commerce licenses most dual-use items, which are items that have both commercial and military applications. Although the Commerce licensing system is the primary vehicle to control dual-use items, some dual-use items–those of such military sensitivity that stronger control is merited–are controlled under the State system.

Commercial communications satellites are intended to facilitate civil communication functions through various media, such as voice, data, and video, but they often carry military data as well. In contrast, military communications satellites are used exclusively to transfer information related to national security and have one or more of nine characteristics that allow the satellites to be used for such purposes as providing real-time battlefield data and relaying intelligence data for specific military needs. There are similarities in the technologies used to integrate a satellite to its launch vehicle and ballistic missiles.

In March 1996, the executive branch announced a change in licensing jurisdiction transferring two items–commercial jet engine hot section technologies and commercial communications satellites–from State to Commerce. In October and November 1996,

Commerce and State published regulations implementing this change, with Commerce defining enhanced export controls to apply when licensing these two items.

KEY ELEMENTS OF EXPORT CONTROL SYSTEM

State and Commerce's export control systems are based on fundamentally different premises. The Arms Export Control Act gives the State Department the authority to use export controls to further national security and foreign policy interests, without regard to economic or commercial interests. In contrast, the Commerce Department, as the overseer of the system created by the Export Administration Act, is charged with weighing U.S. economic and trade interests along with national security and foreign policy interests.

Differences in the underlying purposes of the control system are manifested in the systems' structure. Key differences reflect

- who participates in licensing decisions,
- scope of controls,
- time frame for the decision,
- coverage by sanctions, and
- requirements for congressional notification.

Participants. Commerce's process involves five agencies–the Departments of Commerce, State, Defense, Energy, and the Arms Control and Disarmament Agency. Other agencies can be asked to review specific license applications. For most items, Commerce approves the license if there is no disagreement from reviewing agencies.

When there is a disagreement, the chair of an interagency group known as the Operating Committee, a Commerce official, makes the initial decision after receiving input from the reviewing agencies. This decision can be appealed to the Advisory Committee on Export Policy, a sub-cabinet level group comprised of officials from the same five agencies, and from there to the cabinet-level Export Administration Review Board, and then to the President.

In contrast, the State system commonly involves only Defense and State. While no formal multi-level review process exists, Defense officials stated that license applications for commercial communications satellites are frequently referred to other agencies, such as the Arms Control and Disarmament Agency, the National Security Agency, and the Defense Intelligence Agency. Day-to-day licensing decisions are made by the Director, Office of Defense Trade Controls, but disagreements could be discussed through organizational levels up to the Secretary of State.

This difference in who makes licensing decisions underscores the weight the two systems assign to economic and commercial interests relative to national security concerns. Commerce, as the advocate for commercial interests, is the focal point for the process and makes the initial determination. Under States' system, Commerce is not involved, underscoring the primacy of national security and foreign policy concern.

Scope of Controls. The two systems also differ in the scope of controls. Commerce controls items to specific destinations for specific reasons. Some items are subject to controls targeted to former communist countries while others are controlled to prevent them from reaching countries for reasons that include antiterrorism, regional stability, and nonproliferation. In contrast, munitions items are controlled to all destinations, and State has broad authority to deny a license; it can deny a request simply with the explanation that it is against U.S. national security or foreign policy interests.

Time frames. Commerce's system is more transparent to the license applicant than State's system. Time frames are clearly established, the review process is more predictable, and more information is shared with the exporter on the reasons for denials or conditions on the license.

Congressional Notification. Exports under State's system that exceed certain dollar thresholds (including all satellites) require notification to the Congress. Licenses for Commerce-controlled items are not subject to congressional notification, with the exception of items controlled for antiterrorism.

Sanctions. The applicability of sanctions may also differ under the two export control systems. Commercial communication satellites are subject to two important types of sanctions: (1) Missile Technology Control Regime and (2) Tiananmen Square sanctions. Under Missile Technology sanctions, both State and Commerce are required to deny the export of identified, missile-related goods and technologies. Communication satellites are not so-identified but contain components that are identified as missile-related. When the United States imposed Missile Technology sanctions on China in 1993, exports of communication satellites controlled by State were not approved while exports of satellites controlled by Commerce were permitted.

Under Tiananmen Square sanctions, satellites licensed by State and Commerce have identical treatment. These sanctions prohibit the export of satellites for launch from launch vehicles owned by China. However, the President can waive this prohibition if such a waiver is in the national interest.

EVOLUTION OF EXPORT CONTROLS
FOR COMMERCIAL SATELLITES

Export control of commercial communications satellites has been a matter of contention over the years among U.S. satellite manufacturers and the agencies involved in their export licensing jurisdiction–the Departments of Commerce, Defense, State, and the intelligence community. To put their views in context, I would now like to provide a brief chronology of key events in the transfer of commercial communications satellites to the Commerce Control List.[2]

Origin of Commercial Space Cooperation with China

As the demand for satellite launch capabilities grew, U. S. satellite manufacturers looked abroad to supplement domestic facilities. In 1988, President Reagan proposed that China be allowed to launch U.S.-origin commercial satellites. The United States and China signed an agreement in January 1989 under which China agreed to charge prices for commercial launch services similar to those charged by other competitors for launch services and to launch nine U.S.-built satellites through 1994.

Following the June 1989 crackdown by the Chinese government on peaceful political demonstrations on Tiananmen Square in Beijing, President Bush imposed export sanctions on China. President Bush subsequently waived these sanctions for the export of three U.S.-origin satellites for launch from China. In February 1990, Congress passed the Tiananmen Square sanctions law (P.L. 101-246) to suspend certain programs and

[2]For a chronology and background information on satellite launches from China, see China: Possible Missile Technology Transfers from U.S. Satellite Export Policy–Background and Chronology, by Shirley A. Kan, Congressional Research Service, May 20, 1998 (98-485 F).

activities relating to the Peoples Republic of China. This law also suspends the export of U.S. satellites for launch from Chinese-owned vehicles.

First Transfer of Licensing Jurisdiction

In November 1990, the President ordered the removal of dual-use items from State's munitions list unless significant U.S. national security interests would be jeopardized. This action was designed to bring U.S. controls in line with the industrial (dual-use) list maintained by the Coordinating Committee for Multilateral Export Controls, a multilateral export control arrangement. Commercial communications satellites were contained on the industrial list. Pursuant to this order, State led an interagency review, including officials from Defense, Commerce, and other agencies to determine which dual-use items should be removed from State's munitions list and transferred to Commerce's jurisdiction. The review was conducted between December 1990 and April 1992. As part of this review, a working group identified and established performance parameters for the militarily-sensitive characteristics of communications satellites. During the review period, industry groups supported moving commercial communications satellites, ground stations, and associated technical data to the Commerce Control List.

In October 1992, State issued regulations transferring jurisdiction of some commercial communications satellites to Commerce. These regulations also defined what satellites remained under its control by listing nine militarily sensitive characteristics that, if included in a commercial communication satellite, warranted their control on State's munitions list. (These characteristics are discussed in appendix 1.) The regulations noted that parts, components, accessories, attachments, and associated equipment (including ground support equipment) remained on the munitions list, but could be included on a Commerce license application if the equipment was needed for a specific launch of a commercial communications satellite controlled by Commerce. After the transfer, Commerce noted that this limited transfer only partially fulfilled the President's 1990 directive.

Interagency Groups Consider
Whether to Transfer Additional Satellites

Export controls over commercial communication satellites were again taken up in September 1993. The Trade Promotion Coordinating Committee, an interagency body composed of representatives from most government agencies, issued a report in which it committed the administration to review dual-use items on the munitions list, such as commercial communication satellites, to expedite moving them to the Commerce Control List.

Industry continued to support the move of commercial communications satellites, ground stations, and associated technical data from State to Commerce control. In April 1995, the Chairman of the President's Export Council met with the Secretary of State to discuss issues related to the jurisdiction of commercial communications satellites and the impact of sanctions that affected the export and launch of satellites to China.

Also in April 1995, State formed the Comsat Technical Working Group to examine export controls over commercial communications satellites and to recommend whether the militarily sensitive characteristics of satellites could be more narrowly defined consistent with national security and intelligence interests. This interagency group included representatives from State, Defense, the National Security Agency, Commerce, the National Aeronautics and Space Agency, and the intelligence community. The interagency group reported its findings in October 1995.

Consistent with the findings of the Comsat Technical Working Group and with the input from industry through the Defense Trade Advisory Group, the Secretary of State denied the transfer of commercial communications satellites to Commerce in October 1995 and approved a plan to narrow, but not eliminate, State's jurisdiction over these satellites.

President Overturns State's Decision
to Retain Export Control of Satellites

Unhappy with State's decision to retain jurisdiction of commercial communications satellites, Commerce appealed it to the National Security Council and the President. In March 1996, the President, after additional interagency meetings on this issue, announced the transfer of export control authority for all commercial communications satellites from State to Commerce. A key part of these discussions was the issuance of an executive order in December 1995 that modified Commerce's procedures for processing licenses. This executive order required Commerce to refer all licenses to State, Defense, Energy, and the Arms Control and Disarmament Agency. This change addressed a key shortcoming that we had reported on in several prior reviews.[3]

In response to the concerns of Defense and State officials about this transfer, Commerce agreed to add additional controls to exports of satellites designed to mirror the stronger controls already applied to items on State's munitions list. Changes included the establishment of a new control, the significant item control, for the export of sensitive satellites to all destinations. The policy objective of this control–consistency with U.S. national security and foreign policy interests–is broadly stated. The functioning of the Operating Committee, the interagency group that makes the initial licensing

[3]See Export Controls: Some Controls Over Missile-Related Technology Exports To China Are Weak (GAO/NSIAD-95-82, Apr. 17, 1995) and Export Controls: Concerns Over Stealth-Related Exports (GAO/NSIAD-95-140, May 10, 1995).

determination, was also modified. This change required that the licensing decision for these satellites be made by majority vote of the five agencies, rather than by the chair of the Committee. Satellites were also exempted from other provisions governing the licensing of most items on the Commerce Control List.

In October and November 1996, Commerce and State published changes to their respective regulations, formally transferring licensing jurisdiction for commercial communications satellites with militarily sensitive characteristics from State to Commerce. Additional procedural changes were implemented through an executive order and a presidential decision directive issued in October 1996.

CONCERNS AND ISSUES DEBATED IN THE DECISION

According to Commerce officials, the President's March 1996 decision reflected Commerce's long-held position that all commercial communications satellites should be under its jurisdiction. Commerce argued that these satellites are intended for commercial end use and are therefore not munitions. Commerce maintained that transferring jurisdiction to the dual-use list would also make U.S. controls consistent with treatment of these items under multilateral export control regimes.

Manufacturers of satellites supported the transfer of commercial communications satellites to the Commerce Control List. They believed that such satellites are intended for commercial end use and are therefore not munitions subject to State's licensing process. They also believed that the Commerce process was more responsive to business due to its clearly established time frames and predictability of the licensing process. Under State's jurisdiction, the satellites were subject to Missile Technology sanctions requiring denial of exports and to congressional notifications. Satellite manufacturers also expressed the view that some of the militarily sensitive characteristics of communications satellites are no longer unique to military satellites.

State and Defense point out that the basis for including items on the munitions list is the sensitivity of the item and whether it has been specifically designed for military applications, not how the item will be used. These officials have expressed concern about the potential for improvements in missile capabilities through disclosure of technical data to integrate the satellite with the launch vehicle and the operational capability that specific satellite characteristics could give a potential adversary. The process of planning a satellite launch takes several months, and there is concern that technical discussions between U.S. and foreign representatives may lead to the transfer of information on militarily sensitive components.

Defense and State officials said they were particularly concerned about the technologies to integrate the satellite to the launch vehicle because this technology can also be applied to launch ballistic missiles to improve their performance and reliability. Accelerometers, kick motors, separation mechanisms, and attitude control systems are examples of equipment used in both satellites and ballistic missiles. State officials said that such equipment and technology merit control for national security reasons. They also expressed concern about the operational capability that specific characteristics, in particular antijam capability, crosslinks, and baseband processing, could give a potential adversary.

SAFEGUARDS APPLIED TO COMMERCE AND STATE SATELLITE EXPORTS

No export license application for a satellite launch has been denied under either the State or Commerce systems. Therefore, the conditions attached to the license are particularly significant.

Exports of U.S. satellites for launch in China are governed by a government-to-government agreement addressing technology safeguards. This agreement establishes the basic authorities for the U.S. government to institute controls intended to ensure that sensitive technology is not inadvertently transferred to China. This agreement is one of three government-to-government agreements with China on satellites. The others address pricing and liability issues.

During our 1997 review and in recent discussions, officials pointed to two principal safeguard mechanisms to protect technologies. These safeguard mechanisms include technology transfer control plans and the presence of Defense Department monitors during the launch of the satellites. State or Commerce may choose to include these safeguards as conditions to licenses.

- Technology transfer control plans are prepared by the exporter and approved by Defense. The plans outline the internal control procedures the company will follow to prevent the disclosure of technology except as authorized for the integration and launch of the satellite. These plans typically include requirements for the presence of Defense monitors at technical meetings with Chinese officials as well as procedures to ensure that Defense reviews and clears the release of any technical data provided by the company.

- Defense monitors at the launch help ensure that the physical security over the satellite is maintained and monitor any on-site technical meetings between the company and Chinese officials. Authority for these monitors to perform this work in China is granted under the terms of the government to government safeguards agreement.

Additional government control may be exercised on technology transfers through State's licensing of technical assistance and technical data. State technical assistance agreements detail the types of information that can be provided and give Defense an opportunity to scrutinize the type of information being considered for export. Technical assistance agreements, however, are not always required for satellite exports to China. While such licenses were required for satellites licensed for export by State, Commerce licensed satellites do not have a separate technical assistance licensing requirement.[4]

OBSERVATIONS ON THE CURRENT
EXPORT CONTROL SYSTEM

The addition of new controls over satellites transferred to Commerce's jurisdiction in 1996 addressed some of the key areas where the Commerce procedures are less stringent than those at State. There remain, however, differences in how the export of satellites are controlled under these new procedures.

- Congressional notification requirements no longer apply, although Congress is currently notified because of the Tiananmen waiver process.

---

[4]A Commerce licensed satellite would also require a State technical assistance license if the technical discussions exceed the basic information required to attach the satellite to the rocket, commonly described as "form, fit, and function" data.

- Sanctions do not always apply to items under Commerce's jurisdiction. For example, under the 1993 Missile Technology sanctions, sanctions were not imposed on satellites that included missile-related components.

- Defense's power to influence the decision making process has diminished since the transfer. When under State jurisdiction, State and Defense officials stated that State would routinely defer to the recommendations of Defense if national security concerns are raised. Under Commerce jurisdiction, Defense must now either persuade a majority of other agencies to agree with its position to stop an export or escalate their objection to the cabinet-level Export Administration Review Board, an event that has not occurred in recent years.

- Technical information may not be as clearly controlled under the Commerce system. Unlike State, Commerce does not require a company to obtain an export license to market a satellite. Commerce regulations also do not have a separate export commodity control category for technical data, leaving it unclear how this information is licensed. Commerce has informed one large satellite maker that some of this technical data does not require an individual license. Without clear licensing requirements for technical information, Defense does not have an opportunity to review the need for monitors and safeguards or attend technical meetings to ensure that sensitive information is not inadvertently disclosed.

- The additional controls applied to the militarily sensitive commercial communications satellites transferred to Commerce's control in 1996 were not applied to the satellites transferred in 1993. These satellites are therefore reviewed under the normal interagency process and are subject to more limited controls.

This concludes our statement. We appreciate the opportunity to provide this information for the record of this hearing.

Appendix 1: Militarily Sensitive Characteristics Integrated in Commercial Communications Satellites

Component or Characteristic	Definition	Military Sensitivity of Characteristics Exceeding Certain Performance Parameters
Antijam capability	Antennas and/or antenna systems with the ability to respond to incoming interference by adaptively reducing antenna gain in the direction of the interference.	Ensures that communications remain open during crises.
Antenna	Allows a satellite to receive incoming signals.	An antenna aimed at a spot roughly 200 nautical miles in diameter or less can become a sensitive radio listening device and is very effective against ground-based interception efforts.
Crosslinks	Provide the capability to transmit data from one satellite to another without going through a ground station.	Permits the expansion of regional satellite communication coverage to global coverage and provides source-to-destination connectivity that can span the globe. It is very difficult to intercept and permits very secure communications.
Baseband processing	Allows a satellite to switch from one frequency to another with an on-board processor.	On-board switching can provide resistance to jamming of signals.
Encryption devices	Scramble signals and data transmitted to and from a satellite.	Allows telemetry and control of a satellite, which provides positive control and denies unauthorized access. Certain encryption capabilities have significant intelligence features important to the National Security Agency.
Radiation-hardened devices	Provide protection from natural and man-made radiation environment in space, which can be harmful to electronic circuits.	Permit a satellite to operate in nuclear war environments and may enable its electronic components to survive a nuclear explosion.
Propulsion system	Allows rapid changes when the satellite is on orbit.	Military maneuvers require that a satellite have the capability to accelerate faster than a certain speed to cover new areas of interest.
Pointing accuracy	Provides a low probability that a signal will be intercepted.	High performance pointing capabilities provide superior intelligence-gathering capabilities.
Kick motors	Used to deliver satellites to their proper orbital slots.	If the motors can be restarted, the satellite can execute military maneuvers because it can move to cover new areas.

Testimony of the Honorable John D. Holum
Acting Under Secretary of State for
Arms Control and International Security Affairs

Before the

House International Relations Committee
and National Security Committee

June 18, 1998

Mr. Chairman and members of the Committee, thank you for the opportunity to address U.S. policy regarding the launching of U.S. satellites from China.

I'd like to make several fundamental points to place this issue in context. The first is that nonproliferation of weapons of mass destruction is a cornerstone in U.S. foreign and national security policy. We have placed great emphasis on strengthening international standards - a permanent NPT, completing the comprehensive test ban treaty, entry into force of the Chemical Weapons Convention, efforts to strengthen the Biological Weapons Convention, tighter multilateral restraint such as in the Missile Technology Control Regime. We have pressed better detection - our own technologies, enhanced IAEA safeguards, and others. We have pursued all the tools of diplomacy at all levels.

In her speech last week at the Stimpson Center, Secretary Albright catalogued in some detail our emphasis on nonproliferation and arms control and our broad strategy for pursuing those goals. Trends such as Iran's progress toward a medium range missile capability and the nuclear tests in South Asia make clear that these are not theoretical concerns, but looming threats. There is no disagreement between the Executive Branch and the Congress on the vital importance of these issues.

The second point is that China is indispensable to any effective nonproliferation policy.

China is a nuclear weapon state. Since the 1980s it has had a small but potent arsenal of ballistic missiles, some of which are capable of reaching the United States. It is a potential supplier of weapons, materials and technology to countries of proliferation concern. China's approach can make the crucial difference between success and failure -- whether in negotiating international arms control and non-proliferation agreements, dealing with difficult regional proliferation challenges, or constraining the transfer of potentially destabilizing goods and technologies.

Unquestionably, China has been part of the proliferation problem. Its relationship with Pakistan's nuclear weapons has been a major concern since the 1970's; by 1990, the U.S. could no longer certify the absence of a nuclear weapons program in Pakistan. We also take sharp issue with China's chemical and missile cooperation with Iran. In 1991, the Bush Administration sanctioned two Chinese entities, and in 1993 the Clinton Administration sanctioned eleven Chinese entities, for transferring missile equipment and technology to Pakistan.

At the same time, my third basic point is that although we still have serious concerns, China's approach to nonproliferation has changed markedly in recent years.

Recall that China once advocated proliferation of nuclear weapons as a means of "breaking the hegemony of the superpowers." But in 1992 China joined the Nuclear Non-Proliferation Treaty (NPT) and thereby accepted a legal obligation not to assist others to acquire nuclear arms. As an NPT member, China has proven to be a constructive partner in our efforts to constrain the nuclear ambitions of North Korea. China has ratified the Chemical Weapons Convention and signed the Comprehensive Test Ban Treaty.

In 1994, China committed not to export MTCR-class ground-to-ground missiles. China's exports of missile-related components and technology appear to reflect a narrower understanding of the MTCR Guidelines than ours, but we have no evidence that China has acted inconsistently with its basic 1994 commitment. Similarly, we assess that China continues to abide by its 1996 agreement to end assistance to unsafeguarded nuclear facilities in Pakistan or anywhere else. China is taking steps to improve its export controls, including recent announcement of new dual use nuclear-related controls. And last year, China agreed to conclude its nuclear cooperation with Iran, and also to terminate its export of cruise missiles to that country. Today, it is working with us in efforts to promote stability and prevent an arms race in South Asia.

So the picture is mixed. The progress is substantial, but not enough, especially given the stakes. Therefore, my final broad point is that we must continue to use all the tools at our disposal to make China part of the nonproliferation solution. And we need to fashion the mix of tools in the way best designed not necessarily for psychic satisfaction, but for nonproliferation results.

What are the tools?

- Intensive diplomacy at all levels, including that of President Clinton and his predecessors.

- The front-line, day-to-day work of nonproliferation, with experts sifting through intelligence reports and generating demarches aimed at preventing specific transfers.

- Technical collaboration, such as our expanding cooperation to improve China's export controls.

- Sanctions. China's undertakings on missile restraint in 1992 and 1994 came about through waivers of sanctions then in force. The potential for sanctions was part of the atmosphere in 1996, when China pledged not to assist unsafeguarded nuclear facilities, and in 1997 when China agreed to cease sales of cruise missiles to Iran.

- And positive incentives. Unquestionably, China's recent far-reaching steps on nuclear nonproliferation were motivated, at least in part, by the prospect of civil nuclear cooperation with the United States.

Let me emphasize, however, that there are clear limits to incentives. Specifically, the United States continues our long-standing support for and cooperation with Taiwan, as President Clinton's actions during the 1996 crisis demonstrated. And, of particular relevance to the subject of this hearing, neither this Administration nor its predecessors have been willing to sell China arms or transfer sensitive technologies that could contribute to China's own WMD or missile programs.

* * * *

One aspect of our efforts to persuade China to adopt a more responsible nonproliferation policy, particularly regarding missile transfers, has been the basic policy of three administrations, beginning in 1988, to allow U.S.-made

55-784 99 - 9

satellites and foreign satellites with significant U.S. components and technology to be launched on Chinese rockets. This policy has been used judiciously as a "carrot" to encourage China to enforce strengthened nonproliferation standards.

But, again, the incentive is clearly limited, to exclude transfer of sensitive missile or satellite technology when satellites are licensed for launch from China. The U.S. has a very strict policy, secured in a bilateral technology safeguards agreement between the U.S. and China, designed to prevent the transfer of sensitive missile technology to China that could assist its space launch vehicle program. The agreement specifically precludes U.S. persons from assisting China in any way on the design, development, operation, maintenance, modification, or repair of launch vehicles.

The agreement itself also limits the level of technical data that may be provided to the Chinese to specific interface data related to form, fit and function that are necessary to mate the satellite to the launch vehicle. And the agreement provides for the right of U.S. Government oversight of all stages of a planned Chinese launch, including preparations, satellite transportation, and launch.

The Department of State has delegated the responsibility for implementing this agreement to the Department of Defense. The Defense Technology Security Administration (DTSA) acts as the executive agent for the U.S. side.

Licenses for the export of satellites for launch from China are issued subject to conditions designed to ensure the protection of sensitive technologies, including a requirement that the U.S. licensee conform to the technology safeguards agreement. Under current regulations DTSA representatives attend meetings between manufacturers and Chinese launch service providers to ensure that there is no transfer of unauthorized technology or technical data.

We do not believe that the commercial space launch activities that have been authorized by licenses and monitored under these procedures have benefited China's missile or military satellite capabilities.

* * * *

Against this background, I would like to give you the State Department's perspective on two events that have been the subject of broad reporting and commentary: first, the transfer of jurisdiction from the State Department to the Commerce Department for commercial communications satellites, and second, the most recent waiver, in February 1998, for Loral's Chinasat-8 project.

One unfinished piece of business facing the Clinton Administration when it took office in 1993, was a set of amendments to the International Traffic in Arms Regulations (ITAR) that had been prepared at the end of President Bush's Administration. The ITAR, administered by the State Department, implements the President's authorities under Section 38 of the Arms Export Control Act. The ITAR contains the U.S. Munitions List, which specifies articles and services which require a State Department license before they may be exported or, in certain cases, discussed with a foreign person.

In 1990, Congress inserted specific provisions in reauthorization of the Export Administration Act calling for removal of certain items from the U.S. Munitions List. In his veto message on the bill, President Bush said his Administration nonetheless would act to remove dual-use items from the U.S. Munitions List, except for those warranting the special controls of the ITAR. This led to an inter-agency study, and then draft amendments to the ITAR to remove certain dual use items, including commercial communications satellites.

However, since the conclusion of that study generally coincided with the election of President Clinton, the State Department deferred implementation so the incoming Administration could review the matter.

In July 1993 following further interagency study, the Clinton Administration approved the Bush Administration's proposed ITAR amendments without change. As a result, many commercial communications satellites were removed from the U.S. Munitions List and placed under the export licensing jurisdiction of the Department of Commerce. Commercial satellites remaining on the Munitions List were outlined in Category XV of the list, and cover nine specific performance characteristics such as antennae capabilities, encryption devices, and propulsion systems. Over the next two years, those characteristics continued to define which communications satellites required a U.S. Munitions License and which required approval by the Department of Commerce.

The U.S. aerospace industry continued to press for treatment comparable with other communications trade, such as fiber optics and telephone switching equipment, which were controlled under the Commerce Department's Control List. They pointed out that characteristics once unique to military satellites were now routinely employed on commercial communications satellites. And they argued that the 30-year U.S. lead in building and exporting commercial communications satellites was under challenge from Japan, Europe and Canada, who were promoting the view that American manufacturers were unreliable because of the U.S. Government's restrictive export policies.

Secretary Christopher agreed on the need to ensure that U.S. Munitions List controls in this area are up to date and justified, and requested that an interagency study be undertaken on whether the characteristics specified in the International Traffic in Arms Regulations appropriately identified those communications satellites having significant military or intelligence capability. This inter-agency study was organized by the State Department and included the participation of the Defense Department, the Intelligence Community, the Arms Control and Disarmament Agency, the Department of Commerce, NASA and other interested agencies.

In September 1995, Secretary Christopher received and approved recommendations from the interagency group narrowing, but not eliminating, U.S. Munitions List controls. Those recommendations were supported by the Defense Department and the Intelligence Community. The Commerce Department supported removal of all commercial communications satellites from the U.S. Munitions List. Commerce then exercised its right to seek Presidential review.

This led to additional inter-agency review under the aegis of the National Security Council. As distinct from the earlier, split recommendation, this review produced a common recommendation from State,

Commerce, Defense and the Intelligence Community, with two important parts.

Henceforth, commercial communications satellites would be controlled by Commerce even if they had embedded in them individual munitions list components or technologies; in all other cases, munitions list technologies or components themselves would continue to be controlled on the U.S. Munitions List, subject to State Department licensing.

However, the further shift in control was accompanied by new control procedures and regulations to strengthen safeguards. Interagency review was strengthened, giving State and Defense the right to review all Commerce export license applications. A new foreign policy and national security control was established in Commerce's Export Administration Regulations whereby State and Defense could recommend denial of a satellite export to any destination on the basis of national security or foreign policy interests. Commercial communications satellites were made exempt from the foreign availability requirements of the Export Administration Act.

As Secretary Christopher noted in a recent letter published in the Los Angeles Times, these new features made it possible for the State Department to change its position and support the March 1996 recommendation to the President.

The bottom-line question is, of course, has this change resulted in a degradation of protection for U.S. national security. It was Secretary Christopher's conclusion, and remains the judgment of the Department of State, that the changes made in the Commerce export licensing system in 1996 were sufficient to deal with the national security sensitivities associated with foreign launches of communications satellites. They provide a degree of protection for these items when under Commerce control that approximates the strict controls of the International Traffic in Arms Regulations. Therefore, the State Department was provided with reasonable assurance that U.S. national security would not be adversely affected with the jurisdictional change.

* * * *

Finally, let me report that the waiver of Tiananmen Sanctions earlier this year for Loral's Chinasat-8 project was handled in a normal manner, in accordance with the procedures used in previous waiver requests.

This dealt with the proposed export under a Commerce license of a commercial communications satellite to the China National Postal and Telecommunications Appliances Corporation for launch from China. The satellite, once launched, will provide commercial voice, video and data traffic to China. The State Department received applications to approve technical assistance agreements in conjunction with the export, and these were distributed for interagency review. After all consulted agencies concurred in the issuance of the proposed technical assistance agreements, subject to the normal limitations and conditions, the State Department recommended to the President that he waive Tiananmen sanctions, in accordance with established procedures.

When we recommended the waiver, senior Administration decision-makers were aware that Loral was under criminal investigation for alleged violations of the Arms Export Control Act. But the State Department's long-standing policy has been that, provided the activity proposed for waiver is consistent with U.S. national security and foreign policy, we do not deny export privileges to U.S. firms that are under investigation but have not been indicted. However, if a U.S. firm is indicted, the Department does adopt a denial policy on the basis of the indictment, and does not wait for a conviction.

The State Department recommended the Presidential waiver for Chinasat-8 in line with the established policy that such exports are in our national interest. Permitting the launch of satellites from China is part of our broader engagement policy, which includes a strong basis for U.S.-Chinese cooperation on missile nonproliferation issues.

Of course there were other arguments for the waiver, including the desire to help preserve the U.S. lead in telecommunications industry and jobs for American workers. But these aspects would not outweigh nonproliferation and national security considerations.

* * * *

It is against this backdrop that the United States conducts commercial space launch cooperation with China. We strive to accommodate U.S. commercial and economic interests -- including promoting U.S. satellite exports -- but within our paramount nonproliferation and national security objectives. We have a system, involving the licensing and technical safeguards processes, to deny access to sensitive missile technology by China. At the same time, if any persons violate our laws and regulations in this area, then such violations need to be investigated fully and prosecuted accordingly.

The United States has engaged China at the highest levels regarding its nonproliferation policies and practices. We continually encourage China to strengthen its export controls and to bring its nonproliferation polices and practices more in line with international norms. The prospect of launching U.S. satellites -- under technology safeguards and according to the disciplines of our commercial space agreement -- is an important inducement to a positive evolution in Chinese policy, which, in turn, is indispensable to the containment of proliferation in a dangerous world.

TESTIMONY OF
WILLIAM A. REINSCH
UNDER SECRETARY FOR EXPORT ADMINISTRATION
U.S. DEPARTMENT OF COMMERCE
BEFORE THE
HOUSE COMMITTEE ON NATIONAL SECURITY
AND
HOUSE COMMITTEE ON INTERNATIONAL RELATIONS
SATELLITE EXPORT POLICY TO CHINA
JUNE 18, 1998

Mr. Chairman, I want to thank the subcommittee for this opportunity to testify on the export of commercial communications satellites to China. I believe this Administration's policy on the export of commercial communications satellites to China both protects our national security and facilitates our economic well-being. In allowing China to launch U.S. satellites and transferring licensing jurisdiction for commercial communications satellites to Commerce, this Administration has continued and enhanced the policy of the Reagan and Bush Administrations and has been consistent with Congress' expressed intent.

Our current policy continues the decision by previous administrations to allow China to launch U.S.-built satellites subject to bilateral agreements on price, number of launches and technology safeguards. President Reagan began this policy in light of the growing demand for satellite launches, a demand which could not be met by U.S. launch service providers. His decision was a wise one, and he deserves credit for ensuring that the U.S. would continue its lead of the commercial communications satellite industry. Our view, like that of Presidents Reagan and Bush, is that under the appropriate safeguards these launches need not pose a risk to national security. In a moment I will describe these safeguards as they apply to Commerce licensed commercial communications satellites.

Commerce licensing of commercial communications satellites grew out of a 1990 decision by President Bush to veto a revised Export Administration Act which would, among other things, have moved all communications satellites to Commerce jurisdiction. President Bush's veto was not related to the satellite issue, but in his veto message he directed that State review its control list to determine if a range of items, including communications satellites, could be moved to Commerce jurisdiction, in light of the strong interest expressed by members of both parties in the jurisdictional issue and as the U.S. was the only country in the world to control communications satellites as munitions items. As a result of this directive, the State Department transferred an initial tranche of roughly half of all commercial communications satellites to Commerce licensing jurisdiction at the end of the Bush Administration. President Clinton committed in 1993 to continue the list rationalization exercise begun by President Bush. It is also worth noting that in 1990 and 1992 both houses of Congress passed legislation that would have transferred jurisdiction over commercial communications satellites to Commerce, and in 1994, committees in the House introduced and in the Senate reported bills with this provision. These actions are in addition to the letters the Administration received from a number of Members of Congress urging either the jurisdiction transfer or the export of satellites to China.

The 1992 transfer had established nine technical parameters relating to communications satellites' components and capabilities. Satellites with capabilities above the parameters were controlled by State. Those with lesser capabilities were controlled by Commerce. A 1995 review, undertaken by a working level group of technical experts led by the State Department, attempted to adjust these parameters to see if additional satellites could be moved to Commerce jurisdiction. However, this technical working group was unable to resolve several issues. Rapid technological change in the current generation of commercial communications satellites meant that the nine parameters could not consistently separate "military" from "commercial" communications satellites. These changes made the parameters outdated and illustrate the migration of military technology into the commercial communications satellite sector as the demand for satellite telecommunications services grows.

A cursory review of some of the parameters illustrates the problem. For example, a satellite flown at high earth orbit has a larger "footprint" on the ground than one flown at a low earth orbit. Even though the two satellites could be exactly the same, the one with the larger footprint would be treated as a munition if it flew at a higher orbit. As satellites become lighter, the same propulsion system provides greater acceleration. This means that new satellites, which weigh less than old models, would be controlled as munitions only because of their lighter weight.

Another parameter, cross-link capability, allows satellites to exchange data without going through an intermediary ground station. In 1990, only military satellites needed to "talk," but the development of communications satellites for a global mobile phone network means that civil satellites also needed to be able to ensure global coverage and to avoid the expense of numerous ground stations. Satellite antennae size was also a parameter. Satellite antennas allow the transmission and reception of signals from the ground. The larger the antenna, the better it can receive signals from small transmitters like cell phones. Successful implementation of global mobile telephone systems required that future civil communications satellites use these specialized antennas. Using these parameters to determine licensing jurisdiction means that Iridium, Teledesic and other satellite systems would now be licensed as arms exports.

With respect to the transfer of jurisdiction, Secretary of State Christopher, in a June 8 statement in the Los Angeles Times, makes clear that his 1995 decision was to task State and its industry advisory group to continue the process begun by President Bush in 1990 to determine what additional communications satellites could be moved to Commerce control. It was the Secretary's instruction that led to the eventual transfer of all satellites to Commerce, a decision which all agencies involved -- State, Defense, Commerce -- supported.

As Secretary Christopher noted, this unanimous recommendation was due in part to President Clinton's December 1995 Executive Order 12981, which revised the Commerce licensing process. This executive order provided the participating agencies -- State, Defense, Energy, and ACDA -- with the right to review and make recommendations on any Commerce license application they wished, and it established time lines and rules for interagency review and procedures for dispute resolution and allows any agency the chance to object to proposed exports through a hierarchy of committees from the working level all the way to the President. Moreover, as part of the transfer of jurisdiction, President Clinton amended Executive Order 12981 to specifically give State, Defense, Energy, and ACDA greater ability to object to a proposed export of commercial communications satellites.

In reality, no licenses have ever been sent to either Cabinet Secretaries or the President in this Administration, but disagreements go as high as the assistant secretary level in perhaps five percent of our cases. Every license approved by Commerce for commercial communications satellites since the 1996 transfer has had the unanimous assent of State, Defense, and ACDA, has been subject to the same level of stringent technology safeguards as satellites licensed at State, and has made clear to exporters that no rocket technology could be transferred.

The President's decision applied only to commercial communications satellites and the minimum equipment and technology needed for launch. Space launch vehicles and all detailed design or manufacturing data for space launch vehicles or communications satellites were not transferred to Commerce as part of this decision. Nor was there a transfer of numerous satellite components, if they were being exported separately rather than as part of a single satellite launch package.

The President's decision to transfer jurisdiction required that Commerce impose enhanced controls on satellites. These new controls excluded commercial communications satellites from certain provisions of the Commerce regulations, such as foreign availability, which can be used to release items from licensing. In addition, the enhanced controls created a new and higher standard for reviewing communications satellite licenses which requires a review of every application to determine if the export is consistent with U.S. national security and foreign policy interests. Language from the Arms Export Control Act was used with the intent of providing the same level of control under Commerce regulations as was found under State regulations.

Commerce licenses for communications satellites contain numerous conditions and provisos developed in conjunction with the Departments of Defense and State. Under Commerce licenses, exporters are obliged to comply with the terms of the satellite technology safeguards agreement between the U.S. and China, which requires them to:

-- develop a technology transfer control plan which identifies the level and extent of technical data to be released, and which also includes plans for securing the satellite during its transportation to the launch site;

-- have all technical data under the license reviewed by the Defense Department prior to its release to the launch service provider and have a Defense Department monitor present at technical meetings and launch activities with the Chinese launch service provider;

-- transport the satellite in a sealed container allowing no access to equipment or technical data and with U.S. monitors to accompany the satellite if it is transported on a non-U.S. aircraft;

-- have a separate cryptographic equipment safeguard plan for communications security equipment;

-- limit technology which can be released under the Commerce license to only form, fit and function data used to mate the satellite to the rocket and require the exporter, in the event of a launch failure, to obtain a license from State before releasing any new technical data.

In light of these safeguards, I believe the existing Commerce licensing system fully protects our nationals security and foreign policy concerns. There have been no allegations regarding export control violations of Commerce satellite licenses since the transfer of jurisdiction,

I understand there have been questions raised about an analysis conducted of the 1995 APSTAR II launch failure. After that failure, the company involved conducted an analysis without the participation of the Chinese launch service provider. The analysis was written in order to satisfy insurance requirements. The analysis was reviewed by the Department of Commerce, which determined that it contained only information already authorized for export under the original Commerce license issued in February 1994. The unclassified report was provided first to a consortium of Western insurance companies and later to the Chinese launch service provider.

I would also like to correct some misunderstandings which have arisen regarding the Commerce licensing process as a result of an earlier review by the General Accounting Office. 'In particular, GAO asserted that there are five differences in the treatment of satellite licenses at Commerce and at State. A closer look shows these differences do not affect national security. GAO reported that:

-- *Congressional notification of individual licenses is not required in the Commerce system.* Commerce regularly briefs the Hill, issues annual reports, provides licensing documentation and answers inquiries upon request. We have provided briefings on satellite exports, and we briefed on the transfer of jurisdiction in 1996.. We are not aware that the Congress has ever objected to any satellite export, and the message that Congress has consistently sent is that it wants satellites controlled as dual-use items under the Export Administration Act, which does not generally provide for Congressional review of individual licenses. In the specific case of satellites, of course, there can be no exports to China without a Tiananmen Square waiver which is notified to the Congress.

-- *Sanctions for missile proliferation do not apply to Commerce licenses.* Sanctions do apply to Commerce in cases of Category I violations, and the President generally has flexibility to include dual use export sanctions in other cases. Normally, however, Category II missile sanctions apply only to munitions and dual-use items controlled under the Missile Technology Control Regime. Commercial communications satellites fall into neither category. Congress clearly intended Category II sanctions to be less onerous than Category I sanctions, which do cover dual use items.

-- *Defense's power to influence the decision making process has diminished.* DOD's authority has not diminished in this regard. Commerce has denied licenses when Defense has raised national security concerns found credible by the reviewing agencies, but Executive Order 12981 does not give Defense or any other agency a veto over licensing, which would be contrary to legislative authorities and Congressional intent. It does, however, permit an agency, including DOD, to prevent approval of any license until the President has heard and decided on its objections.

-- *Technical information may not be as clearly controlled under Commerce procedures.* Since Commerce technology conditions are almost identical to those used at State, it is hard to see how the level of control on technology has changed.

-- *The additional controls placed on communications satellites transferred in 1996 do not apply to those transferred in the Bush Administration.* This the assertion misses the point. In practice, all satellite applications subject to Commerce license after the transfer are subject to the same safeguards, and the other agencies have the same review and escalation rights.

Let me also suggest that as a matter of policy there are several reasons why allowing Chinese launches of U.S.-manufactured satellites is in our interest. First, this is a large and important industry that is growing rapidly. U.S. industry revenues last year were $23.1 billion, a 15% increase from the previous year. Employment in 1997 was over 100,000, a 10% increase from the previous year. The industry indicates that it currently has $1.7 billion in launch contracts on Chinese rockets, with 8,000 U.S. aerospace jobs directly supported by those contracts. With over 1200 satellites expected to be launched over the next ten years, it is clear that the U.S. industry will continue to need access to the full range of launch providers if it is to remain the world's leader. And that is a status I think we all support, because it is not only good for our economy, it is good for our military and our national security as well. As the line between military and civilian technology becomes increasingly blurred, what remains clear is that a second class commercial satellite industry means a second class military satellite industry as well -- the same companies make both products, and they depend on exports for their health and for the revenues that allow them to develop the next generation of products.

Second, some of these satellites bring telephone, television, and Internet services to the Chinese people. I believe such services are an integral part of any effort to bring democracy and freedom to China. History has shown that it was the successful example of the West -- not only in military strength but in standard of living and freedom of expression -- that brought the Cold War to an end. Our goal should be to bring not only our products but our ideas and values to China, but we cannot do that if they do not have the technological tools to receive them.

International security since the end of the Cold War poses very real problems for the United States. We are in the midst of a serious debate as to whether we should seek to constructively engage those with whom we have disputes or whether we should simply try to punish them through unilateral embargoes and sanctions. It may make us feel good to impose Cold War-style embargoes on these countries -- even though they rarely work -- but they do not help us achieve our objective of changing the other country's behavior. Those who find it in their interest to exaggerate the threat of trade with China seem incapable of defining our relations with this emerging power in any terms but those of military conflict. However, we believe that treating China as a committed adversary is the quickest way to ensure it becomes one, and we remain convinced that it is better to engage China frankly in dialogue, in trade, and in ideas than it is to seek to isolate them.

STATEMENT OF JAN M. LODAL
PRINCIPAL DEPUTY UNDER SECRETARY OF DEFENSE FOR POLICY

BEFORE A JOINT HEARING OF
THE COMMITTEE ON NATIONAL SECURITY AND THE COMMITTEE ON INTERNATIONAL RELATIONS,
U.S. HOUSE OF REPRESENTATIVES

June 18, 1998

I am pleased to appear today to discuss U.S. policy regarding the export of satellites to China. In my statement I will outline the policy framework and the role of the Department of Defense in implementing this policy.

In September 1988, President Reagan decided to permit the launch of U.S. commercial communication satellites by China. This decision was motivated by the desire to allow commercial relations with China to expand in a more normal manner. The Reagan administration understood the potential risk that such a program could lead to the transfer of missile-related technology to China, but also recognized that China had for many years had the basic technology necessary to develop and deploy effective ballistic missiles, including intercontinental missiles capable of hitting the United States.

To help ensure that no significant missile or satellite technology is transferred to China, the U.S. negotiated a bilateral technology safeguards agreement with the PRC in 1988. This bilateral technology safeguards agreement was renewed with minor modifications in 1993 and remains in force today. The agreement has two important features to protect U.S. national security interests: (1) it prohibits the transfer of specific technical data and assistance by U.S. companies to China and prohibits the Chinese from seeking that technical data or assistance; and (2) requires oversight and monitoring of the launch campaign by the U.S. government. I will address these particulars of this monitoring activity later in my statement

The requirements of the bilateral technology safeguards agreement are implemented through conditions on export licenses. From 1988 to today, the jurisdiction for export licensing of satellites has shifted between the State Department munitions system and the Commerce Department dual-use system. There are three periods of time where the jurisdiction was different: (1) 1988 to 1992; (2) 1993 to 1996; and (3) 1996 to today. I will discuss each of these periods in turn.

From 1988 to the end of 1992, all communications satellites were licensed by the State Department under the International Traffic in Arms Regulations. Up to 1988, State had always had jurisdiction over such satellites and the decision by President Reagan to permit transfers to China did not change this system. In brief, State controlled all of the technical data and technical assistance required to perform a launch of a U.S. built satellite in China. These controls also extended to all design, development, and manufacturing data on communication satellites. Licenses issued by State for satellite transfers to China for launch contained several safeguards including conditions that required USG monitoring of the launch campaign as outlined in the U.S.-China bilateral technology safeguards agreement. During this period, there were four licenses issued by State for the launch of U.S. built satellites in China.

In late 1990, Congress passed the "Omnibus Export Amendments Act" which President Bush ultimately vetoed. One of the bill's provisions was a requirement to transfer to Commerce jurisdiction over those items on the State munitions list that were on the then-COCOM dual-use list. As an administrative matter, President Bush directed a review to accomplish this jurisdictional transfer in a manner that also ensured national security interests would be protected.

One result of this review was a decision by the Bush Administration in 1992 to transfer license jurisdiction for purely commercial communication satellites from the State Department to the Commerce Department. Commerce controls also extended to include "form, fit and function" technical data necessary to mate the satellite to the launch vehicle. Nine technologies were identified as giving a satellite specific military capabilities, and any satellite containing any of these nine technologies continued to require a license from the Department of State. For example, satellites with large antennas, intersatellite relay links, and specialized on-board processing remained under State control as did the "kick motors" necessary to launch satellites into high earth orbits. State also retained control over: (1) all launch vehicles; (2) all technical data beyond "form, fit, and function" that is associated with the integration of satellites with launch vehicles; (3) all design, development, and manufacturing data on satellites; and (5) all technical assistance (e.g., engineering services) that might be provided by U.S. companies to the foreign launch service provider including any analyses of launch failures. The Clinton Administration issued some of the regulations implementing this jurisdictional change shortly after taking office in 1993.

During this period, monitoring by the U.S. government was required in all licenses for launches of satellites that contained one or more of the identified military-related technologies or kick motors, any launch vehicle integration technical data or any technical assistance – that is, in all the licenses issued by the Department of State. There were three launches during this period that were not monitored. These were launches of purely commercial satellites, licensed by Commerce, that did not include DoD monitoring. Monitoring had always been associated with the licenses issued by the State Department, and DoD license review procedures anticipated that there would be at least one State license required for the launch of even these commercial satellites now licensed by Commerce. However, these launches did not require any State licenses. We are not aware of any transfer of technology from these unmonitored launches that contributed to China's missile or military satellite capabilities. Nevertheless, DoD did conclude that full monitoring would be a strong safeguard at relatively low cost to the companies that should be applied to all license cases, even those that did not require Department of State licenses. This was agreed by all agencies and incorporated as a requirement in 1996 when jurisdiction was transferred to Commerce for all commercial communication satellites

The basic approach to implementation of the Bush Administration policy was to follow the established interagency procedures for the review of dual-use export licenses. During this period, it became increasingly clear that these procedures needed reform, not just for satellites, but across the board. The Clinton Administration undertook such a review, which led to the issuance of Executive Order 12981 in December of 1995. This Executive Order established strict timelines for license reviews, and put into place a disciplined dispute resolution process.

In 1996, President Clinton decided to transfer additional jurisdiction for commercial communication satellites from the State Department to the Commerce Department. DoD supported this transfer because the transfer did not involve certain sensitive technology associated with satellites and launch vehicles and because the transfer was accompanied by several changes in procedures that protect DoD's ability to ensure that transfers are consistent with U.S. national security. The system is now the following:

(1) Companies can export complete commercial communication satellites under a Commerce license even if they contain one or more of the individual military technologies that defined State jurisdiction over communication satellites prior to 1996. All of those individual military technologies, however, must still get a State license when not exported as part of a complete communications satellite.

(2) Commerce continues to control certain limited "form, fit, and function" technical data necessary to mate the satellite to the launch vehicle.

(3) State retains control over all launch vehicles, all technical data associated with launch vehicles or the integration of satellite payloads with launch vehicles, all design and manufacturing data for satellites, and all technical assistance that might be provided by U.S. companies to Chinese launch service providers including any launch failure analyses.

In addition, several changes were made to strengthen the Commerce system and the 1995 Executive Order governing interagency reviews of dual-use licenses. The changes in procedures that are now in effect include:

(1) License determinations are subject to majority vote of reviewing agencies with a continuing right of any dissenting agency to escalate the matter up to and including the President.

(2) Licenses can be denied for broad national security reasons to any destination.

(3) Communication satellites are not subject to formal foreign availability determinations under the Export Administration Act.

(4) All communication satellite licenses must include strong safeguards including DoD monitoring and payment of DoD monitoring expenses by the companies.

DOD currently reviews all communication satellite licenses to ensure that the proposed export would be consistent with U.S. national security interests. DOD's recommendations reflect inputs from relevant DoD components such as the

Air Force and the National Security Agency. DoD's recommendations to approve such satellite exports are conditional on strong safeguards including:

•• A requirement that the satellite exporter prepare a Technology Control Plan which must be approved by DOD.

•• The Technology Control Plan must include: (1) a detailed transportation plan for shipping the satellite to ensure that only U.S. personnel have access to the satellite at all times; and (2) a detailed physical and operational security plan including procedures for the supervised mating of the satellite to the launch vehicle.

•• A requirement that technical data that the U.S. company wants to transfer to the Chinese launch service provider is approved in advance by DoD's Defense Technology Security Administration.

•• A requirement that a DOD monitor be present at technical meetings between the U.S. exporter and Chinese launch service personnel to ensure that no information is exchanged that would improve Chinese missile or satellite capabilities. This includes a requirement that DOD monitors be present at the launch site in China to oversee physical site security and launch operations.

Statutory Tiananmen sanctions require that the President issue a "national interest" waiver before a license may be issued for any U.S. satellite export to China. A memorandum with a recommendation is prepared for the President, typically by the State Department. In this connection, DOD has reviewed such memoranda to ensure that the memorandum accurately describes the safeguards and other conditions that DoD has recommended for inclusion on the license.

As the committees know, allegations that Loral Space Systems and Hughes Aircraft Company have committed export violations in connection with a failure of a Chinese launch of a Loral satellite in February 1996 are the subject of an ongoing criminal investigation by the Department of Justice. There is little that I can say about this matter beyond assuring you that DOD is cooperating fully with all Department of Justice inquiries into this matter. DoD was aware of these allegations at the time it was asked to review the export license applications for the launch of Loral's Chinasat-8 satellite in 1998. Those applications were reviewed carefully taking into account all the relevant information available to DoD at that time. DoD's decision to recommend approval of those licenses was based on the facts of those particular cases and on the specific safeguards required by the licenses.

In addition, DOD is cooperating fully with all Congressional inquiries and requests for a broad range of documents. DOD has already provided some documents even as we continue to collect and assemble additional documentation. I should note that the range and scope of those document requests is quite voluminous. We are committed to being fully responsive, but this means taking the time to make sure we are providing all of the materials that you and others have requested.

In summary, DoD takes its overall role in the development and implementation of export control policies very seriously. The case of commercial communication satellites and China presents significant challenges to the U.S. export control system as we seek to ensure that no technology is transferred that would improve China's indigenous missile or satellite capabilities. We believe that the current system protects our national security.

98-485 F

CRS Report for Congress

China: Possible Missile Technology Transfers from U.S. Satellite Export Policy - Background and Chronology

Updated June 12, 1998

Shirley A. Kan
Analyst in Foreign Affairs
Foreign Affairs and National Defense Division

CRS

Congressional Research Service • The Library of Congress

Abstract

Members of Congress are concerned about whether U.S. firms have provided technology or expertise to China for use in its ballistic missile program and whether a series of decisions by the Clinton Administration on satellite exports have facilitated legal or illegal transfers of missile-related technology to China. The *New York Times* reported in April 1998 that the Justice Department is conducting an ongoing criminal investigation into whether Loral Space and Communications (of New York), and Hughes Electronics (of Los Angeles) violated export control laws. The firms are alleged to have shared their findings with China on the cause of a Chinese rocket's explosion while launching a U.S.-origin satellite in February 1996. In sharing their conclusions, the companies are said to have provided expertise that China could use to improve its ballistic missiles, including their guidance systems. This CRS report provides detailed background information and a comprehensive chronology. The events summarized here, based on various open sources and interviews, pertain to various aspects of U.S. foreign and security policy. This report may be updated if there are significant developments.

NOTE

This CRS study was initiated at the request of the Committee on International Relations of the House of Representatives and is adapted for general Congressional use with permission of the Committee.

China: Possible Missile Technology Transfers From U.S. Satellite Export Policy — Background and Chronology

Summary

Members of Congress are concerned about whether U.S. firms have provided expertise to China for use in its ballistic missile program and whether a series of decisions by the Clinton Administration on satellite exports have facilitated legal or illegal transfers of missile-related technology to China. The *New York Times* reported in April 1998 that the Justice Department is conducting an ongoing criminal investigation into whether Loral Space and Communications and Hughes Electronics violated export control laws. The firms are alleged to have shared their findings with China on the cause of a Chinese rocket's explosion while launching a U.S.-origin satellite in February 1996. In sharing their conclusions, the companies are said to have provided expertise that China could use to improve the accuracy and reliability of its ballistic missiles, including their guidance systems. The news report cited a classified report in the Pentagon that concluded in May 1997 that as a result of the transfer of expertise to China's missile program, "United States national security has been harmed." In addition, the press reports alleged that President Clinton in February 1998 issued a waiver for sanctions that undermined the investigation by allowing the issuance of licenses for the export of assistance similar to that in question. Moreover, the *Times* article alleged that political considerations may have influenced the Administration's decision, since Loral's chairman was the largest personal donor to the Democratic National Committee for the 1996 election.

Loral issued a statement on May 18, 1998, denying allegations that it provided missile guidance technology to China. Loral denied that it and Hughes conducted an independent investigation to determine the cause of that launch failure. However, the company acknowledged that it formed a committee of several satellite companies, including Hughes, to review the Chinese investigation and that, contrary to its policies, "the committee provided a report to the Chinese before consulting with State Department export licensing authorities." The Administration argues that export controls and strict security measures prevent inappropriate technology transfers to China's ballistic missile program. Administration officials say that export licensing procedures and strict security measures (including escorts of satellites by U.S. military officers prior to launch) preclude any assistance to the design, development, operation, maintenance, modification, or repair of any launch facility or rocket in China. Moreover, effective export controls on dual-use technology (with military and civilian applications) allow U.S. exporters to compete while protecting U.S. security interests. Officials have publicly disputed that there were objections within the executive branch to allowing recent satellite exports to China.

This CRS report provides detailed background information and a comprehensive chronology. The events summarized below, based on various open sources and interviews, pertain to various aspects of U.S. foreign and security policy, including missile nonproliferation, export controls on technology useful for missiles and/or satellites, and Presidential waivers of sanctions imposed on China after the Tiananmen crackdown.

Contents

China: Possible Missile Technology Transfers From U.S. Satellite Export Policy — Background and Chronology

Introduction

Members of Congress are concerned about allegations that U.S. firms provided expertise to China that could be used in its ballistic missile program and that a series of decisions by the Clinton Administration on satellite exports have facilitated legal or illegal transfers of missile-related technology to China. The *New York Times* reported in April 1998 that the Justice Department is conducting an ongoing criminal investigation into whether Loral Space and Communications (of New York), and Hughes Electronics (of Los Angeles) violated export control laws.[1] The firms are alleged to have shared their findings with China, without approval from the U.S. government, on the cause of a Chinese rocket's explosion while launching a U.S.-origin satellite in February 1996. In sharing their conclusions, the companies are said to have provided expertise that China could use to improve the accuracy and reliability of its ballistic missiles, including their guidance systems. A classified report at the Department of Defense's Defense Technology Security Administration (DTSA) reportedly concluded in May 1997 that Loral and Hughes transferred expertise to China that significantly enhanced the reliability of its nuclear ballistic missiles and "United States national security has been harmed."

In addition, the media reports allege that President Clinton in February 1998 issued a waiver of sanctions that undermined the investigation by allowing the issuance of licenses for the export of technology or expertise similar to that in question — despite "strong opposition" from Justice. Moreover, political considerations are alleged to have influenced the Administration's decision, since Loral's chairman was the largest personal donor to the Democratic Party in 1996.

This CRS report provides detailed background information and a comprehensive chronology. The events summarized below, based on various open sources and interviews, pertain to various aspects of U.S. foreign and security policy:

- Presidential waivers for exports of satellites, including the latest waiver for Chinasat-8 (built by Loral) during an ongoing criminal investigation into alleged assistance by Loral and Hughes to China's missile program; waivers are for sanctions imposed after China's Tiananmen Square crackdown;

---

[1] Gerth, Jeff. "Companies are Investigated for Aid to China on Rockets," and "Aerospace Firms' Ties with China Raise Questions," *New York Times*, April 4 and 13, 1998.

- sanctions imposed for missile proliferation by China's space launch company, China Great Wall Industry Corporation;
- quotas on Chinese launches of satellites;
- controls on exports of U.S.-origin satellites and/or satellite technology;
- export controls to prevent technology transfers that could contribute to China's ballistic missile force and/or military satellites.

Background

China Great Wall Industry Corporation

China Great Wall Industry Corporation (CGWIC) has been China's space launch company since 1986. It is a state-owned corporation in China's defense-related aerospace industry. CGWIC belongs to the China Aerospace Corporation (CASC), which oversees China's space and missile research and development establishment. It develops strategic and tactical ballistic missiles, space launch vehicles, surface-to-air missiles, cruise missiles, and military (reconnaissance, communications, or other) and civilian satellites. CGWIC uses the Long March series of rockets to launch satellites. The Long March boosters are also produced as China's DF-4 and DF-5A intercontinental ballistic missiles (ICBM) deployed in the Second Artillery, the strategic missile force of the People's Liberation Army (PLA), China's military.[2] China's launch facilities, such as the Xichang Satellite Launching Center in southwestern Sichuan province, are at PLA bases.

China reportedly launched its first satellite, Dongfanghong ("East is Red") on April 24, 1970. By the end of 1997, China reportedly had launched 40 domestic satellites: 17 retrievable reconnaissance satellites, 3 meteorological satellites, 8 communications and broadcasting satellites, and 12 "experimental" (possibly military) satellites. China is using the satellites and space technology to enhance its national defense, economy, and international prestige.[3] On April 7, 1990, the China Great Wall Industry Corporation launched a foreign satellite, Asiasat, for the first time.[4] Since then, the company has expanded its foreign business, especially with U.S. firms such as Hughes Electronics, Lockheed Martin, and Loral Space and Communications. China probably seeks foreign capital and technology to apply to its domestic satellite research and development efforts, in part to lessen reliance on purchasing foreign satellites. For example, the president of the Chinese Academy of Space Technology said that the Chinese Dongfanghong (East is Red) satellites match the capacities of advanced satellites built by Hughes, but are backward in satellite navigation and

---

[2]See CRS Report 97-391, *China: Ballistic and Cruise Missiles*, by Shirley Kan and Robert Shuey.

[3]Chou Kuan-wu, "China's Reconnaissance Satellites," *Kuang Chiao Ching* (in Hong Kong), March 16, 1998; translated in FBIS.

[4]For commercial space launches in general, see CRS Issue Brief 93062, *Space Launch Vehicles: Government Requirements and Commercial Competition*, updated regularly, by David P. Radzanowski and Marcia S. Smith.

stabilization technologies. The Academy hopes to sell its satellites at world standards by 2000.[5]

China has experienced a number of embarrassing and costly failed satellite launches. In 1992, a Chinese rocket stalled while attempting to launch the Optus-B1 satellite and another rocket exploded and destroyed the Optus-B2 satellite (both built by Hughes). In 1995, A Long March rocket exploded and destroyed the Apstar-2 satellite (built by Hughes). In 1996, another Chinese rocket exploded and destroyed the Intelsat satellite (built by Loral). Aside from the dramatic explosions, other problems have prevented the Chinese rockets from successfully launching satellites into the correct orbits.

China's aerospace industry has shifted from denying all responsibility in failed launches of foreign satellites to expressing willingness to work with foreign companies in determining the causes of explosions and other failures. This practice may have been a strategy to learn from foreign companies methods to improve China's rockets, satellites, and other related space technology. China may also have been requested by insurance companies and satellite manufacturers to share information about problems in Chinese rockets to show that the problems were being solved.

Missile Technology or Expertise

Critics say that allowing China to launch U.S.-origin satellites indirectly subsidizes and assists Chinese military aerospace research and development efforts, because the technology used in launching satellites can be used in ballistic missiles. Henry Sokolski (Executive Director of the Nonproliferation Policy Education Center and a Defense official in the Bush Administration) writes that "intangible technology" is critical to the timely, reliable, and accurate placement of satellites into space as well as launches of warheads against targets by ballistic missiles. Intangible technologies include: coupling load analysis, guidance data packages, upper-stage solid rocket propellant certification, upper-stage control design validation, lower-stage design validation, and general quality assurance. Also, multi-satellite dispensers, or smart dispensers, can be used as multiple-warhead dispensers, thus assisting China's reported efforts to develop a capability in multiple independently targeted reentry vehicles (MIRVs) for its ICBMs.[6] China has used such dispensers to launch multiple satellites for Iridium. In the case of Loral's alleged assistance to China in 1996, the company may have concurred with or identified problems with the Long March rockets in reviewing the cause of the February 1996 explosion.

Acting Undersecretary of State John Holum confirmed on April 9, 1998, that after the accident in February 1996, the Department of State "became aware that there may have been a violation." The case was referred to the Department of Justice for investigation. He said that there are "strong legal remedies" for violations of export control laws, including a denial of future licenses.

---

[5]Parker, Jeffrey. "China to Expand Rocket Production," *Reuters*, August 25, 1993.

[6]Sokolski, Henry. "US Satellites to China: Unseen Proliferation Concerns," *International Defense Review*, April 1994; "Space Technology Transfers and Missile Proliferation," April 10, 1998.

Loral issued a statement on May 18, 1998, saying that allegations that it provided missile guidance technology to China are false. Loral also says that it did not advise China "on how to fix any problems with the Long March rocket." The company states that "the Chinese alone conducted an independent investigation of the launch failure [in February 1996] and they determined that the problem was a defective solder joint in the wiring — a 'low-tech' matter." Loral denied that it and Hughes conducted an independent investigation to determine the cause of that launch failure. However, at the insistence of insurance companies, which required non-Chinese confirmation of resolutions of problems with Long March rockets, Loral formed a committee of several satellite companies, including Hughes, to review the Chinese investigation. According to Loral, the review committee obtained information from the Chinese and was not formed to help them solve their problems. The review agreed with the Chinese conclusion (that a defective solder joint was responsible), without performing tests or providing any test data to the Chinese. The committee did note that further tests by China would be required to establish certainty. Loral says that, during the review, it discussed the committee's work with U.S. officials. As far as Loral's engineer's can determine, the statement says, "no sensitive information — no significant technology — was conveyed to the Chinese."

Nevertheless, Loral admitted that, contrary to its policies, "the committee provided a report to the Chinese before consulting with State Department export licensing authorities." According to Loral, as soon as its executives found out in May 1996, the company notified the Departments of State and Defense. In June 1996, Loral provided to the U.S. government a detailed, written report concerning all communications with China. Loral adds that it is in full cooperation with the Justice Department in its investigation and with Congressional committees. Loral concludes that based upon its own review, it "does not believe that any of its employees dealing with China acted illegally or damaged U.S. national security." In addition, the statement says that Loral's chairman, Bernard Schwartz, was not personally involved in any aspect of this matter. "No political favors or benefits of any kind were requested or extended, directly or indirectly, by any means whatever." Loral also denies any connection between the launch failure in February 1996 and the Presidential waiver for another Loral-built satellite in February 1998. The export license for the latest launch (for Chinasat-8) "applied the strictest prohibitions on technology transfer and specified that any new launch failure investigation would require a separate license." Loral stresses that it complies strictly with export control laws and regulations.

Administration officials say that export licensing procedures and strict security measures (including escorts of satellites by U.S. military officers prior to launch) preclude any assistance to the design, development, operation, maintenance, modification, or repair of any launch facility or rocket in China. Moreover, Undersecretary of Commerce William Reinsch testified to Congress on April 28, 1998, that effective export controls on dual-use technology (with military and civilian applications) allow U.S. exporters to compete while protecting U.S. security interests. He disputed that there were objections within the executive branch to allowing recent satellite exports to China, saying that since November 1996 (when the licensing jurisdiction was transferred from the Department of State to Commerce), the Commerce Department has issued three export licenses for satellites to be launched from China — with the concurrence of all agencies.

However, it is unclear whether the Justice Department was among those agencies that concurred in issuing the export licenses. The *New York Times* alleges that the Justice Department strongly opposed the President's approval and waiver, in February 1998, for the export of a Loral-built satellite. Finally, there is little public information on the export licenses issued by the State Department for technical assistance agreements (TAAs) concerning the transfer of data needed to mate satellites to launch vehicles, and other technical assistance.

Sanctions

CGWIC has been affected by two categories of sanctions imposed on China: those imposed after the Tiananmen crackdown and those imposed for missile proliferation. In 1990, the United States imposed post-Tiananmen sanctions as required in the Foreign Relations Authorization Act for FY1990 and FY1991 (P.L. 101-246). Sec. 902(a) requires suspensions in programs related to: (1) Overseas Private Investment Corporation, (2) Trade and Development Agency, (3) exports of Munitions List items, (4) exports of crime control equipment, (5) export of satellites for launch by China, (6) nuclear cooperation, and (7) liberalization of export controls. Suspensions (3) and (5) affected export of satellites to China. Sec. 902(b) allows Presidential waivers of those suspensions by reporting that "it is in the national interest" to terminate a suspension.

As for sanctions related to missile proliferation, on April 30, 1991, the Bush Administration denied licenses for the export of U.S. parts for a Chinese satellite, the Dongfanghong-3, citing "serious proliferation concerns." On May 27, 1991, President Bush declared sanctions on China for transferring to Pakistan technology related to the M-11 short-range ballistic missile. These sanctions, required by Sec. 73(a) of the Arms Export Control Act (P.L. 90-629) and Sec. 11B(b)(1) of the Export Administration Act (P.L. 96-72), were intended to enforce the Missile Technology Control Regime (MTCR). These sanctions, which took effect on June 16 and 25, 1991, denied export licenses and waivers of sanctions for: (1) high-speed computers to China, which can be used for missile flight testing; (2) satellites for launch by China; and (3) missile technology or equipment. They affected two Chinese aerospace corporations: CGWIC and China Precision Machinery Import Export Corporation. President Bush waived these sanctions on March 23, 1992, after China agreed to abide by the MTCR guidelines.

The Clinton Administration had to impose similar sanctions on August 24, 1993, after China was again determined to have transferred M-11 related equipment to Pakistan. A total of 11 Chinese defense industrial companies were sanctioned, including CGWIC again. Beginning in 1993, the U.S. aerospace industry and aerospace company executives lobbied against sanctions and for expansion of satellite exports to China. China, on October 4, 1994, agreed not to export ground-to-ground missiles inherently capable of delivering at least 500 kg to at least 300 km — an understanding the U.S. side sought to include the M-11 missiles under the MTCR. On November 1, 1994, the Clinton Administration waived those sanctions.[7]

Waivers[8]

Since sanctions were imposed after the Tiananmen crackdown in 1989, Presidents Bush and Clinton have issued at least 13 waivers for 20 satellite projects (each project may involve more than one satellite), on a case-by-case basis, to allow the export to China of U.S.-origin satellites or components (for foreign satellites) subject to export controls. (See the Table below.) Since they were first granted, Presidential waivers have been increasingly issued for satellites used by China — not just launched from China. Some waivers under P.L. 101-246 have specified whether sections 902(a)(3) and 902(a)(5), on Munitions List items and satellites, applied; other waivers simply referred to section 902 or 902(a). The Administration has considered supporting China as a partner in the MTCR, issuing a blanket waiver of sanctions on satellites, and increasing the quota on the numbers of satellites China is allowed to launch — in return for further cooperation in missile nonproliferation, according to a secret March 12, 1998, National Security Council memo.[9]

Congressional Action

Since the early 1990s, Congress has been concerned about the implications for U.S. national security stemming from satellite exports. After the press reports in April 1998, several Congressional committees — led by House Speaker Newt Gingrich and Senate Majority Leader Trent Lott — began to look into the latest allegations concerning Loral and Hughes and associated actions by the Clinton Administration. These committees include: the Senate Governmental Affairs Subcommittee on International Security, Proliferation, and Federal Services; Senate Foreign Relations Committee; the Joint Economic Committee; and House Committees on Intelligence, International Relations, and National Security. The chairmen of the House International Relations and National Security Committees have requested relevant documents (including the Pentagon's report), but have reportedly faced resistance from the Administration — in part because of concerns about compromising the Justice Department's ongoing criminal investigation.[10] White House spokesperson, Mike McCurry, said on May 11, 1998, that "we're willing to make available whatever

---

[7]See CRS Report 96-767, *Chinese Proliferation of Weapons of Mass Destruction: Background and Analysis*, 1996, by Shirley A. Kan.

[8]For sanctions on China in general, see CRS Report 96-272, *China: U.S. Economic Sanctions*, by Dianne E. Rennack.

[9]See CRS Issue Brief 92056, Chinese Proliferation of Weapons of Mass Destruction: Current Policy Issues, updated regularly, by Shirley A. Kan.

[10] Gerth, Jeff, "Congress Probes Sales of Satellite Technology to China," *New York Times*, April 16, 1998; Juliet Eilperin, "GOP Says U.S. Gave China Nuclear Edge" and "GOP Leaders Demand Satellite Export Data," *Washington Post*, May 6, 1998 and May 12, 1998; Juliet Eilperin, "Gingrich to Create Special Panel to Probe China Technology Deal," *Washington Post*, May 20, 1998.

information they need to satisfy themselves that these decisions have been made on sound national security grounds." In addition, Speaker Gingrich announced on May 19, 1998, that he wants to create a select committee, headed by Congressman Christopher Cox, to investigate the various allegations concerning this case. A vote by the House would be needed, perhaps to be scheduled for early June.

The Joint Economic Committee held a hearing on April 28, 1998, on dual-use exports to China. The Senate Governmental Affairs Subcommittee on International Security, Proliferation, and Federal Services scheduled a hearing for May 21, 1998, to look into the applications of technology or expertise in commercial space launch for foreign satellite and ICBM programs.

The House Defense Authorization Act (**H.R. 3616**) includes amendments that expresses the sense of Congress that the United States should not enter into new agreements with China involving space or missile-related technology, prohibits U.S. participation in investigations of Chinese launch failures, prohibits transfers of missile technology to China, and prohibits the export or re-export of U.S. satellites to China (respectively, by Representatives Spence/Gilman, Bereuter, Hefley, and Hunter on May 19, 1998). The Senate Defense Authorization Act (**S. 2057**) contains several amendments expressing concern over relations with China, technology transfer, and U.S. national security interests. Of relevance to satellite export policy, one of Senator Hutchison's amendments seeks to clarify expectations for presidential waivers regarding Chinese programs.

Table: Waivers for Exports of Satellites or Parts

Satellite Projects (may have multiple satellites per project)	End-User	Manufacturer	Waiver
1. Asiasat-1	Asia Satellite	Hughes	12/19/89
* Asia Satellite Telecommunications is a consortium based in Hong Kong and owned by China International Trust and Investment Corporation (CITIC) of China, Cable and Wireless of Britain, and Hutchison Telecommunications Ltd. Of Hong Kong.			
2. Aussat (Optus-B1 and -B2)	Australia	Hughes	4/30/91
Freja	Sweden		
* In the first waiver, President Bush had waived sanctions for Aussat satellites, but reissued waiver and licenses. He also denied export licenses for U.S. components for a Chinese satellite, Dongfanghong-3 (waived later).			
3. Asiasat-2	Asia Satellite	Martin Marietta	9/11/92
Apsat (or Apstar)	APT Satellite	Hughes	
Intelsat-7A	Intelsat	Martin Marietta	
Starsat			
AfriSat (AfriStar)	Afrispace		
Dongfanghong-3	China	China	
4. Iridium	Iridium	Lockheed	7/2/93
Intelsat-8	Intelsat	Loral	
5. Echostar-1 and -2	Echostar	Martin Marietta	7/13/94
6. Mabuhay	Philippines	Loral	2/6/96
7. Cosat-1 and -2	China	Lockheed Martin	2/6/96
* Also called ChinaStar, used by China Oriental Telecom Satellite Co.			
8. Chinasat-7	China	Hughes	2/6/96
9. Asia Pacific Mobile Telecommunications (APMT)	APT Satellite	Hughes	6/23/96
* China invests in project.			
10. Globalstar	Globalstar	Loral	7/9/96
11. Fengyun 1	China	China	11/19/96
12. SinoSat-1	China	Aerospatiale	11/23/96
* Cooperative product between Daimler-Benz Aerospace and China Aerospace Corp.			
13. Chinasat-8	China	Loral	2/18/98

Chronology[11]

Date	Event
Jan. 1989	The United States and China signed an agreement for six years under which China agreed to charge prices for commercial launch services "on a par" with Western competitors and to allow China to launch nine U.S.-built satellites through 1994.
6/4/89	Crackdown on peaceful, political demonstrators in Beijing.
12/19/89	President Bush waived sanctions for export of Aussat-1, Aussat-2, and Asiasat communications satellites for launch from China, under sec. 610 of the Department of Commerce, Justice, and State, the Judiciary, and Related Agencies Appropriations act 1990 (P.L. 101-162).
1990	
2/16/90	P.L. 101-246 enacted to require post-Tiananmen sanctions, including suspensions in approving exports to China of Munitions List items and satellites.
4/7/90	China Great Wall Industry Corporation, using a LM-3 rocket, launched a foreign satellite, Asiasat (built by Hughes), for the first time.
1991	
4/30/91	President Bush waived sanctions under Sec. 902(b) of P.L. 101-246 to allow exports of Aussat-1 and -2 and Freja satellites for launch from China in part because China was not the end-user. President Bush denied a license to export U.S. satellite components for a Chinese satellite, Dongfanghong-3, citing "serious proliferation concerns." In addition, *Space News* (May 6-12, 1991) reported that President Bush's denial was to punish China for attempting to obtain classified missile-related technology. The license to export parts for Dongfanghong-3 was requested by a German firm, but the U.S. components were produced by M/A-COM, Inc. (Burlington, MA).
6/16/91	The Bush Administration announced sanctions to be imposed on China for transferring missile related technology to Pakistan. The sanctions affected high technology trade with China, covering (1) high performance computers, (2) satellites for launch from China (except for the Freja and Aussat satellites), and (3) sanctions for missile proliferation as required by the Arms Export Control Act and Export

[11] For overview of U.S.-China relations, see CRS Report 97-484, *China-U.S. Relations: Chronology of Developments During the Clinton Administration*, by Kerry Dumbaugh.

Administration Act (imposed on China Great Wall Industry Corp. and China Precision Machinery Import/Export Corp.). The U.S. sanctions were intended to enforce the MTCR.

6/25/91 The sanctions on the two Chinese state-owned companies for missile proliferation in Pakistan took effect.

11/21/91 After Secretary of State James Baker visited Beijing, the Chinese foreign ministry issued a vague statement that China "intends to abide" by the MTCR.

1992

2/1/92 According to the Bush Administration, the Chinese foreign minister sent a secret letter to the U.S. Secretary of State promising to abide by the MTCR.

2/22/92 The Chinese foreign ministry issued a statement saying that "China will act in accordance with the guidelines and parameters of the existing missile and missile technology control regime in its export of missiles and missile technology," after the United States effectively lifts the June 1991 sanctions.

3/22/92 Aborted launch of Aussat (Optus-B1) satellite from China after LM-2E rocket malfunctioned and the rocket stalled on the launch pad. *Beijing Review* (Nov. 2-8, 1992) reported that the rocket's malfunction was caused by a fault in the ignition system which triggered an emergency shut-down.

3/23/92 The Bush Administration effectively waived the sanctions imposed in June 1991 on China for missile proliferation.

8/14/92 China successfully launched the Optus-B1 satellite.

9/11/92 President Bush waived sanctions under P.L. 101-246 to allow exports of five satellites (Asiasat-2, Apsat, Intelsat-7A, Starsat, and AfriStar) for launch from China and parts for China's Dongfanghong-3.

Nov. 1992 According to President Clinton's report to Congress submitted in May 1993, China may have supplied M-11 short-range ballistic missiles or related technology to Pakistan. This transfer may have been taken in retaliation for President Bush's decision in September 1992 to sell F-16 fighters to Taiwan.

12/21/92 A Chinese LM-2E launch vehicle exploded and destroyed the Australian Optus-B2 satellite (built by Hughes) it was carrying. After the explosion, Chinese officials denied that Chinese rockets were responsible, blaming the satellite built by Hughes. *Aviation Week and Space Technology* (Jan. 30, 1995) reported that Hughes and China Great Wall Industry Corp. agreed to declare the cause of that failure to be undetermined. Some experts, however, reportedly identified the premature opening of the launch vehicle's payload fairing as causing the accident.

1993

5/28/93 President Clinton decided to extend most-favored-nation trade status to China with conditions on human rights, but no linkage to weapons proliferation. Nonetheless, after persistent reports that China was continuing to transfer missile components to Pakistan — if not complete M-11 short-range ballistic missiles, the President also reported to Congress that "at present, the greatest concern involves reports that China in November 1992 transferred MTCR-class M-11 missiles or related equipment to Pakistan."

7/2/93 President Clinton waived sanctions under P.L. 101-246 to allow exports to China of Iridium and Intelsat-8 satellites for launch from China.

8/16/93 Hughes and CGWIC issued a joint statement after seven months of "vigorous and cooperative investigation" into the cause of the explosion on 12/21/92. The statement did not identify a cause, with each side denying blame.

8/24/93 The Clinton Administration determined that China had shipped M-11 related equipment (not missiles) to Pakistan and imposed sanctions required by the Arms Export Control Act and Export Administration Act. The sanctions were imposed on Pakistan's Ministry of Defense and 11 Chinese defense industrial aerospace entities, including China Great Wall Industry Corp. The sanctions denied U.S. government contracts and export licenses for missile equipment or technology (items in the MTCR annex) for two years. Exports of satellites were affected because of components such as "kick motors" used on missiles and installed on some satellites to boost them into orbit.

8/26/93 The U.S. aerospace industry lobby, including the Aerospace Industries Association, called on the Clinton Administration to weaken the missile proliferation sanctions.[12]

8/31/93 One week after imposing sanctions, Assistant Secretary of State Winston Lord said that "we're ready at any time to sit down with the Chinese, both to try to find a way to lift the sanctions if they cooperate but also to explain more fully the MTCR and its revised guidelines."

---

[12] "Statement by Don Fuqua, President of the Aerospace Industries Association, on the Imposition of U.S. Economic Sanctions on China," August 26, 1993; Steven Greenhouse, "Aerospace Industry Seeks Weaker Sanctions on China," *New York Times*, August 28, 1993.

9/25/93 National Security Adviser Anthony Lake told the Chinese ambassador that the Clinton Administration was willing to negotiate a waiver of the sanctions, but a more formal and binding Chinese commitment than the one made in November 1991 was needed.

10/20/93 The *Washington Post* reported that top executives of U.S. satellite manufacturers, Martin Marietta Corp. and Hughes Aircraft Co., were lobbying intensively for the Clinton Administration to waive the export ban for satellites. Reportedly due to these objections from private industry (which were supported by the Commerce Department), the National Security Council reviewed the decision to implement the sanctions. In September 1993, Norman R. Augustine, chairman of Martin Marietta, wrote a letter to Vice President Al Gore, arguing that the sanctions "present U.S. companies as an unreliable supplier." Some Members of Congress supported the export of satellites for launch from China.

11/9/93 The chairman and CEO of Hughes Aircraft Company, C. Michael Armstrong, delivered a speech in which he objected to the inclusion in the sanctions of commercial communications satellites. He also said that he "asked the President of the United States to review the situation."

11/19/93 President Clinton met with Chinese President Jiang Zemin at the Asian Pacific Economic Cooperation (APEC) meeting in Seattle. On the eve of the meeting, press reports said that the Administration had formally proposed waiving the sanctions in return for another Chinese promise, in more detail and with more authority, not to export MTCR-class missiles. Senior officials reportedly argued in favor of a broad interpretation of the law to allow the export of two of seven satellites.

1994

1/6/94 The Clinton Administration announced a new policy exempting commercial communication satellites from sanctions for missile proliferation imposed on 8/24/93, facilitating export licenses for one Hughes and two Martin Marietta satellites.

4/2/94 A Chinese weather satellite exploded in a plant.

7/13/94 President Clinton waived sanctions under P.L. 101-246 for the Echostar satellite to be exported for launch from China.

7/21/94 A Chinese LM-3 rocket launched the Apstar-1 satellite (built by Hughes).

8/28/94 A Chinese LM-2E rocket launched Australia's Optus-B3 satellite (built by Hughes).

Sept. 1994 Secretary of Commerce Ron Brown led trade delegation to China, including Bernard Schwartz, Loral's chairman.

10/4/94 Secretary of State Warren Christopher and Foreign Minister Qian Qichen issued a joint statement in which the United States agreed to waive the August 1993 sanctions (for missile proliferation) and China agreed not to export "ground-to-ground missiles" that are "inherently capable" of delivering at least 500 kg to at least 300 km (an important understanding meant in part to include the M-11 missiles under the MTCR guidelines).

11/1/94 The Administration's waiver of the sanctions took effect.

11/30/94 China launched its Dongfanghong-3 satellite, but failed to launch it into the correct position due to a fuel leak.

Dec. 1994 President Clinton selected Armstrong of Hughes to head the Export Council.

1995

1/26/95 A Chinese LM-2E launch vehicle exploded after liftoff, destroying the Apstar-2 satellite (built by Hughes) it was carrying. Hughes and China Great Wall Industry Corporation were reported as planning to determine the cause of the explosion. (*Aviation Week and Space Technology*, Jan. 30, 1995)

2/9/95 The *Wall Street Journal* reported that Chinese aerospace industry officials contradicted an official Chinese newspaper's account that blamed Hughes for the explosion on January 26, 1995. Instead of blaming Hughes, as *Ta Kung Pao* (in Hong Kong) did, officials from China Great Wall Industries Corp. and the China National Space Administration said that the article did not reflect China's official view and that the investigation had not concluded. A spokesman for Hughes said that a thorough investigation into the cause of the explosion would take months to complete.

3/13/95 The United States and China concluded a new agreement for 7 years to allow China to launch up to 11 new satellites to geostationary orbit at prices not less than 15 percent below that charged by Western competitors.

7/21-28/95 The PLA Second Artillery test-fired M-9 short-range ballistic missiles toward Taiwan, after Taiwan's president visited Cornell University in June.

7/25/95 Hughes and CGWIC issued a joint statement on separate findings of six-month investigations into the cause of the explosion on 1/26/95. CGWIC blamed strong winds for shaking Hughes' satellite apart, while Hughes said that severe winds caused the Chinese rocket's fairing to collapse.

10/9/95 Secretary of State Warren Christopher initialed a classified memorandum to retain the State Department's licensing authority over commercial communications satellites (cited in *New York Times*, May 17, 1998).

11/28/95 A Chinese LM-2E rocket launched the Asiasat-2 satellite (built by Lockheed Martin), but the bumpy launch knocked the satellite's antenna-feed horns out of alignment, resulting in a loss of signal power. Asiasat company claimed $58 million in insurance for the damage. (*Flight International*, Oct. 2-8, 1996).

12/28/95 A Chinese LM-2E rocket launched the Echostar-1 satellite (built by Martin Marietta).

1996

2/6/96 President Clinton waived sanctions under P.L. 101-246 for the Chinasat-7 satellite to be exported for launch from China.

2/6/96 President Clinton waived sanctions under P.L. 101-246 for 2 Cosat (later called Chinastar) satellites to be exported for launch from China.

2/6/96 President Clinton waived sanctions under P.L. 101-246 for the Mabuhay satellite to be exported for launch from China.

2/15/96 A LM-3B rocket exploded after liftoff, destroyed the Intelsat satellite (built by Loral), and smashed into a village. The death toll was probably higher than the official report of six deaths and 57 injured.

3/8-15/96 Despite the dramatic explosion of a Chinese rocket one month before, the PLA's Second Artillery again test-fired M-9 ballistic missiles toward targets close to Taiwan's ports, on the eve of Taiwan's first presidential election.

3/10-11/96 In further deterioration of U.S.-China relations, the United States deployed two carrier battle groups to waters off Taiwan, calling China's live-fire exercises "reckless" and "risky."

3/14/96 The Clinton Administration announced a decision to move commercial communications satellites from the Munitions List to the Commerce Control List of dual-use items, so that the export license jurisdiction was moved from the Department of State to the Department of Commerce (implemented in November 1996).[13]

April 1996 Loral headed a commission to study the accident on Feb. 15, 1996.

5/10/96 Loral's commission completed a preliminary review that was shared with China, finding that the cause of the accident was an electrical flaw in the electronic flight control system. The report allegedly discussed weaknesses in the Chinese rocket's guidance and control systems. (*New York Times*, April 13, 1998)

May 1996 Loral's executives found out that the review was shared with the Chinese, and the company notified the Departments of State and Defense.

June 1996 Loral provided to the U.S. government a detailed, written report concerning all communications with China.

6/23/96 President Clinton waived sanctions under P.L. 101-246 for the Asia Pacific Mobile Telecommunications (APMT) satellite to be exported for launch from China.

7/9/96 President Clinton waived sanctions under P.L. 101-246 for a Globalstar satellite to be exported for launch from China.

8/18/96 China failed to launch its Chinasat-7 satellite (built by Hughes) into the correct orbit, after the third stage of the LM-3 rocket shut down early, reported the *Far Eastern Economic Review* (Aug. 29, 1996).

10/21/96 The Bureau of Export Administration of the Department of Commerce issued regulations to implement the transfer of commercial satellites from control under the Munitions List to the Commerce Control List.

11/5/96 The Department of State issued regulations to implement the transfer of commercial satellites from control under the Munitions List to the Commerce Control List, even if the satellites include individual components or technologies on the Munitions List.

11/19/96 President Clinton waived sanctions under P.L. 101-246 for U.S. parts for the Chinese Fengyun-1 (FY-1) meteorological satellite. The waiver cited suspensions under sections 902(a)(3) and 902(a)(5), indicating that technologies controlled under the Munitions List were involved.

---

[13]Also see GAO report GAO/NSIAD-97-24, *Export Controls: Change in Export Licensing Jurisdiction for Two Sensitive Dual-Use Items*, January 1997.

11/23/96 President Clinton waived sanctions under P.L. 101-246 for the Sinosat satellite to be exported for launch from China. The waiver cited suspensions under sections 902(a)(3) and 902(a)(5), indicating that technologies controlled under the Munitions List were involved.

1997

5/12/97 China successfully launched its Dongfanghong-3 communications satellite, built by China Aerospace Corp. on a LM-3A rocket, prompting personal congratulations from top government and military leaders.

May 1997 The U.S. Trade Representative (USTR) reported that China had violated the pricing provisions of a bilateral agreement on the Mabuhay launch.

May 1997 A classified report at the Department of Defense's Defense Technology Security Administration (DTSA) concluded that Loral Space and Communications Co. and Hughes Electronics had transferred expertise to China that significantly enhanced the reliability of its nuclear ballistic missiles and "United States national security has been harmed" (cited in April 13, 1998, *New York Times*).

6/10/97 China successfully launched its Fengyun-2, a second-generation Chinese meteorological satellite, on a LM-3 rocket.

8/19/97 China launched the Agila 2 (formerly called Mabuhay) satellite.

9/1/97 China launched two test satellites for Iridium to demonstrate the technical viability of the new Long March variant, LM-2C/SC.

9/10/97 The *Washington Times*, citing Israeli and U.S. intelligence sources, reported that China Great Wall Industry Corporation was supplying key telemetry equipment (for sending and collecting guidance data during flight tests) to Iran for its development of the Shahab-3 and Shahab-4 medium-range ballistic missiles.

Sept. 1997 Likely prompted by DTSA's report, the Department of Justice began its criminal investigation into allegations that Loral and Hughes illegally passed technical assistance to China. The investigation is still ongoing.

10/17/97 China launched Asia Pacific Telecommunications Satellite (ApStar-2R) (built by Loral) on LM-3B rocket.

10/27/97 The USTR announced that the United States and China agreed on new provisions for the Bilateral Agreement on Space Launch Services (signed in 1995). The new provisions set clear terms for Chinese pricing of launch services to low earth orbit.

Nov. 1997 After a summit in Washington, Chinese President Jiang Zemin visited a Hughes satellite plant in California.

12/8/97 China launched two satellites for Iridium on one Long March 2C/SC rocket to low earth orbit. The rocket had two stages and a smart dispenser on top that deployed the two satellites. Some in the United States are concerned that the technology used to launch more than one satellite at once will help China in its suspected development of multiple-independently targeted re-entry vehicles (MIRVs) for its ICBMs.

1998

2/18/98 President Clinton waived sanctions under P.L. 101-246 for the Chinasat-8 satellite (built by Loral) to be exported to China. Loral says that it is the most powerful satellite that China has ever bought.

3/12/98 Gary Samore, Special Assistant to the President and Senior Director for Nonproliferation and Export Controls in the National Security Council, wrote a Secret memo proposing to support Chinese membership in the MTCR, issue a "blanket waiver" of the post-Tiananmen sanctions to cover all future satellite launches, and increase the number of space launches from China — in return for Chinese cooperation in missile nonproliferation. (The classified memo was printed in the March 23, 1998, *Washington Times*.)

3/16/98 Loral Space and Communications signed an agreement with China Great Wall Industry Corp. to launch five of Loral's communication satellites between March 1998 and March 2002 using Long March-3B rockets.

3/22/98 China Aerospace Corp. kicked off a Quality Promotion Plan to help ensure success in its commercial launch business in research, production, and testing.

3/26/98 China launched two Iridium satellites, built by Lockheed Martin for Motorola, on an improved LM-2C rocket. (According to China, this launch was China's 15th "successful" commercial launch for foreign customers since 1990.)

3/26/98 John Holum, Acting Under Secretary of State for Arms Control and International Security Affairs, concluded his visit to China and confirmed that he discussed increasing the quota on the number of satellite launches from China.

3/29/98 A Hong Kong newspaper owned by the Chinese government reported that China Aerospace Corporation found in its investigations into past failed launches of satellites that all the failures were caused by problems in production and management related to quality control. A previous explosion of an LM-3B rocket (on 2/15/96) was found to have been caused by a defect in a power pack nodal point which caused a short circuit when the rocket ignited, resulting in a malfunction in the inertial platform.

4/3/98 China's official news agency quoted Zhang Haiming, general-manager of a division of Lockheed Martin, as saying that the company is "consulting with the Chinese on satellite manufacturing."

4/4/98 The *New York Times* reported that a Federal grand jury is investigating whether Loral Space and Communications of New York and Hughes Electronics of Los Angeles provided expertise to China that "significantly advanced" the guidance systems of its ballistic missiles in studying the accidental destruction in February 1996 of a satellite built by Loral. Administration officials reportedly said that the Department of Justice, fearing that its criminal investigation would be undermined, opposed the President's February 1998 waiver and approval for export of similar technology to China (for Chinasat-8). Loral's chief executive was reported as the largest personal donor to the Democratic National Committee for the 1996 election.

4/9/98 John Holum, Acting Undersecretary of State for Arms Control and International Security Affairs, stressed that exports of satellites to China for launch occur with an export license and strict security measures to "preclude assistance to the design, development, operation, maintenance, modification or repair of any launch facility or rocket in China, and we monitor that very carefully." He also confirmed that after the accident in February 1996, the Department of State "became aware that there may have been a violation." The case was referred to the Department of Justice for investigation. He said that there are "strong legal remedies" for violations of export control laws, including a denial of future licenses.

4/13/98 The *New York Times* again reported on the criminal investigation of Loral and Hughes, adding that a highly classified Pentagon report concluded in May 1997 that the companies had transferred expertise to China that "significantly improved" the reliability of China's nuclear ballistic missiles.

4/15/98 Loral's president and chief operating officer, Gregory Clark, stated that Loral "did not divulge any information that was inappropriate."

4/16/98 A Chinese Foreign Ministry spokesman stated that "the exchange of technical information about satellite launchings between U.S. companies and the Chinese aerospace department was a normal activity and fell under international rules." He also said that the companies "did not provide technical information about missile technology."

4/21/98 Loral's chairman and CEO, Bernard Schwartz, said that "we have done our own internal investigation, and I'm satisfied that our people acted well — good behavior and in compliance [with U.S. export control regulations]."

4/28/98 Under Secretary of Commerce for Export Administration William Reinsch testified to the Joint Economic Committee that satellite exports to China have shown how effective dual-use export controls allow U.S. exporters to compete and "win without risk to our national security." He said that controls on satellite exports to China are extensive and include measures to "reduce the risk" of illicit technology transfers. Since November 1996 (when the licensing jurisdiction was transferred from the Department of State to Commerce), Commerce issued three export licenses for satellites to be launched from China — "with the concurrence of all agencies."

4/30/98 A spokesman at the State Department, James Foley, denied a *Washington Times* report that the Administration presented China with a draft agreement for space cooperation. He admitted, however, that officials have considered scientific space cooperation as one way to encourage Chinese cooperation in missile non-proliferation. He also stressed that "there still is not any U.S. plan or proposal to offer China access to missile technology."

5/2/98 A Chinese Long March 2C/SD rocket launched two Iridium satellites (built by Lockheed Martin) to low earth orbit.

May 1998 The Justice Department began a preliminary inquiry into whether political donations influenced President Clinton's approval of satellites to China.

5/15/98 The *New York Times* reports that fund-raiser Johnny Chung told the Justice Department that part of his donations to the Democratic Party in the summer of 1996 came from the PLA through Liu Chaoying, a PLA lieutenant colonel (possibly retired) and a senior manager and vice president for China Aerospace International Holdings, Ltd. (a subsidiary of China Aerospace Corporation in Hong Kong). She is also a daughter of retired General Liu Huaqing, formerly a vice chairman of the PLA's command, the Central Military Commission, and formerly a member of the Standing Committee of the Politburo.

5/18/98 Loral issued a statement saying that allegations that it provided missile guidance technology to China are false. The company states that "the Chinese alone conducted an independent investigation of the launch failure [in February 1996] and they determined that the problem was a defective solder joint in the wiring — a 'low-tech' matter." Loral denied that it and Hughes conducted an independent investigation to determine the cause of that launch failure. It was at the insistence of insurance companies, which required non-Chinese confirmation of resolutions of problems with Long March rockets, that Loral formed a committee of several satellite companies, including Hughes, to review the Chinese investigation. However, Loral admitted that, contrary to its policies, "the committee provided a report to the Chinese before consulting with State Department export licensing authorities." Loral adds that it is in full cooperation with the Justice Department in its investigation and with Congressional committees. Loral concludes that based upon its own review, it "does not believe that any of its employees dealing with China acted illegally or damaged U.S. national security." In addition, the statement says that Loral's chairman, Bernard Schwartz, was not personally involved in any aspect of this matter. "No political favors or benefits of any kind were requested or extended, directly or indirectly, by any means whatever." Loral also denies any connection between the launch failure in February 1996 and the Presidential waiver for another Loral-built satellite in February 1998. The export license for the latest launch (for Chinasat-8) "applied the strictest prohibitions on technology transfer and specified that any new launch failure investigation would require a separate license." Loral stresses that it complies strictly with export control laws and regulations.

97-391 F

CRS Report for Congress

China: Ballistic and Cruise Missiles

Updated May 27, 1998

Shirley A. Kan
Analyst in Foreign Affairs

Robert D. Shuey
Specialist in U.S. Foreign Policy & National Defense
Foreign Affairs and National Defense Division

CRS

Congressional Research Service • The Library of Congress

Abstract

China's military, the People's Liberation Army (PLA), is currently believed to have deployed a nuclear-armed missile force consisting of over 100 nuclear warheads on perhaps as many as 43 intercontinental ballistic missiles (ICBMs), over 50 medium-range ballistic missiles (MRBMs), and 12 submarine-launched ballistic missiles (SLBMs) on one nuclear ballistic missile submarine. Some are believed to be targeted on the United States and other countries. There are also short-range ballistic missiles (SRBMs) for tactical operations. Ballistic missiles are deployed in the PLA's Second Artillery Corps, the strategic missile force.

This CRS report contains three parts. The first discusses Chinese ballistic missiles. The second discusses Chinese cruise missiles. The last offers concluding observations. Two tables summarize the discussion on current ballistic and cruise missiles in service or under development. The appendix, prepared by Robert Shuey, discusses China's reported application of global positioning system (GPS) technology to improve the accuracy of its missiles. This report focuses on the status and current developments of China's missile programs, rather than their history, and is based on public sources of information which have been cited.

China: Ballistic and Cruise Missiles

Summary

China's military, the People's Liberation Army (PLA), is currently believed to have deployed a nuclear-armed missile force consisting of over 100 nuclear warheads on perhaps as many as 43 intercontinental ballistic missiles (ICBMs), over 50 medium-range ballistic missiles (MRBMs), and 12 submarine-launched ballistic missiles (SLBMs) on one nuclear ballistic missile submarine. Some are believed to be targeted on the United States and other countries. There are also short-range ballistic missiles (SRBMs) for tactical operations. Ballistic missiles are deployed in the PLA's Second Artillery Corps, the strategic missile force.

Chinese ballistic missiles can be divided into four phases of development and deployment. After developing land-based MRBMs and limited-range ICBMs in the first phase, China turned to diversity and reliability. In the second phase, China added long-range ICBMs, SLBMs, and SRBMs to its ballistic missile force, and achieved mobility and the use of solid fuel. In the third phase, China hopes to deploy after 2000 or 2005 a new MRBM with a large conventional warhead for tactical operations (although this program may have been canceled), a new land-mobile, solid fuel ICBM with a lighter warhead, and a new, longer-range SLBM on a next-generation submarine. For the fourth phase, China plans to deploy a land-mobile, long-range, lighter-warhead ICBM for the 21st century. China may also deploy ICBMs with multiple independently targetable reentry vehicle (MIRV) capability in the next century.

In addition, China has developed and deployed a number of conventionally-armed anti-ship, air-launched, and ground-based, coastal defense cruise missiles, including the supersonic C-301 anti-ship cruise missile. At least two cruise missiles are believed to be under development. China does not have land attack cruise missiles, but may develop one in the future. The reported purchase of Russian Sunburn anti-ship cruise missiles at the end of 1996 indicates that China is now choosing to modernize more rapidly through foreign acquisitions rather than relying solely on its domestic defense industries.

China has worked to improve its ballistic and cruise missiles to provide a greater range of options for warfighting and to increase combat effectiveness. In the future, Chinese ballistic missiles can be expected to include longer-range SLBMs (perhaps deployed on an advanced submarine) and land-mobile, solid fuel ICBMs with smaller nuclear warheads, improved accuracy, MIRV capability, and improved penetrability. Chinese cruise missiles can be expected to have extended ranges and greater accuracy, and include land-attack cruise missiles.

While China continues to modernize its missiles, it is uncertain whether China will substantially expand the size of its missile forces. Chinese leaders face domestic and foreign policy considerations that limit the feasibility of significant increases in the number of missiles.

NOTE

This CRS study was initiated at the request of the Committee on National Security of the House of Representatives and is adapted for general congressional use with permission of the Committee.

Contents

List of Tables

China: Ballistic and Cruise Missiles

Introduction

This CRS report contains three parts. The first part discusses Chinese ballistic missiles. The second discusses Chinese cruise missiles. The last section offers concluding observations. Two tables summarize the discussion on current ballistic and cruise missiles in service or under development. The appendix, prepared by Robert Shuey, discusses China's reported application of global positioning system (GPS) technology to improve the accuracy of its missiles. This report focuses on the status and current developments of China's missile programs, rather than their history. The report is based on public sources of information which have been cited.

Ballistic missiles are self-propelled weapon delivery systems that are powered and guided during a part of their ascent and then follow a ballistic (unguided and unpowered) trajectory over the remainder of the flight. More advanced and accurate missiles also have terminal guidance to direct the weapon to the target. Cruise missiles are continually powered by an air-breathing or rocket engine and are generally guided for their entire flight.[1]

U.S. Classification of Ballistic Missiles

Short-range ballistic missile (SRBM)	70-1000 km	(43-620 mi)
Medium-range ballistic missile (MRBM)	1001-3000 km	(621-1860 mi)
Intermediate-range ballistic missile (IRBM)	3001-5000 km	(1861-3100 mi)
Intercontinental ballistic missile (ICBM)	5001 km +	(3101 mi +)

China is not party to international arms control agreements that limit the development or deployment of its missile forces, such as the Intermediate-Range Nuclear Forces (INF) Treaty signed between the United States and the Soviet Union in 1987. (The INF Treaty bans all intermediate-range and shorter-range ballistic missiles and ground-launched cruise missiles with ranges of 500-5,500 km.) China has promised to adhere to the Missile Technology Control Regime (MTCR), an informal set of guidelines limiting the transfer of ballistic and cruise missiles that have the inherent capability to deliver 500 kg. warheads to 300 km.

[1] CRS Report 95-688, *Ballistic and Cruise Missile Forces of Foreign Countries*, by Robert Shuey.

Table 1. Chinese Ballistic Missiles: Launchers Deployed and Under Development

Type	CH des.	US des.	Deploy-ed	Range/Warhead (km/kg)	CEP (m)	Launch Platform	Fuel	Comments
MRBM	DF-3A	CSS-2	40+	2,800/2,150	1,000	land-mobile	liquid	in service
ICBM	DF-4	CSS-3	10-25	5,500+/2,200	1,370	land-mobile	liquid	in service
ICBM	DF-5A	CSS-4	7-18	13,000/3,200	500	hardened underground silos	liquid	in service
SLBM	JL-1	CSS-N-3	12	1,700/600	?	Xia SSBN	solid	in service
MRBM	DF-21	CSS-5	10	1,800/600	?	mobile TEL	solid	in service
SRBM	DF-15/ M-9	CSS-6	?	600/500	300	mobile TEL	solid	in service; test-fired in 7/95 & 3/96 near Taiwan
SRBM	DF-11/M-11	CSS-7	?	300/500	?	mobile TEL	solid	in service
SRBM	8610/ M-7	CSS-8	?	150/190	?	mobile launcher	solid	modified HQ-2 SAM
MRBM	DF-25			1,700/2,000	?	land-mobile	solid	development
ICBM	DF-31			8,000/700	?	land-mobile	solid	development; after 2000
SLBM	JL-2			8,000/700	?	planned 094 SSBN	solid	development; after 2005
ICBM	DF-41			12,000/800	?	land-mobile	solid	development; after 2010

Chinese Ballistic Missiles[2]

China is currently believed to have deployed a nuclear-armed missile force consisting of over 100 nuclear warheads[3] on perhaps as many as 43 DF-4 and DF-5A ICBMs, over 50 DF-3A and DF-21 MRBMs, and 12 JL-1 SLBMs on one Xia nuclear-armed ballistic missile submarine (SSBN). A number of these are believed to be targeted on the United States and other countries. In an unconfirmed report, the *Washington Times* on May 1, 1998, cited a classified CIA report as saying that 13 of China's 18 strategic missiles have single nuclear warheads aimed at U.S. cities. When asked about this report on the same day, White House spokesman Mike McCurry confirmed that Chinese nuclear missiles are targeted at the United States by replying that "the President has spoken to the issue of 'detargeting' on a number of occasions, making clear there are no *Russian* nuclear missiles targeted on the United States." There are also SRBMs for tactical operations. Ballistic missiles are deployed in the People's Liberation Army (PLA)'s Second Artillery Corps, the strategic missile force. According to the Undersecretary of Defense for Acquisition and Technology, Paul Kaminski, China's strategic ballistic missile capability is "limited" today, but it is "emerging and growing."[4]

Chinese ballistic missiles can be divided into four phases of development and deployment. After developing land-based MRBMs and limited-range ICBMs in the first phase, China sought diversity and reliability. In the second phase, China added long-range ICBMs, SLBMs and SRBMs to its ballistic missile force, and achieved mobility and the use of solid fuel. In the third phase, China hopes to deploy after 2000 or 2005 a new land-mobile, solid fuel ICBM with a lighter warhead and a new, longer-range SLBM on a next-generation submarine. A new MRBM with a large conventional warhead for tactical operations has been in development (but this program may have been canceled). For the fourth phase, China plans to deploy a land-mobile, long-range, lighter-warhead ICBM for the 21st century. China's ballistic missile programs are summarized in the table, "Chinese Ballistic Missiles."

China may also deploy ICBMs with multiple independently targetable reentry vehicles (MIRVs) in the next century. China first decided to develop MIRVs for deployment in 1970. Development was in part stalled, however, by a lack of

---

[2] Sources of information on Chinese ballistic missiles: Institute of International Strategic Studies. *The Military Balance 1996-97*; *Jane's Weapons Systems*; Lewis, John Wilson and Hua Di. "China's Ballistic Missile Programs: Technologies, Strategies, Goals" (*International Security*, Fall 1992); Lewis, John Wilson and Xue Litai. *China's Strategic Seapower* (Stanford: Stanford University Press, 1994); Norris, Robert S., Andrew S. Burrows, and Richard W. Fieldhouse. *Nuclear Weapons Databook: British, French, and Chinese Nuclear Weapons*. Volume V. (Boulder: Westview Press, 1994); Shuey, Robert, *Ballistic and Cruise Missile Forces of Foreign Countries*; Teal Group Corp., "Chinese Tactical Ballistic Missiles," *World Missiles Briefing*, May 1996.

[3] Office of the Secretary of Defense, *Proliferation: Threat and Response*, November 1997.

[4] Testimony at a hearing on ballistic missile defense before the Subcommittee on Military Research and Development, House National Security Committee, March 6, 1997.

capability to miniaturize warheads.[5] The priority for the project on MIRVs was lowered in March 1980, but research and development on MIRVs resumed on November 10, 1983, as part of the DF-5 modification program. Some are concerned that China's use of multi-satellite dispensers, or smart dispensers, to launch more than one satellite on a Long March rocket has contributed to its development of MIRV capability.[6]

Phase 1: DF-3A, DF-4

DF-3A MRBM. The Chinese-designated DF-3A (DF stands for *Dong Feng*, or East Wind) has the U.S. designation of CSS-2. About 40 or more launchers are deployed. The DF-3A MRBM can deliver a 2,150 kg. (4,730 lb.) warhead to 2,800 km. (1,739 mi.) As an indicator of accuracy, the circular-error-probable (CEP) is 1,000 m. (0.6 mi.). The missile is land-mobile and launched from presurveyed above-ground launch sites and bases. It is a single-stage missile using storable liquid propellant and strap-down inertial guidance. It has been deployed since 1971. The DF-3A has a response time (preparation for launch after warning) of one hour and 50 minutes. In 1987, China sold an estimated 36 CSS-2s to Saudi Arabia.[7]

DF-4 ICBM. Ten to 25[8] DF-4 (CSS-3) ICBMs are deployed. The DF-4s can each deliver a 2,200 kg. (4,840 lb.) warhead to more than 5,500 km. (3,416 mi).[9] A CEP of 1,370 m. (0.9 mi.) has been reported. The missiles are based in caves or mountainside tunnels and are land-mobile. They have two stages and use storable liquid propellant and strap-down inertial guidance. The Chinese have deployed DF-4s since 1980. The missiles have also been produced as boosters for the Long March space vehicles that launch satellites. The DF-4s have a response time of two and a half hours.

Phase 2: DF-5A, JL-1, DF-21, DF-15/M-9, DF-11/M-11, 8610/M-7

In the second phase, China achieved a diversified ballistic missile force, with the addition of long-range ICBMs, SLBMs, and SRBMs. The newer missile force also featured greater mobility and solid fuel.

DF-5A ICBM. Development of the DF-5 (CSS-4) ICBMs involved overcoming challenges of new technical requirements before the first full-range test flights into the

---

[5] See also CRS report 97-1022, *Chinese Nuclear Testing and Warhead Development*, November 14, 1997, by Jonathan Medalia.

[6] See also CRS Report 98-485, *China: Possible Missile Technology Transfers From U.S. Satellite Export Policy — Background and Chronology*, by Shirley A. Kan.

[7] For information on Chinese transfers of missiles, see CRS Issue Brief 92056, *Chinese Proliferation of Weapons of Mass Destruction: Current Policy Issues*, by Shirley Kan, updated regularly; and CRS Report 96-767, *Chinese Proliferation of Weapons of Mass Destruction: Background and Analysis*, by Shirley Kan.

[8] Gertz, Bill. "China Targets Nukes at U.S.," *Washington Times*, May 1, 1998.

[9] Office of the Secretary of Defense.

Pacific Ocean in May 1980. The DF-5A ICBMs can each carry a 3,200 kg. (7,040 lb.) warhead to 13,000 km. (8,073 mi.). The missiles have a CEP of 500 m. (0.3 mi.) The DF-5As are two-stage missiles, and use storable liquid fuel and gyroplatform inertial guidance with on-board computers. The missiles are deployed in hardened underground silos and have been deployed since 1981. The missiles are also produced as boosters for Long March vehicles to launch satellites. Previously, four DF-5A launchers were thought to be deployed in the Second Artillery, while there were a number of fake silos to provide deception. *The Military Balance 1995-96*, however, reports that seven DF-5A ICBMs are deployed, while noting that other sources suggest there may be as many as ten. The *Washington Times* (May 1, 1998) cited a new classified CIA report as saying that there are 18 of these long-range strategic missiles, with 13 deployed with nuclear warheads aimed at U.S. cities.

JL-1 SLBM. The JL-1 (JL stands for *Julang*, or Giant Wave) submarine launched ballistic missile (SLBM) is designated CSS-N-3 in the United States. The JL-1 demonstrated a significant technological improvement. It is China's first SLBM which provides the third leg for its nuclear triad and increased nuclear deterrence capability. The JL-1 is also China's first solid-fuel missile, providing greater safety and more rapid response time. 12 JL-1 SLBMs are deployed on China's one Xia nuclear ballistic missile submarine. The JL-1 can deliver 600 kg. (1,320 lb.) to 1,700 km. (1,056 mi.). The missile has two stages and gyroplatform inertial guidance with an on-board computer. The JL-1 has been deployed since 1983. Successful underwater test-firings were conducted in 1988.

DF-21 MRBM. The new technology developed for the JL-1 was applied to other ballistic missiles, most notably its land-based variant, the DF-21 MRBM. An estimated ten DF-21 MRBMs have been deployed. The DF-21 (CSS-5) MRBM can deliver a 600 kg. (1,320 lb.) warhead to 1,800 km. (1,118 mi.). It uses the same fuel and guidance as the JL-1 SLBM. The DF-21's automatic command-control-firing system is said to be the first in the Second Artillery's missile force. The missile is launched from a mobile transporter-erector-launcher (TEL). The Second Artillery's first regiment for firing DF-21s was set up in 1985. In 1996, an engineer from the Beijing Research Institute for Telemetry told an analyst of the Heritage Foundation that China is developing a terminal guidance system for the DF-21. Other sources in Taiwan said that the new guidance system will be radar-based.[10] A modified DF-21 may be named DF-21A.

DF-15/M-9 SRBM. The DF-15, or M-9, SRBM was given the U.S. designation of CSS-6. The M-9 is capable of delivering a conventional or nuclear warhead of at least 500 kg. (1,100 lb.) to at least 600 km. (373 mi.). The missile has been promoted with a CEP of 300 m. (0.2 mi.). It is launched from a mobile TEL with a preparation time of 30 minutes.[11] The M-9 is a single-stage missile with solid fuel. M-9s have been deployed in the Second Artillery since 1990. The M-9 has a strap-down inertial guidance system with an on-board computer. In addition, a miniature propulsion system on the warhead can correct the missile's terminal velocity, reentry attitude, flight trajectory, and range. This control during the descent phase improves the M-9's

---

[10] Fisher, Richard D., Jr. "China's Missile Threat," *Wall Street Journal*, December 30, 1996.

[11] *Jane's Weapons Systems 1988-89*.

accuracy and penetration. In 1996, the engineer from the Beijing Research Institute for Telemetry cited above said that Chinese researchers are integrating GPS technology to improve the accuracy of the M-9 missiles. If true, the CEP of the M-9 could be better than 300 meters, as originally advertised in the 1980s.

In part to show displeasure with Lee Teng-hui's trip to the United States and to intimidate Taiwanese voters in advance of their first direct presidential election, China held military exercises including missile tests in July 1995 and March 1996. In both periods, M-9s were test-fired. In the first missile test, four M-9s were fired. Three landed in a target area about 90 miles north of Taiwan in the East China Sea, but one missed the target area, crashing prematurely. Two M-9s that landed within the closure area were said to have landed on the target's southern edge. The M-9s test-fired in July 1995 displayed a CEP of 3,200 meters.[12] During March 8-15, 1996, the Chinese fired four M-9 missiles into two closure areas about 20 miles off Taiwan's port city of Keelung and 30 miles off the port city of Kaohsiung — closer to Taiwan than the target used in July 1995.[13] According to one report, China was ready to launch 20 to 30 M-9 missiles, although only four were fired.[14]

DF-11/M-11 SRBM. The DF-11 or M-11 (CSS-7) SRBM is similar to the M-9 missile, but has two stages and a shorter range of 300 km. (186 mi.) for a payload of at least 500 kg. (1,100 lb.). It is also considered nuclear or conventional capable. It may have been deployed since 1992. The United States twice imposed sanctions on China and Pakistan for M-11 related missile technology transfers.

8610/M-7 SRBM. The 8610 or M-7 (CSS-8) SRBM is a modified HQ-2 surface-to-air missile (SAM). It is a two-stage missile, believed to have a range of 150 km. (93 mi.) with a warhead of 190 kg. (418 lb.). The 8610 was developed solely for export. Iran has bought the 8610 ballistic missile.

Phase 3: DF-25, DF-31, JL-2

The next group of ballistic missiles, currently under development for planned deployment after the year 2000 or 2005, include a new DF-25 MRBM with a heavy conventional warhead, a new DF-31 ICBM that is land-mobile and has a lighter warhead, and a next-generation JL-2 SLBM on a new ballistic missile submarine.

DF-25 MRBM. The DF-25 MRBM was conceived along with the DF-15 SRBM as a mobile, solid-fueled missile with a conventional warhead. The DF-25 would deliver a 2,000 kg. (4,400 lb.) warhead to 1,700 km. (1,056 mi.). The missile would have two stages, inertial guidance, and solid propellant. The DF-25 would also be land-mobile, with trucks to transport it from semi-hardened storage sites to launch

---

[12] Chen, Kevin. "Chinese Missiles Hit Near Taiwan," *Washington Times*, July 24, 1995; Teal Group Corporation.

[13] Office of Naval Intelligence. *Chinese Exercise Strait 961: 8-25 March 1996*. May 1996; Mufson, Steven. "China Targets Missile Tests On Sites Near Taiwan's Ports," *Washington Post*, March 6, 1996.

[14] Fisher.

sites. The missile was originally intended as a tactical weapon for use in the South China Sea in disputes over the Spratly Islands. However, the priority of this program may have declined. Since mid-1995, the PLA has focused its attention on Taiwan, and there are indications that China has since exercised restraint in its claims over the South China Sea. There have not been reports of test-firings. Indeed, one report said that China has abandoned development of the DF-25.[15]

DF-31 ICBM. The DF-31 would be China's next-generation ICBM. The three-stage missile would be able to deliver a 700 kg. (1,540 lb.) warhead to 8,000 km. (4,968 mi.). It would be fueled by solid propellant and be land-mobile. The DF-31 is expected to be in service after the year 2000. The Chinese may develop MIRVs for deployment on the DF-31 ICBM.

JL-2 SLBM. The JL-2 SLBM would be the submarine-launched version of the DF-31 ICBM. The JL-2 would have a much greater range than the JL-1 (8,000 km. (4,968 mi.) compared to 1,700 km. (1,056 mi.)) and be launched from the planned, next generation 094 ballistic missile submarine. However, development of the land-based DF-31 was given priority over the JL-2 SLBM, perhaps because of the cost and difficulties of building the Type 094 submarine. Speculatively, JL-2s may be deployed after 2005. With the JL-2 SLBMs, Chinese SSBNs will be able to target, for the first time, parts of the continental United States from areas near China. Each Type 094 SSBN is expected to carry 16 JL-2 SLBMs.[16]

Phase 4: DF-41

The fourth phase is represented by the DF-41 ICBM program. The DF-41 ICBM would have three stages and use solid fuel. The new ICBM would be capable of delivering an 800 kg. (1,760 lb.) warhead to 12,000 km. (7,452 mi.). This ICBM would also be road-, rail-, and/or river-mobile. Deployment is planned for the first decade of the 21st century, after 2010 when the DF-5 ICBM will be withdrawn from service. China may develop MIRVs for the DF-41s and is expected to deploy ICBMs with MIRVs after 2010.[17]

---

[15] Arnett, Eric. "Chinese Blow Cold On East Wind Missile Plan," *Jane's Defence Weekly*, December 4, 1996.

[16] Office of Naval Intelligence. *Worldwide Submarine Challenges*. 1997.

[17] Montaperto, Ronald. "China," *Strategic Assessment 1997*. National Defense University.

Table 2. Chinese Cruise Missiles
(with ranges over 80 km.)

Type	CH des.	US des.	Range/Warhead (km/kg)	Guidance	Speed (Mach)	Launch Platform	Comments
Anti-ship	HY-2	CSS-C-2 Silkworm	85/400	active radar	0.9	coastal defense	in service
Anti-ship	HY-2A/ C-201	CSS-C-3 Seersucker	95/513	active radar or IR homing	0.9	Luda destroyer; Jianghu frigate	in service
Air-launched	C-601	CAS-1 Kraken	110/513	active radar or IR homing	0.9	H-6D bomber	in service
Anti-ship	HY-3/ C-301	CSS-C-6 Sawhorse	180/512	active radar	2.0	coastal defense	in service
Anti-ship/ Air-launched	HY-4/ C-201W	CSS-C-7 Sadsack	135/500	active radar	0.8	coastal defense	in service
Anti-ship	YJ-2/ C-802	CSS-C-8 Saccade	120/165	active radar	0.9	Luhu destroyer; Jiangwei & Jianghu frigates	in service
Air-launched	C-611		200/?			H-6D bomber	development; improved C-601
Anti-ship/ Air-launched	XW-41		300/?				development; improved HY-4
Anti-ship		SS-N-22 Sunburn	160/300	active/passive radar homing	2.5	Sovremen-ny destroyer	to be purchased from Russia

Chinese Cruise Missiles[18]

China has developed and deployed a number of anti-ship, air-launched, and ground-based, coastal defense cruise missiles. C-802 anti-ship cruise missiles have been sold to Iran. At least two cruise missiles are believed to be under development. Chinese cruise missiles with ranges of at least 80 km. (50 mi.) are discussed here and are summarized in the table, "Chinese Cruise Missiles." Of these, the C-301 is a supersonic anti-ship cruise missile. China apparently does not have land attack cruise missiles, but may develop one in the future.

Cruise Missiles in Service

HY-2. China's HY-2 (HY stands for *Hai Ying*, or Sea Eagle) anti-ship cruise missile is designated in the United States as the CSS-C-2 Silkworm. The missile can deliver a 400 kg. (880 lb.) warhead to 85 km. (53 mi.). Its guidance uses active radar. The HY-2 travels at a speed of 0.9 Mach. The anti-ship missile is deployed with the Chinese navy's coastal defense forces. In the 1980s, China sold HY-2 missiles to both Iran and Iraq.[19]

HY-2A/C-201. The HY-2A anti-ship missile has the export name of C-201 and a U.S. designation of CSS-C-3 Seersucker. The missile has a range of 95 km. (59 mi.) with a payload of 513 kg. (1,129 lb.). It is an improvement on the HY-2 in that its guidance uses an infra-red homing sensor. The missile cruises at 0.9 Mach. It has been deployed on Luda-class destroyers and Jianghu-class frigates.[20]

C-601. The C-601 cruise missile is the air-launched version of the C-201. The C-601 can deliver a 513 kg. (1,129 lb.) warhead to 110 km. (68 mi.).[21] The C-601 air-launched cruise missile is deployed on H-6D bombers in the naval airforce.

HY-3/C-301. The HY-3/C-301 anti-ship cruise missile has the U.S. designation of CSS-C-6 Sawhorse. This is China's only reported supersonic cruise missile with a range of over 80 km. (50 mi.). (The C-101 also cruises at 2.0 Mach, but has a range of only 50 km. (31 mi.), so it has not been included in this report.) Some sources report that the C-301 has a range of 130 km. (81 mi.) with a payload of 512 kg. (1,126 lb.), and that developments were being made to extend the range to 180 km.

---

[18] Sources include: *The Military Balance 1995-96*; *Jane's Fighting Ships 1995-96*; Shuey; and Teal Group Corporation. "Chinese Anti-Ship Missiles," *World Missiles Briefing*. May 1996.

[19] Bussert, Jim. "HY-2 'Silkworm' Anti-Ship Missile Detailed," *Jane's Soviet Intelligence Review*, March 1991.

[20] Information also provided in: Yang Xinsheng and Luo Jiahong. "China's Hai Ying-2 Shore-to-Ship Missile Weapon System." *Missiles and Spacecraft*, January 1990, translated in *JPRS Report (JPRS-CST-91-016)*, August 1, 1991.

[21] This maximum range is given in: Yang Jingqing and Xu Zimou. "China's C-601 Air-to-Ship Missile Weapon System." *Missiles and Spacecraft*, July 1990, translated in *JPRS Report (JPRS-CST-91-016)*, August 1, 1991.

(112 mi.).[22] According to an engineer of the China Aerospace Corporation, this goal of flying up to 180 km. at twice the speed of sound has been reached.[23] The C-301 uses active radar guidance and a coastal launcher system.

HY-4/C-201W. Efforts to extend the range of HY-2 series missiles resulted in the HY-4 anti-ship cruise missile. It is also called the C-201W missile with the U.S. designation of CSS-C-7 Sadsack. Its air-launched version is also called C-201W. With a turbo-jet engine, the HY-4 has a range of 135 km. (84 mi.) with a 500 kg. (1,100 lb.) warhead. It uses active radar guidance and cruises at 0.8 Mach. It has been deployed with the PLA Navy's coastal defense forces.

YJ-2/C-802. The YJ-2 (YJ stands for *Ying Ji*, or Eagle Strike), or C-802, anti-ship cruise missile is designated as CSS-C-8 Saccade in the United States. It delivers a more compact, 165 kg. (363 lb.) warhead to 120 km. (75 mi.). (The YJ-1/C-801 variant has a range of only 40 km. (25 mi.), so it is not discussed here.) The C-802 missile uses active radar guidance and cruises at 0.9 Mach. The C-802 has been deployed on Luhu-class destroyers, and Jiangwei-class and Jianghu-class frigates. In 1995, China transferred C-802 anti-ship cruise missiles to Iran for use on Chinese and French supplied patrol boats.

China displayed C-802 missiles with improved guidance at the "Airshow China 1996" in November 1996.[24] Also, China is reportedly developing a ground-launched, land-attack version of the C-802 to more accurately hit targets in Taiwan. GPS technology and terrain contour-matching radar systems are being applied.[25]

Cruise Missiles under Development

C-611. An improved version of the C-601 air-launched, anti-ship cruise missile has been named C-611. It would have an increased range of 200 km. (124 mi.), and improved propulsion, electronics, and terminal guidance. The C-611 would use new high-energy fuel. It is also intended for launch by the H-6D bomber.[26]

XW-41. China is developing an improved version of the HY-4 missile, called XW-41. This new missile would have a longer range of 300 km. (186 mi.).

---

[22] Jiang Junliang and Huang Taian. "China's C-301 Supersonic Anti-Ship Missile," *Missiles and Spacecraft*, February 1991, translated in *JPRS Report (JPRS-CST-91-016)*, August 1, 1991; Lennox, Duncan. "China's New Supersonic Anti-ship Missile," *Jane's Intelligence Review*, November 1992.

[23] Opall, Barbara. "Chinese Strive to Boost Range, Aim of Missiles," *Defense News*, December 9-15, 1996.

[24] Opall.

[25] Glashow, Jason and Theresa Hitchens. "China Speeds Development of Missile with Taiwan Range," *Defense News*, March 4-10, 1996.

[26] Yang and Xu.

Cruise Missiles to Be Purchased

SS-N-22. At the end of 1996, Chinese Premier Li Peng visited Russia. Reports say that he concluded an agreement to purchase two Sovremenny-class destroyers to be equipped with SS-N-22 Sunburn anti-ship cruise missiles.[27] Defense Department officials were quoted as saying that the Sunburn missiles were designed to counter U.S. naval warships equipped with the Aegis system. The sea-skimming Sunburn anti-ship cruise missiles are supersonic, travelling at 2.5 Mach. The Sunburns can carry a nuclear or high-explosive warhead weighing 300 kg. (660 lb.) and have a range of 160 km. (99 mi.).[28]

One report says that Israel Military Industries (IMI) is developing a cruise missile with a hard-target penetrating warhead for China. The missile is said to have a range of 397 km. (247 mi.), with guidance that uses GPS and inertial navigation.[29]

---

[27] Gertz, Bill. Pentagon Says Russians Sell Destroyers to China. *Washington Times*, January 10, 1997.

[28] *Jane's Fighting Ships 1995-96.*

[29] "Israel to Equip Chinese Cruise Missile with Penetrator Warhead," *Flight International*, February 5-11, 1997, p. 13.

Conclusion

China has worked to improve its ballistic and cruise missiles to provide a greater range of options for warfighting and increase combat effectiveness. Portending events to come, the June 12, 1995, *Liberation Army Daily* printed an article on the utility of conventional ballistic missiles. Then, in July 1995 and March 1996, Beijing's willingness to use its ballistic missile forces for political purposes, if not actual conflict, was forcefully demonstrated when M-9 SRBMs were test-fired to targets near Taiwan. The missile test-firings were part of a psychological warfare campaign against Taipei. Other rationales driving China's missile modernization programs include technological advances, bureaucratic competition for resources, political objectives, international prestige, and development of the related space program and satellite-launch business.

After developing land-based MRBMs and limited-range ICBMs in the first phase, China pursued diversity and reliability. In the second phase, China added long-range ICBMs, SLBMs, and SRBMs to its ballistic missile force, and achieved mobility and the use of solid fuel. In the third phase, China hopes to deploy after 2000 or 2005 a new land-mobile, solid fuel ICBM with a lighter warhead, and a new, longer-range SLBM on a next-generation submarine. Also in development has been a new MRBM with a large conventional warhead for tactical operations (although this program may have been canceled). For the fourth phase, China plans to deploy a land-mobile, long-range, lighter-warhead ICBM for the 21st century.

China has continued to increase the range, accuracy, survivability, reliability, safety, mobility, and response capability of its missile forces. China now has missiles for conventional and nuclear, strategic and tactical operations, as well as for escalating military tensions short of war. The PLA has strengthened nuclear deterrence through the deployment of ground-, air-, and submarine-launched missiles. Because the reliability of bombers in China's nuclear triad is questioned,[30] ground- and submarine-launched ballistic missiles may be emphasized. China's missiles now have greater mobility, use solid instead of liquid propellent, on-board computers, and GPS technology for greater accuracy. Since 1990, the Second Artillery Corps has deployed a diversified ballistic missile force consisting of short-, medium-, and intermediate-range, and intercontinental ballistic missiles. China's cruise missiles now also include supersonic cruise missiles. However, China does not have land attack cruise missiles. The reported purchase of Russian Sunburn cruise missiles indicates that China is now choosing to modernize more rapidly through foreign acquisitions rather than relying solely on the domestic defense industries.

In the future, Chinese ballistic missiles can be expected to include longer-range SLBMs (perhaps deployed on an advanced submarine) and land-mobile ICBMs with smaller nuclear warheads, improved accuracy, MIRV capability, and improved penetration capability. Chinese cruise missiles can be expected to have extended ranges and greater accuracy, and include land-attack cruise missiles.

---

[30] Lewis and Xue; Arnett.

While clearly modernizing its ballistic and cruise missiles, it is uncertain whether China will substantially expand the size of its missile forces. Chinese leaders face at least two considerations that limit the feasibility of significant increases in the number of missiles.

First, the PLA Air Force, Navy, and Army have obsolete and backward equipment. Limitations in funds mean that the Second Artillery will have to compete with the other services which demand more advanced conventional weapons. There are indications that China has accelerated its arms purchases from Russia to modernize the PLA Air Force and Navy.

Second, a significant increase in China's missile force would send alarms through capitals in Asia, the United States, and Europe, at a time when Chinese leaders have endeavored to persuade all that China will never pose a threat to any other country as it modernizes. A significant increase in China's missile forces would raise further questions about China's nuclear doctrine and military intentions. A dramatic change would also spur calls for meeting Taiwan's missile defense requirements and expansion of arms control agreements to China. Chinese leaders have publicly stated that their current priorities are economic development and promotion of international trade and investment to boost development. Analysts believe that, in the absence of a credible threat or a persuasive ideology, the Chinese leaders recognize that their domestic political standing depends largely on the well-being of the economy and nationalistic appeals. A growing economy can afford, and even be stimulated by, the development of some new weaponry, but there are limits to the resources of capital, technology, and talent that can be beneficially diverted from economically productive enterprises.

If China's leaders decide to expand the size of the ballistic missile force, they could tap the civilian side of the Chinese aerospace industry. The DF-4 and DF-5A missiles have also been produced as boosters for the Long March space vehicles that launch satellites for China and foreign companies.

Appendix: GPS and Chinese Missiles[31]

China is reportedly trying to adapt satellite navigational technology to improve the accuracy of certain ballistic and cruise missiles. Satellites of the U.S. Global Positioning System (GPS) transmit signals that are accessible at no charge to anyone who has a commercially available receiver or a more capable military receiver. The receiver uses the signals to calculate its three-dimensional location on, above, or below the earth's surface. Russia has a similar system — the Global Navigation Satellite System (GLONASS).

The U.S. GPS consists of 24 satellites, ground control and monitoring stations, and receivers that take precise information from four or more satellites on the locations of the satellites and the time the signal was sent. A commercial receiver can determine the length of time it took the Standard Positioning Service signals to arrive from the satellites and calculate its position to within 100 meters. U.S. military receivers receive encrypted Precise Positioning Service signals that establish location within 21 meters. The system can be augmented to provide more accurate information. The most commonly used augmentation is Differential GPS (DGPS) which adds a signal from a base station whose location is precisely known, thereby enhancing the Standard Positioning Service to an accuracy of less than one meter.[32] In mid-1996, a U.S. company offered for sale a unit that receives and compares signals from U.S. and Russian satellites to provide more accurate position data.

If the guidance systems of China's missiles are supplemented by GPS, they can compare the actual position of the missile during flight, as determined by GPS, with the planned position along the flight trajectory and send correction signals to the missile control system. Cruise missiles generally travel in powered, aerodynamic flight all the way to the target, so they may use their control surfaces and propulsion adjustments to adjust course. During the boost phase, ballistic missiles can use control vanes or thrust deflectors to adjust course. Many missiles (but not SCUDs) jettison the rocket motor after the boost phase. In some sophisticated missiles, the reentry vehicle(s) can respond to signals from GPS-assisted guidance systems, using small thrusters to correct their course. Ballistic and cruise missiles can also use terminal guidance that will direct them to a recognized or designated point in the target area. Terminal guidance generally employs radar, imaging infrared, electro-optical, laser, millimeter wave, terrain contour mapping, or digital scene matching area correlation, rather than GPS.[33]

---

[31] This appendix was prepared by Robert Shuey, Specialist in U.S. Foreign Policy and National Defense, Foreign Affairs and National Defense Division.

[32] Grier, Peter. "GPS in Peace and War," in *Air Force*, April 1996, pp. 76-78; "The Global Positioning System: Charting the Future," the National Academy of Public Administration and the National Research Council, May 1995, pp. 2-5.

[33] Ballistic Missile Proliferation. Training Course, TASC, Arlington, VA.

China has access to Standard Positioning Service GPS and commercial receivers. China has reportedly, "refined the commercial GPS capability to provide accuracy of up to 10 meters,"[34] possibly using DGPS or an equivalent Russian augmented system.

China is reportedly trying to develop a land attack cruise missile based on its C-802 anti-ship missile that has a range variously reported from 120 to 180 km. with a payload of 165 kg. GPS-assisted guidance is said to be a key ingredient in the development of this cruise missile. Russia is said to be assisting China in the missile development and in the application of GLONASS to its guidance and control systems.[35]

GPS- and GLONASS-assisted missile guidance are advanced technologies, but probably within China's capabilities. North Korea and Taiwan have much less experience in missile and space technology, but both have GPS-guided missiles.[36]

---

[34] Glashow and Hitchens.

[35] *Ibid.*, p. 1, 36; Lien-Ho Pao (Taipei), April 29, 1996, p. 1, cited in FBIS-CHI-96-086, April 29, 1996, abstracted in the Center for Nonproliferation Studies database, Monterey Institute for International Studies; Gertz, Bill. "China's SS-18 plan underscores nuke-proliferation threat," in *The Washington Times,* May 26, 1996, p. 4. The Taiwanese source that cited Russian technical assistance said China had probably bought from Russia and deployed cruise missiles capable of traveling 100 to 3,000 km with an accuracy of 10 meters. Russia may have such a system (perhaps the KH-65SE/KH-101), but K. Scott McMahon and Dennis M. Gormley report, "there have been no confirmed Russian transfers of advanced cruise missiles...." See their "Cruise Missile Proliferation: Threat and Response 1995-1996," Pacific-Sierra Research Corporation, October 14, 1996, p. 5.

[36] National Ballistic Missile Briefings: North Korea, Center for Defense and International Studies, cited in Center for Nonproliferation Studies database; Lu Chao-Lung, "National Army to Test Surface-to-Surface Missiles this Year," in *Chung-Kuo Shih-Pao* (Taipei), Sept. 11, 1996, FBIS-CHI-96-180, cited in Center for Nonproliferation Studies database.

Order Code IB93062

CRS Issue Brief

Space Launch Vehicles: Government Requirements and Commercial Competition

Updated June 12, 1998

Marcia S. Smith
Science, Technology, and Medicine Division

CRS

Congressional Research Service • The Library of Congress

CONTENTS

55-784 99 - 11

Space Launch Vehicles: Government Requirements and Commercial Competition

SUMMARY

The 1986 space shuttle *Challenger* accident, coupled with the failures of two other major U.S. launch vehicles, initiated a reassessment of U.S. launch vehicle capabilities and needs for placing military, civilian, and commercial satellites into orbit. The U.S. has two types of launch vehicles: those that can be used only once ("expendable launch vehicles" or ELVs) and the reusable space shuttle. There currently are five families of operational U.S. ELVs; from smallest to largest they are Pegasus/Taurus (Orbital Sciences Corporation), Athena (Lockheed Martin), Delta (Boeing), and Atlas and Titan (Lockheed Martin). The U.S. space shuttle is the only operational reusable launch vehicle in the world.

In the 1970s, the United States decided that once the space shuttle was operational, all the ELVs except Loral Vought's Scout would be phased out of use on the basis that the shuttle would be more economical than ELVs. In the early 1980s, production lines for the Delta and Atlas began to close down, although in 1984 the Air Force decided to develop a new version of Titan to "complement" the space shuttle and assure that the Department of Defense could place satellites in orbit in the event the space shuttle was grounded.

DOD's concerns proved well founded following the *Challenger* tragedy in 1986, which ended shuttle flights for 32 months. In the wake of *Challenger*, U.S. policy changed to support a "mixed fleet" approach using both the shuttle and ELVs. DOD initiated a "Space Recovery Plan" for getting its satellites in orbit as expeditiously as possible, and assuring access to space in the future. In addition to the new version of Titan, it ordered new versions of the Delta and Atlas. Both DOD and NASA continue to study the need for new launch vehicles using more advanced technology and/or capable of lifting heavier objects into space. NASA is undertaking the Reusable Launch Vehicle (RLV) program and DOD is pursuing the Evolved Expendable Launch Vehicle (EELV) program.

Interest in facilitating the emergence of a U.S. commercial space launch industry, begun before *Challenger*, blossomed with the change in launch policy. Congress and the White House have taken steps to help commercial space launch services companies, for example, by requiring NASA to procure commercial launch services rather than the rockets themselves, which then are launched by the commercial company rather than the government, and by indemnifying launches.

The main competitor to U.S. companies for commercial launch services is Europe's Arianespace, which launches the Ariane ELV. China, Russia, Ukraine, and Japan also offer commercial launch services. Deciding whether to allow U.S.-made satellites to be exported to other countries for launch intermixes technology transfer, trade, and ballistic missile proliferation issues.

Among the issues facing the 105th Congress are: whether export licenses were violated by U.S. companies who build satellites that are launched by China (the "Loral/Hughes issue"), what else could or should be done to facilitate the U.S. commercial ELV industry, the development of future U.S. launch systems and technology, and space shuttle safety as the shuttle transitions to a single prime contractor.

MOST RECENT DEVELOPMENTS

On June 11, the Senate Appropriations Committee completed markup of its version of the FY1999 VA-HUD-Independent Agencies appropriations bill that includes NASA (no bill number, report number, or text available yet). The committee increased NASA's budget by $150 million for the space station and other critical NASA programs. The Senate is expected to consider the NASA authorization bill (S. 1250) soon. The Senate bill covers FY1998-2000, while the House version, H.R. 1275 (passed by the House last year), covers FY1998-1999. As reported from the Senate Commerce Committee (S.Rept. 105-195), S. 1250 would authorize $13.465 billion for FY1999, the same as the request.

The Senate Appropriations Committee also has reported the FY1999 DOD appropriations bill (S. 2132, S.Rept., 105-200). It is expected to consider that bill and resume consideration of the FY1999 DOD authorization bill (S. 2057) soon. The House passed its version of the DOD authorization bill (H.R. 3616) in May. The House passed several amendments to the bill related to allegations that two U.S. satellite manufacturing companies, Hughes and Loral, may have transferred technology to China that improved the reliability of Chinese nuclear missiles (see ***"The Loral/Hughes Issue"*** *below). The amendments would, inter alia, prohibit the export of U.S. satellites to China. Similar amendments may be offered during Senate floor debate. Several congressional committees are holding hearings on the issue. The House Appropriations Committee is marking up its version of the FY1999 DOD appropriations bill.*

The Senate Commerce Committee reported the Commercial Space Act on June 2 (H.R. 1702, S.Rept. 105-198). Separately, on May 22, Senator Breaux introduced S. 2121, the Space Launch Cost Reduction Act to create mechanisms for ensuring the availability of financing for private sector development of low-cost commercial space launch vehicles.

BACKGROUND AND ANALYSIS

U.S. Launch Vehicle Policy

The National Aeronautics and Space Administration (NASA) was created in 1958 to conduct U.S. civilian space activities. Military space activities are conducted by the Department of Defense (DOD). During the early part of the space program, both agencies developed launch vehicles to satisfy their requirements. DOD developed the Atlas, Delta, and Titan families of launch vehicles from ballistic missile technology; NASA developed Scout and Saturn (both are no longer produced). Private companies developed Pegasus (and a variant of it called Taurus), and Athena (formerly called the Lockheed Launch Vehicle). All these launch vehicles can only be used once and, thus, are called Expendable Launch Vehicles (ELVs). Government and commercial users utilize whatever vehicle they need based on the size, weight, and destination of the spacecraft. From smallest to largest, the operational U.S. ELVs are Pegasus, Taurus, Athena, Delta, Atlas, and Titan.

From "Shuttle-Only" to "Mixed Fleet"

In 1972, President Nixon approved NASA's plan to create the first reusable launch vehicle, called the space shuttle, and directed that it become the nation's primary launch vehicle, replacing all the ELVs except Scout (later discontinued for unrelated reasons). This would have made NASA and DOD dependent on a single launch vehicle, but the resulting high launch rate was expected to significantly reduce the cost per flight. The shuttle was first launched in 1981, and was declared operational in 1982. The phase-out of the ELVs began, but in 1984 the Air Force successfully argued that it needed a "complementary" ELV as a backup to the shuttle for "assured access to space" and initiated what is now known as the Titan IV program. Production lines for the Delta and Atlas began to close down, and it was expected that only the shuttle, Scouts, and Titan IVs would be in use by the end of the 1980s.

Everything changed on January 28, 1986, however, when the space shuttle *Challenger* exploded 73 seconds after launch. The space shuttle program had enjoyed 24 successful missions prior to *Challenger*. Apart from the human tragedy, the *Challenger* accident deeply affected U.S. space launch policy, demonstrating the vulnerability of relying too heavily on a single system. Many military and civilian satellites had been designed specifically to be launched on the shuttle, and could not have been transferred to ELVs even if the ELVs were not already being phased out. The few remaining ELVs had their own problems in 1986. A Titan exploded in April and a Delta failed in May, which also grounded Atlas because of design similarities. As a result of these failures, U.S. policy was significantly revised from primary dependence on the shuttle to a "mixed fleet" approach. The country once again has a wide variety of launch vehicles from which to choose. The shuttle is used principally for missions that require crew interaction, while ELVs are used for other spacecraft.

As part of the change in policy, President Reagan also decided that commercial payloads could be flown on the shuttle only if they were "shuttle-unique" (capable of being launched only by the shuttle or requiring unique shuttle capabilities such as crew interaction) or if special foreign policy considerations were involved. This action facilitated the emergence of a U.S. commercial space launch industry whose participants had long argued that they could not compete against what they felt were government-subsidized shuttle launch prices. The White House and Congress had taken steps beginning in 1983 to assist in developing a commercial space launch services business, including President Reagan's 1983 designation of the Department of Transportation as the agency responsible for facilitating and regulating the commercial space launch sector, and passage of the 1984 Commercial Space Launch Act (P.L. 98- 575). But removing the shuttle as a competitor was the major factor in fostering the U.S. launch businesses. Other governmental actions, such as passage of the Commercial Space Launch Act Amendments of 1988 (P.L. 100-657), and the Launch Services Purchase Act of 1990 (P.L. 101-611), also have helped. (See below for recent congressional activities related to the U.S. commercial space launch business.)

Clinton Administration Launch Vehicle Policy

On August 5, 1994, the Clinton Administration released its National Space Transportation Policy. The policy has a two-track strategy of maintaining and upgrading the current fleet of ELVs, and investing R&D resources in developing and demonstrating next-generation reusable space launch systems to replace the space shuttle. Under the policy, DOD has lead responsibility for improvement and evolution of current ELVs. NASA is responsible for upgrading the space shuttle system and for technology development and demonstration of new reusable space transportation systems.

The policy also sets guidelines for the use of foreign launch systems and components, the use of excess ballistic missile assets for space launch, and encourages an expanded private sector role in space transportation R&D. Unless exempted by the President or his designee, U.S. government payloads will be launched by U.S. manufactured launch vehicles. The policy does not apply to the use of foreign launch vehicles on a no-exchange-of-funds basis to support scientific missions and government-to-government cooperative programs. The U.S. government, however, may take advantage of foreign components or technologies in upgrading U.S. launch systems or developing next generation systems. U.S. excess ballistic missiles eliminated under the START agreements are to be retained for government use or destroyed. Agency use of these missiles for launching payloads into orbit is permitted if the payload supports the agency's mission, the use is consistent with international obligations, and results in cost savings to the government relative to the use of available commercial launch services. Requests to use these excess assets are considered by DOD on a case by case basis. Finally, the policy promotes private sector involvement in and private sector financing of the development of new space launch capabilities.

On September 19, 1996, the Clinton Administration released its first comprehensive space policy, covering civil, military and commercial space activities. The policy includes a goal of moving commercial launch service competition with other countries from "negotiated trade" to a "trade environment characterized by the free and open interaction of market economies" once the existing launch services agreements expire. The policy also maintains the goals and guidelines of the 1994 National Space Transportation Policy.

U.S. Launch Vehicle Programs and Issues

NASA's Space Shuttle Program

The space shuttle is the sole U.S. means for launching humans into orbit. The 1986 *Challenger* accident and occasional shuttle launch delays have led many to question the reliability of the shuttle system. *Challenger*, however, is the only failure so far in more than 80 launches. Concerns about shuttle launch delays and the risk of another shuttle accident nonetheless continue to raise questions about future planned NASA programs that rely on the shuttle, such as assembly and servicing of the International Space Station (see CRS Issue Brief 93017, *Space Stations*). Another issue is the development of future vehicles to replace the shuttle system for launching and returning humans to and from orbit. Furthermore, there are concerns about the 26% budget cut the shuttle program has experienced over the past 5 years. Some worry that such a cut and associated personnel reductions could affect the safety of the shuttle program. Turning the shuttle program over to a "single prime contractor" and discussions of eventual privatization of the program also raise issues.

The shuttle consists of four parts: the orbiter (built by Boeing) for crew and cargo; the large cylindrical external tank that contains fuel for the orbiter's main engines (Lockheed Martin); and two solid rocket boosters (SRBs), one on either side of the external tank (Thiokol), for additional thrust during the first two and a half minutes of launch. The SRBs and the orbiter are reusable; the external tank is jettisoned and not recovered. After the *Challenger* accident, the space shuttle returned to flight on September 29, 1988. With a fleet of four orbiters, NASA had planned to build up to a flight rate of 10 flights per year; however, budget constraints have limited the projected annual flight rate to 7-8 per year.

Launch manifests for the shuttle are dynamic, changing to reflect schedule delays and other circumstances. Eight shuttle flights occurred in 1997.

On May 19, 1995, NASA released the results of its Zero Base Review with recommendations of how to achieve a $5 billion reduction in the agency's budget over 5 years. One of the recommendations was to restructure the space shuttle program by consolidating major shuttle contracts so that the program is run by a single prime contractor. This would be the initial step in preparing the program for potential privatization. Moving to a single prime contractor had also been recommended in February 1995 by an independent shuttle review panel headed by Christopher Kraft, former director of Johnson Space Center.

On November 7, 1995, NASA announced that it had non-competitively selected United Space Alliance (USA), a joint venture of Rockwell International (now Boeing) and Lockheed Martin, as the single prime contractor for space shuttle operations. The agency asserted there was no other company that could meet the safety, manifest, and schedule requirements for the program. On September 26, 1996, NASA signed a $7 billion, 6-year Space Flight Operations Contract (SFOC) with USA for space shuttle operations. The con-tract initially consolidates 12 major existing contracts into one. It also includes two, 2-year extension options that could bring the value of the contract to $12 billion over 10 years.

For FY1998, Congress appropriated $2.93 billion for the space shuttle program. The program has experienced a 26% cut in its budget over the past 5 years. Some shuttle observers are concerned that such cuts and associated personnel reductions could affect the shuttle safety. In FY1997, NASA took $190 million from the space shuttle program and $10 million from payload and utilization operations to provide additional funding for the space station program. NASA insists the FY1997 reduction did not affect shuttle safety and a GAO study of the transfer came to the same conclusion. In FY1998, NASA transferred $55 million from the shuttle and $22 million from payload and utilization operations to the space station. The FY1999 request for shuttle is $3.06 billion.

On April 24, 1997, the House passed H.R. 1275, the FY1998-1999 Civilian Space Authorization Act that, *inter alia,* directs NASA to conduct a study of the potential privatization of the space shuttle. The Senate version of the bill, S. 1250, as reported from committee (S.Rept. 105-195) also requires such a study.

Space Shuttle Safety and Upgrades

NASA also is embarking on several shuttle system upgrades to increase reliability and extend the life of the shuttle system. The shuttle program plans to maintain the capability to fly a maximum of 7 to 8 flights per year

On May 31, 1996, NASA announced that, at the request of the White House, it had asked the independent Aerospace Safety Advisory Panel (ASAP) to undertake a focused review of the space shuttle program, concentrating on the safety of the shuttle in light of management changes, planned shuttle upgrades, and flight rates to build and support the International Space Station. On December 13, 1996, ASAP announced its conclusion that efforts to streamline the space shuttle program have not increased risks. The panel, however, made 22 recommendations to NASA, many of which address a need for the agency to take steps to ensure the availability of a skilled and experienced workforce in sufficient numbers to meet ongoing needs. The panel also stated that "NASA should not be misled by the apparent initial success of all of the transition efforts." ASAP asserts that shuttle flight safety could be jeopardized as the agency scales back its safety inspections and hands over more responsibility for running the shuttle program to USA. NASA Administrator Goldin concurred with ASAP's recommendations and instructed that they be implemented.

Two significant shuttle upgrade programs include the Alternate Turbopump (ATP) program and the superlightweight external tank. The most challenging and potentially troublesome components of the shuttle's main engines are the high pressure turbopumps. The ATP program is addressing the shortcomings of the current turbopumps and includes development of an ATP liquid oxygen (LOX) pump and the ATP fuel pump. The July 1995 flight of *Discovery* included the new LOX pump in an upgraded main engine. The new LOX pumps are designed to need detailed inspection every 10 flights versus every flight for the old pumps. The new fuel pump was scheduled to first fly on the shuttle in FY1997, but technical problems have delayed it to late 1998. Development of a superlightweight external tank began in FY1994 to enable the shuttle to take sufficient payload weight to the orbit that will be used for the International Space Station (ISS). Prior to Russia joining the ISS program, it would have been placed in an orbit easily accessible from Cape Canaveral (inclined 28.5 degrees to the equator). Attaining the new orbit, inclined at 51.6 degrees, requires more fuel. Thus, weight had to reduced elsewhere on the shuttle and, as one measure, NASA decided to develop a lighter version of the external tank using an aluminum-lithium alloy. The new tank saves 8,000 pounds. The first flight of the tank took place on June 2, 1998 on STS-91, the final shuttle-Mir docking mission.

Future Launch Vehicle Development Programs

Despite the hopes of the 1970s that the space shuttle would reduce the cost of reaching orbit, operation of U.S. launch systems remains expensive and less efficient and reliable than DOD and NASA would like. In addition, these systems have been losing market share in the international commercial launch arena. Thus, efforts continue to find a way to reduce costs for both expendable and reusable U.S. space launch systems. DOD and NASA initiated several efforts in the late 1980s and early 1990s to develop a new ELV system (see below), but each was terminated in turn because Congress or the agencies themselves were not convinced that the required investment had sufficient priority in a time of constrained budgets. Other government efforts to develop cutting-edge technology for new launch concepts have also been discontinued (see below).

In response to the Administration's August 1994 National Space Transportation Policy, DOD and NASA initiated two new launch vehicle R&D programs. DOD began the Evolved Expendable Launch Vehicle (EELV) program and NASA initiated a Reusable Launch Vehicle (RLV) program. Both agencies also are continuing some work on hypersonic technologies for future high-speed cruise aircraft and launch vehicles.

DOD's Evolved Expendable Launch Vehicle (EELV) Program. The most recent effort to develop advanced ELV systems, the Evolved Expendable Launch Vehicle (EELV) program, was initiated in FY1995.

As noted, EELV is the successor to several failed attempts. President Bush's 1991 National Space Launch Strategy began a joint DOD/NASA ELV development program known as the National Launch System (NLS). The NLS concept was based on a series of DOD studies initiated in 1985 as the Space Transportation Architecture Study. This study was supplemented by studies of a "Shuttle-C" derivative of the space shuttle performed by NASA from 1987 to 1990, and a joint DOD/NASA Advanced Launch System (ALS) program. ALS was originally conceived as an evolutionary family of increasingly capable launch vehicles to be available in the early 1990s, but changed into a technology development program (called the Advanced Launcher Development Program — ALDP) with no specific launch vehicle. When the NLS program began in 1991, it consisted of plans for three launch vehicles (NLS-1, NLS-2, and NLS-3) and supporting launch infrastructure. DOD and NASA estimated NLS costs through the year 2005 at $15 billion. In its FY1993 budget deliberations, however, Congress directed DOD to terminate the NLS program.

A launch capability study released by the Bush Administration's National Space Council in November 1992 recommended that the Air Force instead develop a new medium-sized ELV, dubbed "Spacelifter." This new launch vehicle was similar in capability to the envisioned NLS-3 vehicle. However, in the spring of 1993, DOD underwent a "Bottom-Up Review" to evaluate its programs in order to meet long-range budget reduction targets. The review did not support starting the acquisition process for Spacelifter because of the estimated $6-10 billion cost of the program and the availability of current ELVs that already met DOD's needs.

During its FY1994 budget deliberations, Congress did not provide any funds to begin Spacelifter. Nevertheless, the FY1994 DOD Authorization Act (**P.L. 103-160**) required DOD to develop a Space Launch Modernization Plan to upgrade its space launch capabilities. Known as the Moorman study, it was submitted to Congress in April 1994. Later in 1994, the Administration's National Space Transportation Policy directed DOD to assume the lead responsibility for improvement and evolution of current expendable launch vehicles.

For FY1995, Congress authorized $50 million to initiate a competitive "evolved" ELV program to replace existing ELV launch capabilities. The FY1995 DOD Appropriations Act (**P.L. 103-335**) appropriated $30 million to begin development of a new family of medium- and heavy-lift launch vehicles evolved from existing technologies. For FY1996, in response to the FY1995 congressional direction and the Administration's Space Transportation Policy, DOD requested funding for what it calls the Evolved Expendable Launch Vehicle (EELV) program. The Air Force awarded four $30 million contracts on August 24, 1995, for 15-month studies of EELV concepts to Lockheed Martin, Boeing, Alliant Systems, and McDonnell Douglas. On December 20, 1996, the Air Force announced that Lockheed Martin and McDonnell Douglas (later bought by Boeing) each won 17-month pre-engineer-ing and manufacturing development contracts worth $60 million as part of the EELV program. According to original Air Force plans, one of these companies would have been selected in 1998 to develop the EELV with the first test of an medium-lift vehicle planned for 2001. Revised Air Force plans have both contractors being chosen to develop two separate EELV systems.

DOD requested and received $44.46 million for EELV in FY1997. For FY1998, $91.64 million is available for the program (**P.L. 105-56**). For FY1999, the Air Force is requesting $283.6 million for the program.

NASA's Reusable Launch Vehicle (RLV) Program. The Clinton Administration's National Space Transportation Policy gave NASA the lead responsibility for technology development and demonstration for a next-generation reusable space transportation system, such as the single-stage-to-orbit (SSTO) concept. In response, NASA initiated the Reusable Launch Vehicle (RLV) program to develop experimental RLVs and associated technology to form the basis for next-generation vehicles to replace the space shuttle system and ELVs. The SSTO concept involves a rocket that can attain orbit with only one stage (instead of two or more as is common today) carrying a crew or cargo. The goal is to have this vehicle capable of returning to Earth, being serviced quickly, and flying again within a very short time. Proponents believe that RLV, particularly SSTO, technology could dramatically lower the cost of accessing space. Lower launch costs could increase the future competitiveness of the U.S. commercial launch industry.

One feature of the RLV program is that NASA wants industry, not government, to eventually lead this program. NASA is trying to establish new forms of cooperation with private industry by sharing the costs of developing RLV technology, with the intent that industry will take over development, operation, and financing of the operational vehicle after the year 2000. The RLV program includes ground-based development and testing of reusable launch technologies and experimental flight test vehicles known as the X-33 and the X-34. NASA tests of another vehicle, DC-XA, have been discontinued (see below). The RLV program is estimated to cost NASA $1.25 billion over 5 years.

Originally, the RLV program that became known as DC-XA was conducted by DOD's Ballistic Missile Defense Organization (BMDO). Under contract to DOD, McDonnell Douglas built a 1/3 scale suborbital prototype rocket called the Delta Clipper Experimental (DC-X) that took off and landed vertically. Several flight tests were conducted successfully in 1993. During a flight test in June 1994, the rocket was damaged by an explosion but landed safely. Two test flights of the repaired DC-X succeeded. Because of NASA's designated role as lead agency for reusable launch vehicle technology and BMDO's reluctance to fund continued testing because it did not fit BMDO's role and mission, the DC-X was transferred to NASA. After finishing its test flights in 1995, the DC-X was upgraded to begin a NASA-sponsored flight test program and became known as DC-XA (later renamed Clipper Graham). Four test flights occurred in 1996. On July 30, 1996, the Clipper Graham was severely damaged when it tipped over and burned upon landing. On January 7, 1997, NASA released an investigative report concluding that the accident's primary cause was that the break line on the helium pneumatic system for landing gear 2 was not connected and therefore that landing gear did not extend. As a result the vehicle toppled onto its side during landing, exploded, and burned. The report further stated that failure to consider operations as part of McDonnell Douglas's rapid-prototyping guidelines, and a management approach that empowered ground personnel to make decisions but without putting an individual in charge of the total process led to the accident. NASA spent $31 million on the DC-XA program.

In January 1995, NASA released a Cooperative Agreement Notice (CAN) inviting industry to develop plans for a two-phase technology development project called X-33 for a large reusable launch vehicle (a companion X-34 program for a smaller vehicle is discussed below). As a technology demonstrator, the X-33 test vehicle will not need to demonstrate on-orbit performance. For phase-1 of the project, NASA signed three cooperative agreements with industry to design the X-33 on March 29, 1995. NASA provided approximately $7 million each to the participants during this design phase, with each company investing approximately a matching sum. A CAN for phase-2 was released April 1, 1996, and on July 2, 1996, Vice President Gore announced that NASA had selected Lockheed Martin to build the X-33 test vehicle. Under the agreement, Lockheed Martin is to design, build, and conduct the first test flight of the X-33 by March 1999 and conduct at least 15 suborbital test flights by December 1999 (as noted below, the schedule for the first flight has been delayed). The Lockheed Martin design is based on a lifting body shape and uses a new "aerospike" engine. The vehicle would be launched vertically like a rocket and land horizontally like an airplane. NASA has budgeted $941 million for the project through 1999. Lockheed Martin is to invest $220 million in its X-33 design.

If the X-33 technology development program is successful technically and economically, the goal is then to build a small fleet of operational vehicles to replace the space shuttle in the 2007-2012 time frame. Lockheed Martin calls this operational vehicle *VentureStar*. The announced goal is for the private sector to pay for development of operational vehicles, not government, but some are skeptical that industry will provide that level of investment. Whether to proceed with an operational vehicle or instead to make major upgrades to the space shuttle is a decision NASA expects to make at the turn of the century.

The progress of the X-33 program has come under scrutiny. In June 1997, Lockheed Martin announced that it would postpone the Critical Design Review (CDR) a few months because of technical problems in the program. As a result, test flights were postponed from March 1999 until the summer of 1999, but NASA is hopeful that all 15 tests still can be completed by the end of 1999. The CDR was successfully completed October 31.

NASA also released a CAN in January 1995 for a smaller RLV demonstrator called the X-34. In large measure because the X-34 is a two-stage-to-orbit demonstrator (which is less technically complex than SSTO), the X-34 was on a faster track to completion than the X-33. On March 30, 1995, NASA signed a cooperative agreement with Orbital Sciences Corp. (OSC), teamed with Rockwell International Corp., to jointly develop the X-34. OSC, however, issued a stop work order to its contractors on January 24, 1996 because of design problems and weight and cost growth. On February 15, 1996, OSC announced that it was formally dropping out of the X-34 project. One week earlier, Rockwell had decided to leave the partnership. NASA officials decided to continue the X-34 project by using the $65 million in remaining funds for the project and build a scaled-down version of the X-34. A NASA Research Announcement (NRA) was released in March 1996 for a revised X-34 demonstrator that would fly in 1998. On June 10, 1996, NASA announced that it had selected OSC to undertake the revised X-34 project. The contract was finalized August 30, 1996. The 30-month contract is valued at $50 million with an additional $10 million that will be spent by NASA in direct support of the project. Two flight tests were scheduled to begin in the fall of 1998 at White Sands Missile Range in New Mexico, but the tests have been delayed until March 1999. NASA may exercise a contract option for up to 25 additional test flights per year. The X-34 project originally included only one test vehicle, but NASA modified its contract with OSC in January 1998 to now include two test vehicles.

In FY1997, Congress appropriated the full request (P.L. 104-204), $266 million, for RLV. For FY1998, NASA requested $353.5 million for the RLV program ($333.5 million for the X-33 and $20 million for the X-34). On April 24, 1997, the House passed H.R. 1275, the FY1998-1999 Civilian Space Authorization Act, authorizing the full request for the RLV program. The bill also provides an unrequested $300 million authorization for FY1998 and

$425 million in FY1999 for a new experimental demonstration vehicle that would be a "complementary follow-on" to the X-33. The additional funding was not included in FY1998 VA-HUD-IA appropriations act (P.L. 105-65) or the Senate version of the authorization bill (S. 1250, S.Rept. 105-195). NASA's request for the RLV program in FY1999 is $338.8 million.

Since NASA is the lead agency for RLV development, no DOD funds have been requested for the program. Congress nevertheless appropriated $25 million for Air Force involvement in the RLV program in FY1996 (P.L. 104-61) and $10 million in FY1997 (P.L. 104-208 — the FY1997 DOD authorization act, P.L. 104-201, had authorized $25 million). None was added for FY1998.

U.S. Commercial Launch Services Industry

Congressional Interest

As discussed earlier (see **U.S. Launch Vehicle Policy**), Congress has passed three laws intended to assist the commercial spaces launch services industry: the 1984 Commercial Space Launch Act, the 1988 Commercial Space Launch Act Amendments, and the 1990 Launch Services Purchase Act. Congress continues to be interested in the issue, although no new laws have been passed since 1990. Two bills were introduced in the 104th Congress pertaining to the commercial launch industry. Neither was enacted. (H.R. 1953, to provide tax incentives for individuals and firms investing in commercial space activities; and H.R. 3936, containing provisions regarding licensing of re-entry vehicles, requiring the federal government to procure commercial launch services, and establishing requirements for the use of excess intercontinental ballistic missiles as space launch vehicles. H.R. 3936 was passed by the House September 17, 1996. No further action was taken.) The House included language in the Omnibus Civilian Science Authorization Act of 1996 (H.R. 3322) to amend the Commercial Space Launch Act, *inter alia*, to give regulatory jurisdiction over the reentry (in addition to launch) of spacecraft to the Department of Transportation (DOT). The bill did not pass the Senate. The Senate included similar language in its version of the FY1997 FAA authorization act (S. 1994, P.L. 104-264), but it was deleted during conference (H.Rept. 104-848). (FAA is the part of DOT that regulates commercial space launch activities.)

H.R. 1702, the Commercial Space Act, was introduced on May 22, 1997, and passed the House November 4, 1997. The bill includes, *inter alia*, amendments to the Commercial Space Launch Act dealing with re-entry vehicles and provisions on federal acquisition of space launch services. H.R. 1702 was reported by the Senate Commerce Committee June 2, 1998. S. 1473 was introduced by Senator Bob Graham on November 8, 1997. Some provisions of its provisions that were not in the House version of H.R. 1702 were incorporated into the Senate-reported version of H.R. 1702. Concerns about whether the private sector will be able to finance space launch vehicle projects led to introduction of S. 2121 by Senator Breaux on May 22, 1998. The bill would create a $400 million federal loan guarantee program administered by NASA for companies building low cost commercial space launch vehicles (but not the EELV program).

Foreign Competition

Congress and the Administration have taken many steps to facilitate the success of U.S. commercial launch services companies, including attempts to limit foreign competition and/or ensure a level playing field in the international launch services market. Europe, China, Russia, Ukraine and Japan offer commercial launch services in competition with U.S. companies. Most satellites requiring launch are made by U.S. companies or include U.S. components, and hence require export licenses if their destination is a country like China, Russia or Ukraine. This gives the United States considerable influence over whether these countries can participate in the commercial launch services market. The United States has negotiated trade agreements with China, Russia and Ukraine on "rules of the road" for participating in this market. The September 1996 Clinton National Space Policy states that once those agreements expire, it is a U.S. goal to move from negotiated trade to a "trade environment characterized by the free and open interaction of market economies." How this will affect countries that continue to have non-market economies at that time is unclear.

Europe. The European Space Agency (ESA) developed the Ariane family of launch vehicles. Ariane was first test-launched in 1979, and began operational launches in 1982. ESA continues to develop new versions of Ariane. ESA conducted a successful test launch of the new Ariane 5 on October 30, 1997 (the first test, in 1996, was a failure). Operational launches are conducted by the French company Arianespace, which is owned by the French space agency (CNES) and 51 European aerospace companies and banks. Arianespace is the main competitor to U.S. launch services companies. Ariane 4 is the only version currently operational. Separately, Arianespace markets Russia's Soyuz launch vehicle as part of a French-Russian joint venture called Starsem.

In 1985, a U.S. company (Transpace Carriers Inc.) filed an unfair trade practices complaint against Arianespace, asserting that European governments were unfairly subsidizing Ariane. The Office of the U.S. Trade Representative (USTR) investigated and found that Europe was not behaving differently from the United States with regard to pricing commercial launch services (then offered primarily on the government-owned space shuttle). The incident did raise the question of what "rules of the road" to follow in pricing launch services. In the fall of 1990, USTR and Europe began holding trade talks to establish such rules of the road and to look at how the United States and Europe should respond to the entry of non-market economies into the launch services business. The only formal negotiating session was held in February 1991.

Each side has been concerned about how much the respective governments subsidize commercial launch operations, but another controversial topic (not formally part of the talks) is whether Arianespace should be able to bid for launches of U.S. government satellites, which now must be launched on U.S. launch vehicles as a matter of U.S. policy. Arianespace wants that restriction lifted. France and other European governments do not have written policies requiring the use of Ariane for their government satellites. However, to support the Ariane program, the member governments of ESA agreed from the beginning to pay a surcharge of as much as 15- 20% if they did choose Ariane. In 1992, the ESA member governments also passed a resolution strongly encouraging their governments to use Ariane (but not *requiring* its use). Nevertheless, the surcharge led some cost-conscious European governments to buy launch services from other (notably U.S.) suppliers. In the fall of 1995, ESA's member governments reached agreement with Arianespace to reduce the surcharge to encourage use of Ariane. (ESA itself does give preference to Ariane.)

China. The People's Republic of China offers several versions of its Long March launch vehicles commercially. China poses special issues because of its non-market economy, meaning it can offer comparatively low prices on the basis that its labor costs are low. Technology transfer and political concerns have arisen as well. Launch services are offered through China Great Wall Industry Corp. (CGWIC).

In September 1988, the U.S. government agreed to grant three export licenses for satellites manufactured by Hughes to be launched by CGWIC. Two were Optus communications satellites (formerly called AUSSAT) built for Australia and the third was Asiasat 1, owned by the Hong Kong-based Asiasat Co. (of which China's International Trust and Investment Corp. is a one-third owner). The Reagan Administration granted the export licenses on the conditions that China sign three international treaties related to liability for satellite launches and other subjects; agree to price its launch services "on a par" with Western companies; and establish a government-to-government level regime for protecting technology from possible misuse or diversion. China met the conditions and the two countries signed a 6-year agreement in January 1989. The now-defunct Coordinating Committee on Multilateral Export Controls (COCOM) approved the licenses that March.

On June 5, 1989, after the Tiananmen Square uprising, President Bush suspended all military exports to China. At the time, exports of communications satellites were governed by the State Department's Munitions List. The 3 satellites therefore counted as military exports and the licenses were suspended. Then Congress passed language in the FY1990 Commerce, Justice, State and Judiciary appropriations (P.L. 101-162) and the 1990-91 Foreign Relations Authorization Act (P.L. 101-246, Section 902) prohibiting the export of U.S.-built satellites to China unless the President reported to Congress that (1) China had achieved certain political and human rights reforms, or (2) it was in the national interest of the United States. In December 1989, President Bush notified Congress that export of the satellites was in the national interest and the licenses were reinstated. Asiasat-1 became the first commercial launch by China in April 1990. Final export approval for Optus 1 and 2 was granted in April 1991 (they were launched in 1992, but Optus 2 did not reach orbit intact).

A different issue arose in 1990. China signed a contract to launch an Arabsat Consortium satellite for $25 million, much less than what many consider "on a par" with Western companies. The main competitor was Arianespace, which turned to both the French and U.S. governments to prohibit export of the satellite (the prime contractor for the satellite was French and it included American components). No formal action was taken by the United States. In 1991, the Arabsat Consortium terminated the contract with the Chinese and signed an agreement with Arianespace, so the case became moot, but the issue of what constituted "on a par" remained. China argued that because their costs are so low, they could offer lower prices and still adhere to international norms as to what costs are included in setting the price.

Yet another issue arose in 1991 — linkage of satellite export licenses with U.S. concern over China's ballistic missile proliferation policies. On April 30, the Bush Administration approved final export licenses for Optus 1 and 2, and for U.S. components of a Swedish satellite called Freja (launched by China in October 1992). To emphasize its concern about

Chinese missile proliferation, however, the White House disapproved export of U.S. components for a satellite China itself was building (Dong Fang Hong 3). Then, on June 16, the White House announced that it would be "inappropriate for the United States to approve any further export licenses for commercial satellite launches at this time." On July 17, the State Department identified CGWIC as one of two Chinese entities engaged in missile technology proliferation activities that require the imposition of trade sanctions in accordance with the Arms Export Control Act, including denial of license applications for export items covered by the Missile Technology Control Regime (MTCR). Although the MTCR does not cover satellites (only satellite launch vehicles, which are close cousins of ballistic missiles), the identification of CGWIC as a cause of concern to the U.S. Government complicated China's marketing plans. China then agreed to adhere to the MTCR, and the sanctions were lifted on February 21, 1992.

China's fortunes improved. In May 1992, the International Telecommunications Satellite Organization (Intelsat) agreed to launch at least one of its satellites on a Chinese launch vehicle. On September 11, 1992, the State Department notified Congress that it was waiving legislative restrictions on U.S. exports for six satellite projects with China: APSAT, Asiasat-2, Intelsat 7A, STARSAT, AfriStar, and Dong Fang Hong 3. The first five were satellites China wanted to launch; the sixth was the Dong Fang Hong 3 components for which export was disapproved in April 1991. (The satellite was launched in 1994, but failed once it was in orbit). Many observers saw the move as a conciliatory gesture in the wake of the U.S. decision to sell F-16s to Taiwan.

On August 25, 1993, however, the U.S. government again imposed sanctions against China for ballistic missile proliferation activities, and the State Department said that satellite exports would not be permitted. The State Department announced October 4, 1994 it would lift the sanctions after China pledged to abide by the MTCR. During this period of time, tensions were particularly acute in the United States between those who view the sanctions as harmful to U.S. business interests (notably satellite manufacturers Hughes and Lockheed Martin), and those who want to prevent sensitive technology from reaching China and/or to punish China for MTCR infractions. The debate centered on whether the satellites should continue to be governed by export guidelines of the State Department (Munitions List) or the Commerce Department (Commerce Control List). Following extensive debate (including a House Science, Space and Technology hearings on September 29, 1994), in October 1996 the Clinton Administration transferred primary responsibility for export of commercial communications satellites from the State Department to the Commerce Department.

The 6-year agreement signed in 1989 expired at the end of 1994. A new 7-year agreement was signed on March 13, 1995 allowing China up to 11 new launches for international customers to geostationary orbit (in addition to launch contracts already signed). China agreed to price launches no less than 15% below what Western companies charge or a U.S. review of the price would be triggered. For low Earth orbit (LEO) launches, it was agreed that as long as China and Russia combined do not win contracts to launch more than 50% of any particular LEO satellite constellation (such as Motorola's Iridium system), China will be assumed to be in compliance with the agreement. President Clinton approved the export of four satellites to China for launch (2 COSAT satellites , Chinasat 7, and Mabuhay) on February 6, 1996, despite concerns about China exporting nuclear weapons-related equipment to Pakistan. The COSAT satellites have been renamed Chinastar and are built by Lockheed Martin. Chinasat 7 was built by Hughes, and Mabuhay (later renamed Agila 2) by

Loral. On February 14, 1996, a Long March 3B rocket carrying an Intelsat communications satellite built by Loral malfunctioned seconds after liftoff, breaking apart in flight, then impacting the ground and spreading debris and toxic fumes over the launch site and a nearby village. The Chinese reported 6 dead and 57 injured, but other reports suggested a higher figure. This was the second Chinese launch failure involving fatalities. The first, of the Hughes-built Apstar-2 satellite in January 1995, killed 6 and injured 23. Some customers, including Intelsat, subsequently canceled contracts with China, but what long term impact these accidents will have is unknown. On June 25, 1997, the President notified Congress that it is in the national interest to export another satellite (Asia Pacific Mobile Telecommunication, built by Hughes) to China.

In May 1997, USTR stated that it believed China violated the pricing provisions of the bilateral agreement for launching a Loral-built satellite for the Philippines (Agila 2, formerly called Mabuhay). Chinese officials disagreed. On September 10, 1997, the Washington *Times* published a story that Chinese and Russian entities (including CGWIC) are selling missile technology to Iran. China and Russia deny the allegations. On October 27, the United States and China agreed to pricing provisions for launches of satellites to low Earth orbit (LEO). Only general language had been included in the 1995 agreement. China will price its LEO launches on a par with U.S. and European companies, allowing for adjustments for certain differences. On February 18, 1998, the President notified Congress that it was in the national interest to export another satellite (Chinasat 8, built by Loral) to China. China successfully launched the Lockheed Martin-built Chinastar-1 on May 30, 1998.

The "Loral/Hughes" Issue. On April 4, 1998, the New York *Times* reported that a 1997 classified DOD report alleges that two U.S. satellite manufacturing companies, Space Systems/Loral (part of Loral Space & Communications) and Hughes Electronics (a subsidiary of General Motors), provided technical information to China that improved the reliability of Chinese nuclear missiles. The assistance was provided in the wake of the February 1996 Intelsat launch failure (see above). That Intelsat satellite was built by Loral, which participated in an inquiry into the accident at the request of insurance companies seeking assurances that the Chinese had correctly diagnosed and solved the cause of the failure. Loral formed a review committee that, according to Loral, included representatives of other satellite companies, including Hughes. According to Loral, the review committee did not itself investigate the accident, but listened to Chinese officials explain their investigation and then wrote a report. Loral concedes that a copy of the report was given to the Chinese before it was provided to the U.S. State Department in violation of Loral's internal policies. Loral says it notified the State Department when it learned that the Chinese had been given a copy. According to media sources, DOD's 1997 report says that the companies provided technical information in violation of the export license that allowed the export of the satellite to China for launch. The companies insist they did nothing that violated the export license. The Justice Department is investigating the allegations. (CRS Report 98-485 provides a chronology of activities related to satellite exports to China in the Loral/Hughes context.)

Several congressional committees are holding hearings on the issue: the Joint Economic Committee, April 28; Senate Governmental Affairs May 21; Senate Intelligence Committee June 4, June 5 and June 10; and Senate Armed Services June 11. Senate Majority leader Lott has established a task force to coordinate Senate hearings on the matter by the Armed Services, Foreign Relations, Governmental Affairs, and Intelligence committees. The is establishing a special committee chaired by Representative Cox.

On May 20, the House adopted four amendments to the FY1999 DOD authorization bill (H.R. 3616) affecting satellite exports to China. The Spence amendment (adopted 417-4) states, *inter alia,* that U.S. business interests should not be placed above national security interests; that at the planned June 1998 summit, President Clinton should not support China's membership in the MTCR, agree to a blanket waiver of suspensions regarding the export of U.S. origin satellites to China, agree to increase the number of launches to geostationary orbit by China, support any cooperative project with China to design or manufacture satellites, or enter into new or amend existing scientific or other agreements related to space or missile-related technology; that the February 1998 decision to export another Loral satellite to China was not in the national interest considering the ongoing Justice Department investigation; that the executive branch should ensure that U.S. export laws regarding satellite exports to China are enforced and the Justice Department investigation proceeds with all due dispatch; and the President should indefinitely suspend satellite exports to China including those licensed in February 1998. The Bereuter amendment (adopted 414-7) prohibits U.S. persons from participating in investigations of Chinese launch failures of U.S. origin satellites. The Hefley amendment (adopted 412-6) prohibits exports of missile equipment and technology to China. The Hunter amendment (adopted 364-54) prohibits export and reexport of U.S. origin satellites to China that as of the date of enactment of the Act have not been exported or reexported whether or not an export license has been approved, and also prohibits export or reexport of information, equipment, or technology that could be used in the acquisition, design, development, or production of any satellite or launch vehicle.

On May 21, the House adopted a Spence en bloc amendment (voice vote) that included, *inter alia,* a provision that would return commercial communications satellites to jurisdiction of the State Department's Munitions List. As noted, in 1996 the Administration transferred responsibility for those licenses from the State Department to the Commerce Department. A January 1997 GAO report (GAO/NSIAD-97-24) examines that decision. Critics of the decision now assert that the Departments of State and Defense objected to the transfer of responsibility because of national security concerns. White House spokesman Michael McCurry has characterized the objections as an "interagency bureaucratic squabble."

Russia. Russia, Ukraine, and Kazakstan are the three most important former Soviet republics involved in space activities: the two major launch sites, Plesetsk and Tyuratam (also called Baikonur), are located in Russia and Kazakstan, respectively; a major rocket production facility is in Ukraine (and Ukraine itself offers commercial launch services).

Following the collapse of the Soviet Union, interest developed in the possibility of loosening U.S. policy to permit export of U.S.-made satellites for launch on Russian launch vehicles (previously, U.S. policy flatly prohibited the export of satellites to the Soviet Union for launch). In June 1992, President Bush opened the door by announcing that he would not oppose Russia bidding for launch of an International Maritime Satellite Organization (Inmarsat) satellite; Inmarsat subsequently signed an agreement for such a launch for $36 million (the launch took place in 1996). He also stated that the United States would begin negotiations with Russia over "rules of the road" for future commercial launches. Discussions were held in the fall of 1992, and negotiations were held in Moscow May 5-6, 1993 where the U.S. and Russian governments reached agreement in principle whereby Russia could contract for up to eight launches to geostationary orbit (the largest segment of the market today) through the year 2000, plus three launches to low Earth orbit (potentially a large market in the future). Russia agreed to price its launch services no more than 7.5% lower than Western prices, or a U.S. review of the price would be triggered.

The U.S. government stressed that it would not sign the agreement unless Russia agreed to comply with the MTCR, specifically in a case involving a Russian company, Glavkosmos, that announced in 1992 that it would sell rocket engine technology to the Indian Space Research Organization (ISRO). The United States declared it was a violation of the MTCR and imposed 2-year sanctions against Glavkosmos and ISRO (the sanctions expired in May 1994). In June 1993, the United States threatened to impose sanctions against other Russian companies that do business with Glavkosmos. The two countries finally agreed that Russia would cease transferring rocket engine technology to India (the sale of engines themselves is not an issue, but rather the technology that would allow India to build its own such engines), paving the way for the agreement on launch services which was signed in September 1993. As noted, on September 10, 1997, the Washington *Times* published a story that Russian and Chinese entities (including the Russian Space Agency) are selling missile technology to Iran. Russia and China deny the allegations. At a January 30, 1998 press conference in Washington, Russian Space Agency Director Yuri Koptev stated that the Russian government takes its commitment to the MTCR seriously, and of 13 instances where concerns had been raised about possible MTCR violations, only 2 were found by the Russian government to have merit and both were stopped before any technology was transferred.

On January 30, 1996, the United States agreed to increase from 9 to 16 the number of geostationary launches Russia can conduct, putting Russia on a more even footing with China and Ukraine, and the price Russia can charge (like China) is now 15% below Western prices without triggering a review. There also is the potential to add 4 more launches to the 16 total if the right market conditions exist. Lockheed Martin has a joint venture with two Russian companies to market Russia's Proton launch vehicle. The first Russian launch of a Western commercial satellite (Astra) occurred in April 1996.

Ukraine. Ukraine has a major rocket production facility (though some of the rocket engines are produced in Russia) and offers commercial launch services. The United States commenced negotiations with Ukraine in April 1995 and an agreement was signed in February 1996 that permits up to 20 Ukrainian commercial space launches to geostationary orbit through 2001. Since Ukraine has no launch sites, a joint venture called Sea Launch was created by Boeing, Ukraine's Yuzhnoye, Russia's Energomash, and Norway's Kvaerner to launch Ukrainian Zenit boosters from a mobile ocean oil rig that would be based in the Los Angeles area and towed out into the Pacific for launches. Of the up to 20 permitted launches, Sea Launch would get 11, with 3 more if the market grows. Five other launches of Zenit or another Ukrainian vehicle, Cyclone, are allowed, with one more if the market grows. Separately, Ukraine signed an agreement with the U.S. company Globalstar for three launches of Globalstar low Earth orbit (LEO) communications satellites on Zenit. These launches are scheduled to take place from Baikonur. Also, Rockwell (now Boeing) signed an agreement to market Ukrainian Cyclone launchers, which are used for LEO launches. The market for LEO launches is still evolving and the Ukrainian agreement contains only general guidelines regarding LEO launches, not numerical limits.

Japan. On February 3, 1994, Japan successfully conducted the first launch of its H-2 launch vehicle, the first all-Japanese rocket capable of putting satellites in geostationary orbit. Previous rockets launched by Japan for this purpose were based on U.S. technology and a

1969 U.S.-Japan agreement prohibited Japan from launching for third parties without U.S. consent. With the entry into service of the H-2 rocket, Japan is freed from that constraint and thus may compete in the launch services market. Japan created a private launch vehicle company in 1990 (Rocket Systems Corp., or RSC) to offer commercial launch services. In June 1997, the Japanese government reached agreement with the fishing industry to allow more opportunities for launches from the Tanegashima launch site (used for the H-2). Fishermen must evacuate the area near the launch site during launches. The new agreement extends from 90 to 190 the number of days per year that launches may be conducted, and permits up to eight launches a year instead of only two. The Japanese do not consider the current version of the H-2 to be cost competitive, and are working on a new version, the H2A. RSC has signed contracts with two U.S. satellite builders, Hughes and Loral, for 10 launches each using the H2A between 2000 and 2005.

LEGISLATION

H.R. 1275 (Rohrabacher)/S. 1250 (Frist)

Civilian Space Authorization Act. H.R. 1275 (for FY1998-1999) reported from Science Committee April 21, 1997 (H.Rept. 105-65); passed House April 24. S. 1250 (for FY1998-2000) reported from the Senate Commerce Committee May 22, 1998 (S.Rept. 105-195).

H.R. 1702 (Sensenbrenner)/S. 1473 (Graham)

Commercial Space Act of 1997. H.R. 1702 reported from Science Committee (H.Rept. 105-347) October 24, 1997; passed House November 4, 1997; reported by Senate Commerce Committee June 2, 1998 (S.Rept. 105-198). S. 1473 introduced and referred to Committee on Commerce November 8, 1997.

H.R. 3616 (Spence)/S. 2057 and S. 2060 (Thurmond)

FY1999 Department of Defense authorization act. H.R. 3616 reported from House National Security Committee May 12 (H.Rept. 105-532); passed House May 21. S. 2057 reported from the Senate Armed Services Committee May 11 without written report. It authorizes funding for military activities of the Department of Defense, military construction, defense activities of the Department of Energy, and prescribes personnel strengths for the Armed Forces. Programs in S. 2057 are also split into three other bills: S. 2058 (DOE defense activities), S. 2059 (military construction), and S. 2060 (DOD military activities and personnel strengths). S. 2060 contains the programs discussed in this report and was reported from Senate Armed Services Committee May 11 (S.Rept. 105-189). The Senate began consideration of the omnibus bill, S. 2057, on May 14.

S. 2121 (Breaux)

Space Launch Cost Reduction Act. Introduced May 22, 1998; referred to Committee on Commerce, Science, and Transportation.

S. 2132 (Stevens)

FY1999 DOD appropriations act. Reported from Senate Appropriations Committee June 4 (S.Rept. 105-200).

SELLING CHINA THE ROPE . . .

Clinton Didn't Start It, But He Sure Made It Worse

By Henry Sokolski

Presidential spokesman Mike McCurry last week justified the Clinton administration policy that allowed the transfer of satellite technology to the Chinese military with the hoary "they started it" defense. "This administration," said McCurry, "has pursued the exact same policy pursued by the Bush administration."

This is not really a defense of the policy, of course, but is it true? Republican officials, as we shall see, were not without sin. But you might say that they worried enough to go to confession: They tried to control against the leaking of sensitive technology in their dealings with China by at least monitoring and limiting the transactions. Not so the Clinton administration, which from 1993 on not only showed contempt for enforcing existing satellite controls but loosened them so as to make it all but impossible to know whether they were being violated. You might say they not only skipped confession, but burned the church down.

Today's controversy surrounds what the Chinese have managed to learn through launching satellites made by two American companies, Loral Space and Communications and Hughes Electronics. Details of a federal grand-jury investigation have been leaked to *New York Times* reporter Jeff Gerth and others that make this much clear: In February 1996 a Chinese Long March rocket carrying a Loral-made satellite blew up shortly after liftoff. In an effort to clarify to insurers who was to blame for this accident, analysis done by Loral and Hughes was presented to the Chinese, which the U.S. Defense Department later determined could help China perfect more reliable, accurate, long-range ballistic missiles. (According to a CIA report leaked this spring, 13 Long March missiles with nuclear warheads are aimed at American cities.) The federal grand jury is now trying to determine what, if any, U.S. export-control laws may have been broken.

Focusing on money is exciting, but probably misses the point when it comes to assessing the potential damage done to national security.

This story has exploded because of the tandem revelations that the Chinese military may have made illegal campaign donations to aid Clinton's reelection and that Loral's CEO is a top donor to the Democratic party. Despite Justice Department warnings that he might undermine the grand-jury investigation of Loral, the president went ahead earlier this year and allowed the company to transfer an additional satellite to China. Eager to connect the dots of the scandal, the House last week voted 364 to 54 to suspend all transfers of U.S. satellites to China.

Focusing on the money is exciting, but probably misses the point when it comes to assessing the potential damage done to national security. In fact, not just Loral and Hughes, but Lockheed Martin, Motorola, and Martin Marietta have all worked closely with the Chinese launch industry—work which began not in 1996, but nearly a decade ago in 1989. And all of this history (not just the 1996 Loral-Hughes case) bears investigating. There is no way to judge the administra-

Henry Sokolski, executive director of the Nonproliferation Policy Education Center, was the top official for nonproliferation issues in the Bush Defense Department.

The Nonproliferation Policy Education Center
1718 M Street, N.W., Suite 244, Washington, D.C. 20036 ; phone: 202-466-4406 / fax: 202-659-5429, E-mail: npec@ix.netcom.com

tion's performance in the Loral-Hughes matter without knowing what was attempted by prior administrations.

It was Ronald Reagan, after all, who first allowed the launch of U.S.-made satellites on Chinese rockets, after the *Challenger* space shuttle crash in 1986 deprived the satellite industry of launch alternatives. And it was George Bush who waived Tiananmen Square sanctions to allow the Chinese launch of up to five U.S.-made satellites, three of which—all made by Hughes—were launched before he left office.

If this larger record is examined, three points emerge. First, *all* of our satellite transfers have helped China perfect its military rocketry. China's launching of U.S.-made satellites—worth up to a half-billion dollars in revenue to date—has helped finance China's own missile-modernization efforts and missile exports to nations like Pakistan and Iran. It also has given the Chinese access to U.S. rocket know-how. U.S. contractors have a natural inclination to tutor the Chinese on what they should do to make their crude rockets precise and reliable (they don't want to lose their satellites, which are worth up to 10 times the value of the launcher). Anticipating this, State and Defense officials drew up strict rules in the late 1980s covering precisely what information companies could share with the Chinese. These rules required monitoring of all contractor-Chinese exchanges (including discussions) by a U.S. government rocket-engineer enforcement agent.

Did this prevent militarily useful information from being conveyed to the Chinese? No. But because all exchanges were monitored, there was a clear record of what was conveyed and a concerted effort to keep such transfers to a minimum. Were there infractions? Yes, but when they were reported, senior officials in the Defense and State departments reprimanded the contractors and got them to stop. Yet despite these enforcement measures, a number of key technologies were transferred before 1993. Clean-rooms were constructed in China to assure Hughes's sensitive communications satellites wouldn't be ruined by dust, humidity, or major temperature changes before they were launched. And clean-room technology, as it happens, is also crucial in preparing any advanced system for launch, including reconnaissance satellites and complex warhead packages.

In an attempt to clear up liability for two launch failures in 1992, U.S. contractors also discussed how to improve Chinese payload farings (the nose cone at the rocket's top that shields the satellite) and attitude and engine controls, which fire the rocket's stages and keep them and the payload (either military or civilian) at the precise angles required for proper functioning. Finally, each launch of a Chinese Long March vehicle helped improve the reliability of China's intercontinental ballistic missile fleet, since the rockets are the same.

Republican officials, then, had a spotty record, with the advantage that they worried about it and tried to enforce the law. By the end of the Bush administration, proposals were made to loosen controls over satellite transfers. Whether they would have succeeded no one can know, because the 1992 elections intervened.

The industry, however, correctly sensed that with Clinton's election the time for pushing for decontrol was ripe. Their first step came in late 1993 when they asked the Commerce Department to persuade the White House to drop government monitoring of contractors' discussions with the Chinese. They wanted to share, unimpeded by monitors, a key technology known as "coupling load analysis." The crude Chinese rockets were originally designed to be so rigid that vibration from the rocket's separating stages and engines risked shattering delicate satellites of the sort the U.S. companies would want to launch (and the Chinese would want to develop later on their own). Using coupling load analysis, the Chinese could "soften" their launchers, allowing them to carry more sensitive payloads—be it satellites or the latest in highly accurate, multiple-warhead systems.

The space industry was so eager to share this technology, it lobbied Congress and the executive branch throughout 1993 to be given a free hand to do so. Meanwhile, government monitors continued to file compliance reports on a host of issues. Now, however, their concerns were handled differently: Where before senior State and Defense officials took action, now little or nothing happened. Word got out: Increasingly, industry officials disobeyed government guidance, shared their know-how with the Chinese, and discovered that contempt for the law paid off.

By 1995, the satellites being launched by the Chinese were more sophisticated. One of these, AsiaSat 2, a communications satellite made by Martin Marietta, was to be placed in its orbit with a Chinese solid-rocket kick motor—a final rocket stage strapped to the satellite itself. This kick motor's propellant had to be configured with extreme precision to ensure that it would propel the satellite to an exact point in space and no further and that it would do so without shattering the satellite through vibration or jolts of acceleration.

Martin Marietta and its Hong Kong customers were concerned that the Chinese kick motor might not be capable of such precision. They asked State if they could witness a Chinese test-firing of the motor. Their wish was granted. What's unknown is what, if anything, was then said to the Chinese engineers by the company's foreign staff, who are not bound by U.S. restrictions. Were they briefed by the contractor? Did they speak with the Chinese or otherwise convey U.S. solid-rocket propulsion know-how? We don't know. Why might it matter? Perfecting kick motors can also help in China's development of a warhead-delivery system known among experts as a "post-boost vehicle"—which is designed to penetrate missile defenses. Boosting a satellite up into a precise position in space with a kick motor is little different from blasting warheads off their predictable course down through space and the atmosphere.

The good news in this case is we may have a clue whether this technology was leaked: Industry's campaign to do away with monitoring didn't fully bear fruit until 1996. In 1995, U.S. law still required government monitoring agents, and compliance reports were still being filed. This paper trail and government monitoring work didn't grind to a halt until 1996. That's when President Clinton quietly removed virtually all commercial satellites and related technology from State Department munitions controls (which required official monitors). The responsibility was transferred to the Commerce Department, which (no surprise) trusts industry to monitor itself.

In his defense of the Clinton policy last week, Mike McCurry cited this transfer to Commerce as the one change that distinguished the Clinton administration's policy from Bush administration practices. But the transfer to Commerce was no simple "change." It was tantamount to a complete overthrow of the old export-control regime.

It was under Commerce "controls" that Motorola and Lockheed worked with the Chinese to launch a series of small communications satellites known as Iridium. Two of these satellites at a time were successfully launched on a Long March rocket with a multiple-satellite dispenser of Chinese design. A host of issues about the satellite dispenser were somehow addressed—from proper mounting and release of the satellites to coupling load analysis and attitude control. And all were resolved. The result? China now has mastered a technology virtually interchangeable with that of multiple independently targetable warhead vehicles (MIRV), a delivery system used on America's most advanced intercontinental ballistic missiles. Indeed, the MIRV system that our military uses today was borrowed from dispensers that the commercial-satellite industry first developed.

What's unknown is what, if anything, was said to the Chinese engineers. Were they somehow able to learn about U.S. solid-rocket propulsion know-how?

One could go into greater detail on the potential military significance of our satellite transfers to China. But this much is already abundantly clear: Our national security demands that Congress learn all the facts. This will require going beyond the narrow legal question of whether Loral and Hughes broke the law in 1996. Indeed, allegations of influence peddling by the Chinese and the contractors should not divert attention from the crucial questions raised by a decade of U.S. satellite commerce with China.

Among them are these: Have we already given the Chinese everything of value (in which case, continued satellite commerce could hardly do much harm)? Or is there more that they need or want that we should control and protect? What, if anything, should be done to improve enforcement of controls and assure effective executive-branch backing? Finally, is the spread of missile technology so tied up in the transfer of satellites that we delude ourselves in trying to control their transfer? Would it make more sense to accept this connection and expand such trade, or in the case of China, cut it off entirely?

To get at these questions, Congress will have to hold its own hearings—but it will need the time and depth and expertise that can only come with the creation of an independent commission. The commission and Congress, moreover, are unlikely to get anywhere if U.S. contractors are unwilling to speak freely. Only they know what has actually been transferred to the Chinese since 1996. To encourage them to be forthcoming, Congress and the executive branch should grant contractors immunity from prosecution. Meantime, a moratorium should be placed on further transfers of satellites to China until the commission and

Congress get the answers they need. This will hurt industry only to the extent that it drags its heels in providing information about past transfers.

Certainly, given the seriousness of these matters, it would be shortsighted of Congress to focus exclusively on the political and legal issues surrounding the 1996 Loral case. There is, after all, a broader set of concerns at stake. The president is duty bound to provide for the common defense. Not until we know the truth about the U.S. role in China's missile program can we know whether the Clinton administration has met this most basic obligation. ♦

OFFICE OF THE ASSISTANT SECRETARY OF DEFENSE

WASHINGTON, D.C. 20301-2400

INTERNATIONAL SECURITY AFFAIRS

20 NOV 1991

MEMORANDUM FOR DEPUTY ASSISTANT SECRETARY OF STATE FOR PROLIFERATION

SUBJECT: Position on COMSAT Class Action CJ

I understand that the Space Technical Working Group from the USML/IML Rationalization Exercise is proposing a Class Action Commodities Jurisdiction (CJ) that would transfer licensing of COMSATs launched outside the United States from the Department of State to the Department of Commerce. This action will undermine MTCR controls.

The technology to place satellites in orbit is directly related and can be applied to development of ballistic missiles. Our experience with China has shown that the subsystems used on their LM2E space launched vehicle are similar, and in some cases identical, to those on the M series missile launcher. They have demonstrated with the M series launches how they propose to fix a LM2E failure.

Space is one arena that encompasses more than just the satellite. In order to place a satellite into orbit and maximize his orbital life time, the accuracy and efficiency of the launch vehicle is critical. Any time a satellite is integrated and launched, technical information has to be exchanged that can be used to upgrade launch vehicles and launch processes. Analysis and feedback required to satisfy insurance underwriters for launch of Western satellites results in the transfer of technical knowledge in the areas of increased payload, accuracy and reliability to foreign launch vehicle designers.

To insure technical oversight and stringent MTCR controls, it is our position that State should retain license authority for all satellites launched outside the US. This will provide a clear distinction between satellites sold on orbit and those launched outside the US.

Henry D. Sokolski
Deputy for Non-Proliferation Policy

cc:
Brad Gordon, DoS ACDA/NPP
Richard Clarke, DoS ASS(PM)
Jim LeMunyon, DoC DAS/EA
Jeanne Mintz, OUSD(A)/IP
Peter Sullivan, DTSA

NSA Damage Assessment - Lost Encryption Board from Crash of Loral Satellite - 14 Feb 1996

If the encryption board was reverse-engineered, the knowledge gained could be used to strengthen adversaries' knowledge of COMSEC engineering principles. However, since that crash, we have adopted new space encryption algorithms. Satellites with the older algorithms that are still in orbit use keys that are unique to each vehicle; compromised keys cannot be employed to gain access to or control these satellites.

Department of Defense Damage Assessment
(released after the NSA version)

LAUNCH FAILURE – RECOVERY OF ENCRYPTION DEVICES

Q: There have been reports that sensitive encryption devices were lost during several of the launch failures of US satellites in China? Is this true, and if so, what is the impact?

A: Since 1990, there have been four failures of Chinese launches of US build satellites. Two of the satellites did not have any encryption devices on board. In one of the two cases (the Optus B-2 failure in December 1992), the devices were recovered and returned to the US. In the other case (the Intelsat 708 failure in February 1996), the command processor boxes as a whole were recovered, but not the circuit cards which contained the encryption.

In both cases, the encryption involved embedded, single chip devices. There is some chance that a third-party could reverse-engineer such devices to gain some knowledge about the engineering principles for making such chips. However, any loss of the chips and associated encryption algorithms would have had only minimal impact on national security because the older satellite in Intelsat case used older algorithms for which the keys are unique.

The New York Times

THURSDAY, JUNE 18, 1998

U.S. Rethinking a Satellite Deal Over Links to Chinese Military

Pentagon and State Dept. Raising New Questions

By JEFF GERTH

WASHINGTON, June 17 — Faced with growing criticism of its satellite exports to China, the Clinton Administration is rethinking whether to allow one of the biggest sales to date, a $650 million deal President Clinton quietly approved two years ago.

Government officials said the Pentagon and State Department were raising new questions about whether a Chinese-controlled company with close ties to China's military should be allowed to buy the satellites, which contain some of the United States' most sophisticated communications equipment.

The satellites are the cornerstone of a commercial mobile phone network planned for China and 21 other Asian nations. American officials said their design included a powerful antenna that could eavesdrop on mobile phone calls in China or other countries in the region. It could also be used by the Chinese military to transmit messages through hand-held phones to remote parts of China.

Antennas of these dimensions are a mainstay of the United States' and Russia's eavesdropping satellites and have not previously been exported to China, though a sale to the United Arab Emirates is pending. They also can be used to extend the range of mobile phones.

Mr. Clinton leaves next week for China, and the Administration had hoped to use the trip to showcase a variety of business deals and agreements, including cooperation on civilian satellite and rocket projects. Meanwhile, the House continued investigating the export of space technology today. [Page A10.]

Administration officials said concerns about the pending satellite sale had been deepened by American intelligence reports about Shen Rongjun, the Chinese Army general who oversees his country's military satellite programs. The reports quote the general as saying he planned to emphasize the role of satellites in gathering information.

In an unusual arrangement, Hughes Space and Communications hired General Shen's son, a dual citizen of Canada and China, to work on the project as a manager. The company said it was aware of his familial ties; it is not clear whether the Clinton Administration knew.

Father and son were both directly involved in the project, and American officials said the intelligence reports said the general was pressing his son to move it forward.

The New York Times reported last week that the Chinese military was sending many of its coded messages through American-made commercial satellites sold to Asian companies. China's military satellite network collapsed in 1996, when its first satellites wore out and the replacements failed to work as planned.

President Clinton approved the Hughes project on June 23, 1996, after advisers assured him the communications satellite technology was readily available from European suppliers and would not contribute to Chinese military capabilities.

China already has a burgeoning cellular telephone system, which relies on ground-based transmitters. There are almost 1.5 million cellular phones in Beijing and Shanghai, but the system is less developed in the country's more remote areas, industry officials say.

Donald O'Neal, a spokesman for Hughes, said the satellites were "inherently dual use," meaning that they have both civilian and military potential. "The satellite is not designed for military application," Mr. O'Neal said. "But I don't know how you can prevent it."

The Federal Government could still stop the deal. Mr. O'Neal said Hughes, which is part of Hughes Electronics, a subsidiary of the General Motors Corporation, was waiting for the Commerce Department to review its application to sell the satellite to the Asian consortium, A.P.M.T. or Asia-Pacific Mobile Telecommunications.

Liu Tsun Kie, a spokesman for the consortium, said in a telephone interview from Singapore that the satellite network would be marketed to civilians by regional telecommunications operators. It would be up to Chinese Government regulators, Mr. Liu said, to decide if China's military could use the satellites.

Mr. Liu predicted that the Clinton Administration would eventually approve the deal. "In view of the improving Sino-American relationship, as well as the close rapport established between the U.S. satellite industry and major industry leaders in China and the Asia Pacific," he said, "we are confident that A.P.M.T. will obtain all the necessary approval and export license to insure no delay in satellite launch."

Mr. Liu said the project would attract more than 200,000 mobile phone customers in China within its first two years.

The Two Crucial Steps In a Satellite Sale

Making a satellite sale to China involves two crucial steps that occur simultaneously. Aerospace manufacturers must persuade the President to sign a waiver of the sanctions imposed on Beijing after the Tiananmen Square killings in 1989. Each project requires a separate waiver.

At the same time, companies apply to Federal Government agencies for permission to export specific technologies used in the satellites. Satellite exports to the Chinese military are banned, but sales to Chinese companies are generally allowed, unless they would advance military development in areas like intelligence gathering and nuclear weapons.

Mr. Clinton granted the waiver for the Hughes project two years ago and the company obtained the necessary export licenses. Since then, however, Hughes has changed the design to enhance the satellite's capabilities, requiring it to return to the Government for a new license.

That decision is now before a Government panel headed by the Commerce Department and including officials from the Pentagon, State Department, the Arms Control and Disarmament Agency and the Department of Energy. Each department casts a single vote, with the decision made by majority rule. A dissenting agency can appeal to the President, but that has never happened.

A Commerce Department spokesman declined to discuss the case, saying it involved confidential business information.

Privately, Commerce Department officials are arguing that the deal should go forward because the design approved in 1996 is substantially the same as the current configuration, Administration and Congressional officials said.

But some Pentagon and State Department officials believe the license should face more scrutiny in light of the new information about General Shen and the capabilities of the satellite. Administration officials also said that the increased scrutiny by Congress of the Chinese military and American satellites has prompted officials to pay closer attention to exports to China.

Several Congressional committees are investigating whether the policies on technology exports hurt the national security.

Technical Questions Determine Fate of Deal

The issue turns on highly technical questions. An Administration official who disagrees with the Commerce Department's analysis said the Hughes design is substantially different from what was approved two years ago.

"The antenna sent up the flags," the official said. "It is more powerful than what we have licensed before."

The antenna allows the satellite to receive incoming signals. But a sophisticated antenna, like the one currently under review, can become a listening device that is very effective against ground-based interception efforts, Government reports show.

Before 1996, the Pentagon could easily have stopped the license, because satellites were treated as military items and subject to State Department authority. That year President Clinton shifted jurisdiction to the Commerce Department, easing the controls and lessening the influence of the Pentagon, a senior Government auditor told Congress earlier this month.

A.P.M.T. was organized in the early 1990's. Most of its stock was held by five Chinese state-owned entities: China Satellite Launch and Tracking Control, a unit of Costind, the scientific and research arm of the Chinese military, the China Aerospace Corporation, part of the defense-industrial complex, China Resource Holdings, a trading company that owns a bank in Hong Kong with the Riady family of Indonesia, and subsidiaries of Chinese electronics and telecommunications ministries. A small stake was held by a Singapore company.

In February 1996, the consortium authorized Hughes to proceed with the design and construction of a sweeping mobile satellite telecommunication network that would span 22 countries in Asia and the Pacific, from Pakistan to Indonesia.

China's own space program — both rockets and satellites — was then under severe strain.

A Chinese rocket exploded shortly after liftoff in February. Two months later, engineers from Hughes and Loral Space and Communications were brought in by insurers and China Aerospace to help figure out what went wrong.

The conversations that ensued between the companies and Chinese technicians are now the subject of a criminal investigation, which is seeking to determine whether American export laws were violated. Both companies deny wrongdoing.

While China was trying to repair its rocket program, its satellites began to fail. The first domestically produced satellites, launched by the Chinese military in the early 1990's, were wearing out, and the first replacement, built in cooperation with the German company Daimler-Benz, had failed to achieve proper orbit after its 1994 launch.

In early 1996, all this led China's most senior military official, Gen. Liu Huaqing, to discuss his concern with General Shen, who until a recent reorganization was a senior Costind official and oversees China's satellite and rocket launching programs, American officials said.

General Shen and General Liu have publicly promoted satellite technology as crucial to the future development of China's military capabilities. General Shen has privately assured his colleagues about his ability to fix China's satellite problems and improve the military's surveillance and intelligence-gathering capabilities, American officials said.

At about the same time, there were concerns within Hughes and A.P.M.T. over the how long it was taking President Clinton to make a decision about the deal, Mr. O'Neal and American officials said.

Commercial satellite exports to China have been banned since the killings in Tiananmen Square in 1989, but the President can waive the prohibition, which Presidents George Bush and Clinton have done 20 times.

'Expedited Handling' Of Waiver Was Sought

Hughes officials wanted "an expedited handling" of the waiver in order to meet a contractual deadline, Mr. O'Neal said. And recently released White House documents show that the company hoped to have the President sign off on the deal before Hughes' chairman left China on June 19, 1996.

The staff memorandum that the President relied on to approve the deal made no mention of the Loral-Hughes help for China's rocket program. Three weeks before the memorandum to the President, the State Department had alleged, in a letter to Hughes, that there had been a violation of the arms export control law during the rocket accident review.

The President granted the waiver on June 23.

Soon after the Presidential action, Hughes received a license to export a satellite. Later that summer, Hughes applied for another export license that would allow Shen Jun, the son of General Shen, to work on projects subject to United States export controls, including the A.P.M.T. project, Mr. O'Neal said.

"We applied for and received an export license that allowed him to participate as a translator in the A.P.M.T. preliminary design review," Mr. O'Neal said.

Mr. Shen was hired in 1994 by Hughes for his computer expertise, though the company was also aware of his family ties before he joined the company, Mr. O'Neal said.

General Shen has been involved in the A.P.M.T. project as the overseer of the Chinese launch and tracking company and his son has given Hughes marketing advice about China and technical advice about mobile telephone networks, Mr. Liu and a Hughes executive said.

Mr. O'Neal said he had no comment on the Shen family discussions because "anything he said to his dad is personal."

Despite all the flurry of activity in mid-1996 between Hughes and A.P.M.T., the deal bogged down amid internal squabbles. But by this year the pace had picked up again and last month the consortium reorganized itself and signed another deal with Hughes for an upgraded satellite.

The new satellite will have greater power to transmit and receive signals. Its payload includes a large-scale antenna reflector and a digital-on board processor, Mr. Liu and Mr. O'Neal said.

The antenna and processor enable the consortium's network to pinpoint low-power hand-held phones and simultaneously handle 16,000 phone conversations. Mr. Liu said that the regional affiliates "will be able to intercept calls if required by local authorities" but the consortium will not be able to intercept.

As a result of the recent reorganization, the consortium is now two-thirds owned by its Chinese affiliate, China A.P.M.T., said Mr. Liu, the consortium's deputy president. China A.P.M.T., in turn, is owned by the same five Chinese entities, including the Costind unit, and it will be the local A.P.M.T. franchise in China.

The president of A.P.M.T. and China A.P.M.T. is Li Baoming and A.P.M.T.'s chief engineer is Feng Ruming. Mr. Liu said both men have senior posts with the China Satellite Launch and Tracking Control Corporation, the unit of Costind overseen by General Shen. American intelligence reports say Mr. Feng and Mr. Li are top military officers, according to Administration officials.

Mr. O'Neal said that Hughes was "not aware" of A.P.M.T.'s military ties and while "there could be" some, it was up to the Federal Government to vet those connections. That is precisely what is now happening.

Administration Details Defense Of Controls on Satellite Exports

3 Top Aides Deny That Transfers Helped China

NYT 6-19-98, A1

By ERIC SCHMITT

WASHINGTON, June 18 — The Clinton Administration mounted the most detailed defense of its satellite export control policy today. Three top officials said that China's military had not benefited from transfers of sensitive American technology and that safeguards were in place to prevent any such leakage.

The spirited rebuttals by senior officials from the Departments of Commerce, State and Defense marked a counteroffensive against a growing Republican-led attack on the Administration's handling of satellite exports to China, a policy that Republicans have charged threatens American national security.

"We do not believe that the commercial space launch activities that have been authorized by licenses and monitored under these procedures have benefited China's missile or military satellite companies," John D. Holum, an acting undersecretary of state, told a House hearing today.

And Jan M. Lodal, the principal deputy undersecretary of defense for policy, said, "The current system protects national security."

The Administration officials said intelligence agencies have found no proof that any sensitive technology was transferred to the Chinese military through normal licensing channels. They asserted that adequate safeguards were in place to protect against that. But they did not rule out that there might have been unauthorized transfers. That is a question the Justice Department is looking into in the case of two American satellite companies.

But critics say it is very hard to tell whether there was a transfer of technology because the Chinese missiles that launch American satellites are very similar to the Chinese missiles that carry nuclear warheads. Improving the reliability of the commercial missiles might help the Chinese military, even inadvertently, with its nuclear program.

Despite the united front at today's hearings, cracks of dissension within the Administration's ranks and among usually loyal Democrats broke through.

The Government's arms control agency warned last year that waiving a ban on satellite exports to China to entice Beijing to curb missile sales had accomplished "nothing."

The warning by the Arms Control and Disarmament Agency, contained in a memorandum apparently sent to State Department officials, was made public at the House hearing. The memo undercut a fundamental reason for granting waivers to allow satellite exports to China despite a ban imposed after the 1989 massacre in Tiananmen Square.

"History has proven that the only time we have gotten movement from the Chinese on missile proliferation has been in the face of a penalty being imposed," the memo said. "Carrots have gotten us nothing."

But the agency's advisory was not

Continued on Page A10

Continued From Page A1

included in the State Department's official recommendation to the White House earlier this year. President Clinton approved the waiver last February, despite objections from the Justice Department that it would undermine an ongoing criminal inquiry into whether two satellite makers divulged sensitive information to Chinese rocket scientists.

Mr. Holum, who also heads the arms control agency, told an unusual joint hearing of two House committees, National Security and International Relations, that the warning was not adopted because "it would have been very disruptive" to broader American commercial interests.

In an afternoon Senate hearing, Senator Max W. Cleland of Georgia became one of the first Democrats to openly criticize the White House's shift of licensing authority for all commercial satellites to the Commerce Department from the State Department in 1996. "I've got more and more concern about Commerce becoming the lead dog here," said Senator Cleland. "I'd rather hedge my bets and put national security first and commerce second."

The back-to-back hearings over seven hours featuring the Administration witnesses capped a frenzied day on Capitol Hill over the wide-ranging China accusations, and came just a week before Mr. Clinton visits China.

The Director of Central Intelligence, George J. Tenet, updated the House Intelligence Committee in a closed, two-and-a-half hour briefing on a series of conflicting Government assessments. They involve whether two satellite makers who are the focus of the Justice inquiry — Loral Space & Communications and Hughes Electronics — jeopardized national security by discussing results of their independent review of a failed launch of a satellite on a Chinese Long March missile in February 1996.

With little dissension, the House approved a resolution, 409 to 10, to create a special select committee to lead its inquiry on the China accusations. The 10 who voted against the resolution were Democrats.

Despite bipartisan backing for the resolution, Democrats criticized Republicans during the debate for spending too much money on duplicate investigations of President Clinton and Democratic fund-raising practices. And they expressed concern that the work of the select committee would fall captive to what they labeled a Republican witchhunt to tarnish Mr. Clinton and curry favor with conservative voters.

At the House and Senate hearings, Administration officials said allowing American satellites to be launched on Chinese rockets was a "wise" policy initially set by Republican Presidents Ronald Reagan and George Bush to help make up for a shortage of American launching services.

And by transferring commercial satellite licensing authority to the Commerce Department, the Clinton aides said, the Administration was completing a shift that President Bush started in 1992. In a nutshell, the State Department considers national security implications for sensitive exports, while the Commerce Department also weighs American commercial interests.

Or, as William A. Reinsch, undersecretary of commerce for export administration, told Senators: "Commerce is an open, transparent, time-sensitive approach. State's is not. We've developed a 'process-friendly' " approach.

The General Accounting Office said the shift in licensing authority had loosened controls of satellite exports. But Mr. Reinsch disagreed, saying the Administration had added safeguards, including a broader role for the Pentagon and intelligence agencies to review export applications. Critics say those protections still fall far short of the State Department's regime.

"The existing Commerce licensing system fully protects our national security and foreign policy concerns," Mr. Reinsch said.

THE WALL STREET JOURNAL INTERACTIVE EDITION

FROM THE ARCHIVES

The Wall Street Journal Interactive Edition -- June 23, 1998

Pentagon Didn't Track Some Launches Of U.S. Satellites on Chinese Rockets

By BRIAN DUFFY
Staff Reporter of THE WALL STREET JOURNAL

WASHINGTON -- Pentagon officials didn't monitor the launches of U.S. satellites on three Chinese rockets, because of bureaucratic confusion.

The unmonitored launches occurred between 1994 and 1996, after President [redacted] had transferred licensing authority for export of some U.S. commercial satellites to the Commerce Department from the State Department. In addition to the launches, Pentagon personnel also must oversee all preparations and discussions between the U.S. and officials of foreign countries preparing to launch U.S. satellites abroad.

Business Deals Take Back Seat at U.S.-China Summit Talks

Clinton Hopes Trip Will Sway U.S. Opinion Toward China

"I believe there was a question there on the part of the people doing the licensing," said Clayton Lowry, director of the Satellite Industry Association, a trade group. "I believe that there was perhaps some confusion."

Republican Sen. Thad Cochran of Mississippi, whose Committee on Governmental Affairs is investigating U.S. satellite exports, took a stronger view. "It's just a laxness that has permeated the whole system" of satellite export licenses, he said. "And this is evidence of that."

The disclosure came late last week in testimony before the committee by Jan Lodal, principal deputy undersecretary of defense for policy. Mr. Lodal said that export-licensing regulations for commercial satellites "had anticipated" that both Commerce and State Department approval would be needed for foreign launches of U.S. commercial satellites. A State Department license automatically requires assignment of Pentagon monitors.

Republicans have accused the Clinton administration of permitting the export of sensitive technology and data to **China**, allowing Beijing to improve its missile and military technology. Mr. Clinton has called the allegations groundless.

Between 1994 and 1996, the Commerce Department approved licenses for the export of commercial satellites by four U.S. companies: Martin Marietta Corp., Hughes Electronics Corp., Loral Corp. and Lockheed. The Commerce and State departments concurred in recommending that Mr. Clinton waive sanctions on Beijing so the launches could be carried out. It couldn't be learned which of the launches weren't monitored.

"We are not aware of any transfer of technology from these unmonitored launches that contributed to China's missile or military satellite capabilities," Mr. Lodal testified last week.

The apparent confusion began during Mr. Clinton's first year as president. Following on Mr. Bush's decision to transfer licensing authority to Commerce, the Clinton administration issued regulations implementing the change.

Under the new rules, Commerce could license export of satellites containing nine sensitive technologies, such as intersatellite relay links. To gain approval, however, those technologies had to be "embedded" in the satellite in such a way that foreign technicians couldn't examine them. But the rules don't mention Defense Department oversight of launches.

Government officials now say the unmonitored launches raised concerns, particularly at the Defense Department, about the possible transfer of prohibited technology and technical data overseas. When Mr. Clinton transferred additional export-licensing authority to the Commerce Department in March 1996, he explicitly required that Pentagon personnel monitor all future launches of U.S. satellites abroad and their preparations.

Return to top of page | Format for printing

CONFIDENTIAL

UNITED STATES DEPARTMENT OF COMMERCE
The Under Secretary for Export Administration
Washington, D.C. 20230

July 6, 1998

House National Security Committee
Control No. 98-234
Copy 1 of 1

The Honorable Floyd Spence
Chairman, Committee on National Security
House of Representatives
Washington, D. C. 20515

Dear Mr. Chairman:

During the joint International Relations Committee / National Security Committee hearing on June 18, a number of members asked questions to which I agreed to respond in writing. I have the answers to all but one question and, rather than delay my answer, I wanted to respond to the Committees with the other information.

Congressman Gibbons referred to three satellite waivers dated October 24, 1995, December 4, 1995, and December 8, 1995, that I recommended be signed, and he implied incorrectly that they related to the 1996 Loral launch which exploded. I would like to correct that misconception. They did not. The waiver authorizing the 1996 Loral launch was signed by President Bush on September 11, 1992. The waiver recommendations that I signed were for different satellite exports. Copies of those recommendations are attached. All three are classified and contain information subject to Section 12(c) of the Export Administration Act, and may not be included in the printed hearing report.

Congressman Burton asked about the Yuchai America case. As I stated at the hearing, I and other BXA officials discussed this case extensively with his Governmental Affairs Committee staff over a year ago; however, I committed at the hearing to review the matter. I have done so and determined that the license was properly issued. The export license application was handled through the established licensing process, a full interagency review was conducted, and the licence was issued by interagency consensus. I am attaching a chronology of that case. The information contained therein is subject to the provisions of Section 12(c) of the Export Administration Act . It should not appear in the printed hearing report. Section 12(c) provides that this information may not be further disclosed unless the full Committee determines that withholding of the information is contrary to the national interest. In providing this information to you, I would underscore the importance of protecting it from further disclosure. The information provides details of U.S. exporters' business transactions. Among other concerns, release of the information could discourage exporters from participating, or being fully candid, in the export licensing process.

Congresswoman Fowler asked about a March 21, 1996, license to China Electronic Systems Engineering Corp., which she believed authorized the export of ground stations. A careful search of our licensing records shows no such license being issued, nor any exports to China Systems

CONFIDENTIAL
This Document Is Automatically
Declassified When Classified
Enclosures Are Removed

CONFIDENTIAL

Engineering Corp. I would also point out that in March 1996 the export of a ground station which contained encryption would have been under the jurisdiction of the State Department.

Congressman Hunter asked whether Hughes met with the Chinese as part of the 1995 launch failure investigation. Hughes undertook a launch failure analysis in order to satisfy insurance requirements, which it provided to the Commerce Department. Commerce determined that analysis did not contain information that exceeded the scope of the approved Commerce license. Commerce authorized release of the analysis, which subsequently was provided to a consortium of Western insurance companies and to the Chinese launch service provider.

Hughes worked with the Chinese launch service provider to assemble data on the failure but has indicated it prepared the final analysis independently. In that analysis, Hughes concluded that the satellite itself was not responsible for the launch failure. The data in the analysis was not detailed design or manufacturing data for the satellite or the launch vehicle and was focused on showing that the satellite did not cause the failure.

The outstanding issue is Congressman Gibbons' request for a list of companies and chief executive officers of the Computer Systems Policy Project, when the group's meetings at the White House occurred, and who attended them. The Department of Commerce does not have all of this information and must obtain it from other sources. We are assembling the information and will provide it to the Committees once we have it.

Sincerely,

William A. Reinsch

Enclosures

CONFIDENTIAL

This Document Is Automatically Declassified When Classified Enclosures Are Removed

Computer Systems Policy Project

Apple Computer

Eckhard Pfeiffer
Compaq Computer Corporation

Michael Dell
Dell Computer Corporation

Ronald L. Skates
Data General Corporation

Lewis E. Platt
Hewlett-Packard Company

Louis V. Gerstner. Jr.
IBM Corporation

Andrew S. Grove
Intel Corporation

Lars Nyberg
NCR Corporation

Richard E. Belluzzo
Silicon Graphics

Bruce I. Sachs
Stratus Computer

Scott G. McNealy
Sun Microsystems

Lawrence A. Weinbach
Unisys Corporation

Unit: Equipment in number; systems, components, parts and accessories in $ value.

Related Controls: (1) The corresponding EU list number controls space launch vehicles (not including their payloads) and other ``spacecraft'' (not identified in this CCL entry). These items are subject to the export licensing authority of the U.S. Department of State, Office of Defense Trade Controls (See 22 CFR part 121, Category XV). For the control status of products contained in ``spacecraft'' payloads, see the appropriate categories of the U.S. Munitions List (USML). (2) For the control status of items contained in ``spacecraft'' payloads subject to the EAR, see the appropriate entries on the CCL.

Related Definition: Transferring registration or operational control to any foreign person of any commercial communications satellite controlled by this entry must be authorized on a license issued by the Bureau of Export Administration. This requirement applies whether the commercial communications satellite is physically located in the United States or abroad.

List of Items Controlled

a. Commercial communication satellites;

Technical Note: Commercial communications satellites are subject to Commerce licensing jurisdiction even if they include the individual munitions list systems, components, or parts identified in Category XV(f) of the USML. In all other cases, these systems, components, or parts remain on the USML, except that non-embedded, solid propellant orbit transfer engines (``kick motors'') are subject to Commerce licensing jurisdiction (and not controlled under the USML) when they are to be utilized for the specific commercial communications satellite launch, provided the solid propellant ``kick motor'' being utilized is not specifically designed or modified for military use or capable of being restarted after achievement of mission orbit (such orbit transfer engines are always controlled under Category IV of the USML). Technical data (as defined in Sec. 120.21 of the International Traffic in Arms Regulations (ITAR)) and defense services (as defined in Sec. 120.8 of the ITAR) related to the systems, components, or parts referred to in Category XV(f) of the USML are always controlled under the USML, even when the satellite itself is licensed by the Department of Commerce.

Note: Military communications satellites or multi-mission satellites, including commercial communications satellites having additional non-communication mission(s) or payload(s) are under the jurisdiction of the Department of State.

b. [Reserved]

c. Other ``spacecraft'' not subject to the export licensing authority of the U.S. Department of State, Office of Defense Trade Controls under 22 CFR part 121, Category XV.

Notes: 1. ECCN 9A004.c includes the international space station being developed, launched and operated under the supervision of the U.S. National Aeronautics and Space Administration. Exporters requesting a license from the Department of Commerce for spacecraft other than the international space station or a commercial communications satellite specified in 9A004 must provide a statement from the Department of State, Office of Defense Trade Controls, verifying that the item intended for export is under the licensing jurisdiction of the Department of Commerce.

2. All other spacecraft, including all other satellites not controlled under 9A004 and components, parts, accessories,

attachments, associated equipment, and ground support equipment therefor are subject to the export licensing authority of the Department of State.

3. Items on Category XV(f) of the USML that are included in a commercial communications satellite to be exported under a Commerce license must be specifically listed on the Commerce license application. Such items when not included in a specific commercial communications satellite are under the jurisdiction of the Department of State.

4. Technical data provided to the launch provider (form, fit, function, mass, electrical, mechanical, dynamic/environmental, telemetry, safety, facility, launch pad access, and launch parameters) for commercial communications satellites that describe the interfaces for mating of the satellite to the launch vehicle and parameters for launch (e.g. orbit, timing) of the satellite, are under Commerce jurisdiction. Other technical data and all defense services and technical assistance for satellite and/or launch vehicles, including compatibility, integration, or processing data are controlled and subject to licensing by the Department of State, in accordance with 22 CFR parts 120 through 130. Approval for such technical assistance will require a Technical Assistance Agreement (TAA) and may require U.S. Government oversight.

5. Once a satellite is launched, items remaining unlaunched are required to be returned immediately to the United States. If the satellite launch is canceled or unduly delayed, the satellite and all support equipment must be returned immediately to the United States.

6. Detailed design, development, production, or manufacturing data for all spacecraft, including satellites, regardless of which agency has jurisdiction over the export, and all systems components, parts, accessories, attachments, and associated equipment (including ground support equipment) specifically designed or modified for articles under Category XV on the United States Munitions List (including software source code and operating algorithms) are subject to licensing by the Department of State. This does not include that level of technical data (including marketing data) necessary and reasonable for a purchaser to have assurance that a U.S.-built item intended to operate in space has been designed, manufactured and tested in conformance with specified contract requirements (e.g., operational performance, reliability, lifetime, product quality, or delivery expectations) as well as data necessary for normal in-orbit satellite operations, to evaluate in-orbit anomalies, and to operate and maintain associated ground station equipment (except encryption hardware).

* * * * *

9E003 Other ``technology''.

License Requirements

Reason for Control: NS, AT, SI.

Control(s)	Country chart
NS applies to entire entry...............	NS Column 1.
AT applies to entire entry...............	AT Column 1.

SI applies to 9E003. a.1. through a.12 and f. See Sec. 742.14 of the EAR for additional information.

License Exceptions

CIV: N/A

CHARTS SUBMITTED FOR THE RECORD

JUNE 17, 18, 23, 1998

Last uodated: May 31, 1998

Long March Launch Vehicle Overview

(FB launcher is also included in all tables in this page)

Models: 15
Total launches: 60
Orbited launches: 53
Successful launches: 49
Failed launches: 7
Partial failed launches: 4
Commercial launches: 17
Payload launched: 64
GEO payload: 18
SSO payload: 2

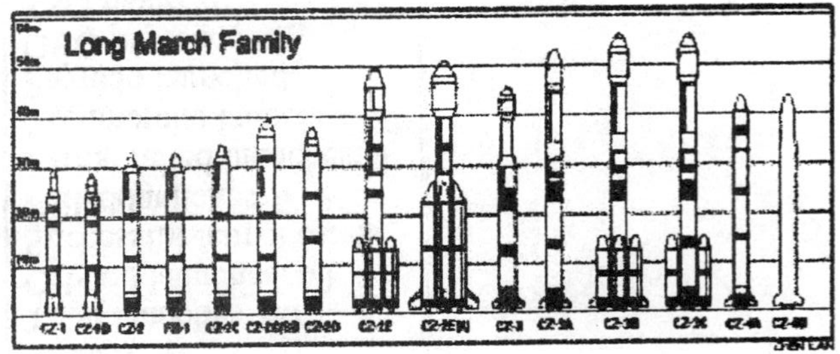

Long March Specification

Model	Configuration	Length/ Diameter (m)	Mass/ Thrust (t/kN)	Capacity (kg)	Flights/ fails
CZ-1	1: YF-2 (NA27S/UDMH) 2: YF-3 (NA27S/UDMH) 3: GF-02 (Solid)	29.86 2.25	81.6 1020	LEO 300	2/0
CZ-1D	1: YF-2A (NA27S/UDMH) 2: 2xYF-40 (N_2O_4/UDMH) 3: GF-36 (Solid)	28.22 2.25	81.1 1101	LEO 930	0/0
CZ-2	1: YF-21 (N_2O_4/UDMH) 2: YF-22 (N_2O_4/UDMH)	31.17 3.35	190 2786	LEO 1800	4/1
CZ-2C	1: YF-21 (N_2O_4/UDMH) 2: YF-22 (N_2O_4/UDMH)	35.1 3.35	192 2786	LEO 2400	11/0
FB-1	1: YF-21 (N_2O_4/UDMH) 2: YF-22 (N_2O_4/UDMH)	32 ? 3.35		LEO 1200	8/4
CZ-2C/SD	1: YF-21 (N_2O_4/UDMH) 2: YF-22 (N_2O_4/UDMH) 3: Solid (HTPB)	39.93 3.35	213 2962	LEO 2800	4/0

Last Updated: May 31, 1998

Overview of Chinese Satellite

(Foreign built satellites and launch failure not included)

Tatal satellites launched: 44
Communication satellite: 10
Meteorological satellite: 3
Earth Resource Satellite: 0
Recoverable Satellite: 17
Technical/Engineering Test Satellite: 7
Science Satellite: 7

Communication Satellite

#	Model	Launch Date	Mass (kg)	Dimension (m)	Orbit (km,degree)	Payload	Comment
1	DFH-1	04/24/1970	173	D=1.0	439/2384, 68.5	Radio transmitter	
2	DFH-2	01/29/1984	900 (GTO)	D=2.1 H=3.1	360/6480, 36	4 C-Band	
3	DFH-2	04/08/1984			125 E		
4	DFH-2A	02/01/1986			103 E		
5	DFH-2A	03/07/1988			87.5 E		
6	DFH-2A	12/22/1988			110.5 E		
7	DFH-2A	02/04/1990			105 E		
8	DFH-2A	12/28/1991					
9	DFH-3	11/30/1994	2266 (GTO)	2.2x2.2x1.7 L=18.1		24 C-Band	
10	DFH-3	05/12/1997			125 E		

DFH: Dong Fang Hong (Red is East)

Meteorological Satellite

#	Model	Launch Date	Mass (kg)	Dimension (m)	Orbit (km, degree)	Payload	Comment
1	FY-1A	04/24/1970	750	1.4x1.4x1.2	832/937, 99.2	3 VIS, 2 IR	Worked 39 days
2	FY-1B	09/03/1990	880	1.4x1.4x1.8	875/897, 98.8	3 VIS, 2 IR	
3	FY-2B	06/10/1997	593	D=2.1, H=1.6	105 E	1 VIS, 2 IR	
4	FY-1C	1998			870, 99	4 VIS, 6 IR	

FY: Feng Yun (Wind and Cloud); VIS: visible channel; IR: infrared channel

Earth Resource Satellite

#	Model	Launch Date	Mass (kg)	Dimension (m)	Orbit (km,degree)	Payload	Comment
1	ZY-1	1998	1450	2x3.3x8.3	778, 98.5	IRC,IR-MSS,WFI	

ZY: Zi Yuan (Resource); IRC: infrared CCD camera;
IR-MSS: IR multi spectural scanner; WFI: wide field imager

Recoverable Satellite

#	Model	Launch Date	Mass (kg)	Dimension (m)	Orbit (km,degree)	Payload	Comment
1	FSW-I	11/05/1975	1790	D=2.1 H=3.14	177/479, 63	O	3 days in orbit Parachute failed
2	FSW-I	12/07/1976	1812		159/489, 59.4	O	3 days in orbit
3	FSW-I	01/26/1978	1810		167/509, 57	O	3 days in orbit
4	FSW-I	09/091982	1783		177/410, 63	O	5 days in orbit
5	FSW-I	08/19/1983	1842		175/410, 63.3	O	5 days in orbit
6	FSW-I	09/12/1984	1809		178/414, 68	O	5 days in orbit
7	FSW-I	10/21/1985	1809		175/409, 63	O	5 days in orbit
8	FSW-I	10/06/1986	1800		176/402, 63	O	5 days in orbit
9	FSW-I	08/05/1987	1819		175/400, 70	O,M(French)	5 days in orbit
10	FSW-II	09/09/1987	2076	D=2.2 H=4.6	208/323, 63	O,M	8 days in orbit
11	FSW-II	08/05/1988	2109		208/313, 63	O.M(Germany)	8 days in orbit
12	FSW-II	10/05/1990	2080		211/311, 57	O,M,B	7 days in orbit
13	FSW-III	08/09/1992	2592		176/351, 63	O,M	15 days in orbit
14	FSW-II	10/06/1992	2080		210/329, 63	O,M	7 days in orbit
15	FSW-II	10/08/1993	2099		209/300, 57	O,M	Failed to retuen
16	FSW-III	07/03/1994	2755		176/359, 64	O,M	15 days in orbit
17	FSW-III	10/20/1996				O,M(Japan)	15 days in orbit

FSW: Fanhui Shi Weixing (Recoverable Satellite)
O: earth observation; M: material processing; B: Biological

Technical/Engineering Test Satellite

#	Model	Launch Date	Mass (kg)	Dimension (m)	Orbit (km,degree)	Payload	Comment
1	JSSW	07/26/1975	1107		186/464, 69		
2	JSSW	12/16/1975	1109		184/387, 69		
3	JSSW	08/30/1976	1108		191/2145, 70		
4	Optus MFS	07/16/1990	7600				
5	DFH-3 MFS	02/08/1994			211/32578, 27.5		
6	Iridium MFS	09/01/1997	680				
7	Iridium MFS	09/01/1997	680				

JSSW: Ji Shu Shiyan Weixing (Technical Test Satellite);
DFH: Dong Fang Hong (Red is East)
MFS: Mass Freqeuncy Simulator

Science Satellite

#	Model	Launch Date	Mass (kg)	Dimension (m)	Orbit (km, degree)	Payload	Comment
1	SJ-1	03/03/1971	221	D=1	266/1826, 69.9	particle detectors	
2	SJ-2	09/20/1981			232/1598, 59.5	Ionosphere study	
3	SJ-2A	09/20/1981			232/1615, 59.4		
4	SJ-2B	09/20/1981			232/1608, 59.4	Atomshere study	
5	DQ-1A	09/03/1990			789/811, 99	Atomshere study	
6	DQ-1B	09/03/1990			596/629, 99	Atomshere study	
7	SJ-4	02/08/1994	400		212/36092, 28.2	particle detectors	

SJ: Shi Jian (Practice); DQ: Da Qi (Atomsphere)

Taiwan's Science Satellite

#	Model	Launch Date	Mass (kg)	Dimension (m)	Orbit (km,degree)	Payload	Comment
1	Rocsat-1	1998	400	2.1x7x1.1	600, 35	OCI, IPEI, ECP	

OCI: Ocean Color Imager
IPEI: Ionosphere and Electrondynamics Instrument
ECP: Experimental Communication Payload

Foreign Built Commercial Satellites

(Including Taiwan and Hong Kong's satellites)

Name	Owner	Model (Transponders)	Launch Date	Launcher	Comment
Chinasat-5	Chinasat	GE 3000 (24)	05/23/1984	Ariane	Former Spacenet-1
Chinasat-7	Chinasat	HS-376 (24)	08/18/1996	CZ-3	Launch failed
Chinasat-8	Chinasat	SS/Loral FS1300 (52)	1998	CZ-3B	
Chinastar-1	China Orient Telecom	LockMart A2100 (48)	05/30/1998	CZ-3B	
Sinosat-1	Sinosat	Aerospatiale (38)	1998	CZ-3B	
Asiasat-1	AST Holdings	HS-376 (24)	04/07/1990	CZ-3	
Asiasat-2	AST Holdings	GE 7000 (33)	11/28/1995	CZ-2E	
Asiasat-3	AST Holdings	HS-601 (44)	12/28/1997	Proton	Launch failed
Asiasat-3S	AST Holdings	HS-601 (44)	1999	Proton	
Asiasat-G	AST Holdings	Russia built (7)	05/20/1994	Proton	Former Gorizont-30
Apstar-1	APS Ltd	HS-376 (24)	07/21/1994	CZ-3	
Apstar-2	APS Ltd	HS-601 (34)	01/25/1995	CZ-2E	Launch failed
Apstar-1A	APS Ltd	HS-376 (24)	07/03/1997	CZ-3	
Apstar-2R	APS Ltd	SS/Loral FS1300 (44)	10/16/1997	CZ-3B	
Gals-1	Land Group	Russia built (3)	01/20/1994	Proton	
Gals-2	Land Group	Russia built (3)	11/17/1995	Proton	
ST-1	Taiwan Telecom/ S'pore Telecom	Matra Marconi (30)	1998	Ariane 4	
APMT-1	APMT	HS GM	2000	CZ-3B	

Red: Launch failure
Blue: Planned

HOME | EMAIL

CZ-2D	1: YF-21B (N_2O_4/UDMH) 2: DaYF-20 (N_2O_4/UDMH)	37.73–41.30 3.35	237 2962	LEO 3100	3/0
CZ-2E	0: 4xYF-20B (N_2O_4/UDMH) 1: YF-21B (N_2O_4/UDMH) 2: YF-24B (N_2O_4/UDMH)	49.7 3.35	462 5923	LEO 9200 GTO 3500 (EPKM)	7/2
CZ-2E(A)	?	?	?	LEO 12000	0/0
CZ-3	1: YF-21 (N_2O_4/UDMH) 2: YF-22 (N_2O_4/UDMH) 3: YF-73 (N_2O_4/LH2)	43.8 3.35	202 2800	LEO 5000 GTO 1400	12/3
CZ-3A	1: YF-21 (N_2O_4/UDMH) 2: YF-22 (N_2O_4/UDMH) 3: YF-75 (LOX/LH2)	52.3 3.35	240 2962	GTO 2500	3/0
CZ-3B	0: 4xYF-20 (N_2O_4/UDMH) 1: YF-21 (N_2O_4/UDMH) 2: YF-22 (N_2O_4/UDMH) 3: YF-75 (LOX/LH2)	54.84 3.35	425.5 5924	GTO 5000	4/1
CZ-3C	0: 2xYF-20 (N_2O_4/UDMH) 1: YF-21 (N_2O_4/UDMH) 2: YF-22 (N_2O_4/UDMH) 3: YF-75 (LOX/LH2)	?	?	GTO 3700	0/0
CZ-4A	1: YF-21B (N_2O_4/UDMH) 2: DaYF-20 (N_2O_4/UDMH) 3: YF-40 (N_2O_4/UDMH)	44.16-45.78 3.35	249 2962	SSO 1650	2/0
CZ-4B	?	?	?	SSO 2800 (?)	0/0
YF-2 = 2xYF-1; YF-21 = 4xYF-20; YF-21B = 4xYF-20B NA27S: Nitric Acid 27S; UDMH: Unsymmetrical Di-Methyl Hydrazine					

Long March Launch Log

#1	#2	#3	Date	ID	Model	Site	Payload	Comment
1	1	1	04/24/1970	70034	CZ-1	Jiuquan	DFH-1	
2	2	2	03/03/1971	71018	CZ-1	Jiuquan	SJ 1	
3	1	1	09/18/1973		FB-1	Jiuquan	JSSW ?	

4	2	2	07/14/1974		FB-1	Jiuquan	JSSW ?	2nd stage engine lost thrust
5	3	1	11/05/1974		CZ-2	Jiuquan	FSW ?	
6	3	3	07/26/1975	75070	FB-1	Jiuquan	JSSW-1	
7	4	2	11/26/1975	75111	CZ-2	Jiuquan	FSW-1	
8	4	4	12/16/1975	75119	FB-1	Jiuquan	JSSW-2	
9	5	5	08/30/1976	76087	FB-1	Jiuquan	JSSW-3'	
10	5	3	12/07/1976	76116	CZ-2	Jiuquan	FSW-2	
11	6	6	11/10/1976		FB-1	Jiuquan	JSSW ?	
12	6	4	01/26/1978	78011	CZ-2	Jiuquan	FSW-3	
13	7	7	07/27/1979		FB-1	Jiuquan	SJ ?	
14	8	8	09/19/1981	81093	FB-1	Jiuquan	SJ 2 SJ 2A SJ 2B	
15	7	1	09/09/1982	82090	CZ-2C	Jiuquan	FSW-4	
16	8	2	08/19/1983	83086	CZ-2C	Jiuquan	FSW-5	
17	9	1	01/29/1984	84008	CZ-3	Xichang	DFH-2-0	3rd Stage early shutdown
18	10	2	04/08/1984	84035	CZ-3	Xichang	DFH-2-1	
19	11	3	09/12/1984	84098	CZ-2C	Jiuquan	FSW-6	
20	12	4	10/21/1985	85096	CZ-2C	Jiuquan	FSW-7	
21	13	3	02/01/1986	86010	CZ-3	Xichang	DFH-2-2	
22	14	5	10/06/1986	86076	CZ-2C	Jiuquan	FSW-8	
23	15	6	08/05/1987	87067	CZ-2C	Jiuquan	FSW-9	
24	16	7	09/09/1987	87075	CZ-2C	Jiuquan	FSW-10	
25	17	4	03/071988	88014	CZ-3	Xichang	DFH-2-3	
26	18	8	08/05/1988	88067	CZ-2C	Jiuquan	FSW-11	
27	19	1	09/06/1988	88080	CZ-4A	Taiyuan	FY-1A	
28	20	5	12/22/1988	88111	CZ-3	Xichang	DFH-2-4	
29	21	6	02/04/1990	90011	CZ-3	Xichang	DFH-2-5	
30	22	7	04/07/1990	90030	CZ-3	Xichang	Asiasat 1	First commercial launch

31	23	1	07/16/1990	90059	CZ-2E	Xichang	Optus Model Badr A	
32	24	2	09/03/1990	90081	CZ-4A	Taiyuan	FY-1B DQ-1A DQ-1B	
33	25	9	10/05/1990	90089	CZ-2C	Jiuquan	FSW-12	
34	26	8	12/28/1991	91088	CZ-3	Xichang	DFH-2-6	3rd Stage early shutdown
35	27	1	08/09/1992	92051	CZ-2D	Jiuquan	FSW-13	
36	28	2	08/13/1992	92054	CZ-2E	Xichang	Optus B1	
37	29	10	10/06/1992	92064	CZ-2C	Jiuquan	FSW-14 Freja	
38	30	3	12/21/1992	92090	CZ-2E	Xichang	Optus B2	payload fairing collapsed
39	31	11	10/08/1993	93063	CZ-2C	Jiuquan	FSW-15	
40	32	1	02/08/1994	94010	CZ-3A	Xichang	DHF-3 Model SJ-4	
41	33	2	07/03/1994	94037	CZ-2D	Jiuquan	FSW-16	
42	34	9	07/21/1994	94043	CZ-3	Xichang	Apstar 1	
43	35	4	08/27/1994	94055	CZ-2E	Xichang	Optus B3	
44	36	2	11/29/1994	94080	CZ-3A	Xichang	DFH-3-1	
45	37	5	01/25/1995		CZ-2E	Xichang	Apstar 2	destroyed due to wind shear
46	38	6	11/28/1995	95064	CZ-2E	Xichang	Asiastar 2	
47	39	7	12/28/1995	95073	CZ-2E	Xichang	Echostar 1	
48	40	1	02/14/1996		CZ-3B	Xichang	Intelsat 708	attitude control lost
49	41	10	07/03/1996	96039	CZ-3	Xichang	Apstar 1A	Foreign Built
50	42	11	08/18/1996	96048	CZ-3	Xichang	Chinasat 7	3rd Stage early shutdown
51	43	3	10/20/1996	96059	CZ-2D	Jiuquan	FSW-17	Chinese Spy Sat
52	44	3	05/11/1997	97021	CZ-3A	Xichang	DFH-3-2	Chinese Com sat
53	45	12	06/10/1997	97029	CZ-3	Xichang	FY-2B	Chinese Met sat
54	46	2	08/19/1997	97042	CZ-3B	Xichang	Agila 2	
55	47	1	09/01/1997	97048	CZ-2C/SD	Taiyuan	Iridium MFS 1 Iridium MFS 2	

56	48	3	10/16/1997	97062	CZ-3B	Xichang	Apstar 2R	Foreign Built	W Commercial
57	49	2	12/08/1997	97077	CZ-2C/SD	Taiyuan	Iridium 42 Iridium 44		W
58	50	3	03/25/1998	98018	CZ-2C/SD	Taiyuan	Iridium 51 Iridium 61		W
59	51	4	05/02/1998	98026	CZ-2C/SD	Taiyuan	Iridium 69 Iridium 71		W
60	52	4	05/30/1998	98033	CZ-3B	Xichang	Chinastar-1	Foreign Built	W
#1: all launches #2: all launches of CZ or FB #3: launches of specific model									

Red : Failure or partial failure

CZ: Chang Zheng (Long March)
FB: Feng Bao (Storm)
DFH: Dong Fang Hong (Red is East)
SJ: Shi Jian (Practice)
FY: Feng Yun (Wind and Cloud)
DQ: Da Qi (Atmosphere)
FSW: Fanhui Shi Weixing (Recoverable Satellite)
JSSW: Ji Shu Shiyan Weixing (Technical Test Satellite)

HOME | EMAIL

CHART A

CRS-7

Presidential Waivers of Post-Tiananmen Sanctions for Exports of Satellites or Parts to China

Satellite Projects (may have multiple satellites per project)	End-User	Manufacturer	Waiver
Asiasat-1	Asia Satellite	Hughes	12/19/89
Aussat (Optus-B1 and -B2)	Australia	Hughes	4/30/91
Freja	Sweden		
Asiasat-2	Asia Satellite	Martin Marietta	9/11/92
Apsat (or Apstar)	APT Satellite	Hughes	
Intelsat-7A	Intelsat	Martin Marietta	
Starsat			
AfriSat (AfriStar)	Afrispace		
Dongfanghong-3	China	China	
Iridium	Iridium	Lockheed	7/2/93
Intelsat-8	Intelsat	Loral	
Echostar-1 and -2	Echostar	Martin Marietta	7/13/94
Mabuhay	Philippines	Loral	2/6/96
ChinaStar-1 and -2	China	Lockheed Martin	2/6/96
Chinasat-7	China	Hughes	2/6/96
Asia Pacific Mobile Telecommunications (APMT)	APT Satellite	Hughes	6/23/96
Globalstar	Globalstar	Loral	7/9/96
Fengyun 1	China	China	11/19/96
SinoSat-1	China	Aerospatiale	11/23/96
Chinasat-8	China	Loral	2/18/98

Source: CRS Report 98-485, *China: Possible Missile Technology Transfers From U.S. Satellite Export Policy — Background and Chronology*.

CHART B

CRS-8

Chinese Ballistic Missiles: Launchers Deployed and Under Development

Type	CH des.	US des.	Deploy-ed	Range/Warhead (km/kg)	CEP (m)	Launch Platform	Fuel	Comments
MRBM	DF-3A	CSS-2	40+	2,800/2,150	1,000	land-mobile	liquid	in service
ICBM	DF-4	CSS-3	10-25	5,500+/2,200	1,370	land-mobile	liquid	in service
ICBM	DF-5A	CSS-4	7-18	13,000/3,200	500	hardened underground silos	liquid	in service
SLBM	JL-1	CSS-N-3	12	1,700/600	?	Xia SSBN	solid	in service
MRBM	DF-21	CSS-5	10	1,800/600	?	mobile TEL	solid	in service
SRBM	DF-15/ M-9	CSS-6	?	600/500	300	mobile TEL	solid	in service; test-fired in 7/95 & 3/96 near Taiwan
SRBM	DF-11/M-11	CSS-7	?	300/500	?	mobile TEL	solid	in service
SRBM	8610/ M-7	CSS-8	?	150/190	?	mobile launcher	solid	modified HQ-2 SAM
MRBM	DF-25			1,700/2,000	?	land-mobile	solid	development
ICBM	DF-31			8,000/700	?	land-mobile	solid	development; after 2000
SLBM	JL-2			8,000/700	?	planned 094 SSBN	solid	development; after 2005
ICBM	DF-41			12,000/800	?	land-mobile	solid	development; after 2010

Source: CRS Report 97-391, *China: Ballistic and Cruise Missiles*. This information has been derived from various open press and government reports.

CHART C

INCREASED CHINESE ROCKET RELIABILTY:HOW MUCH ARE US SATELLITE TECHNOLOGY TRANSFERS HELPING?

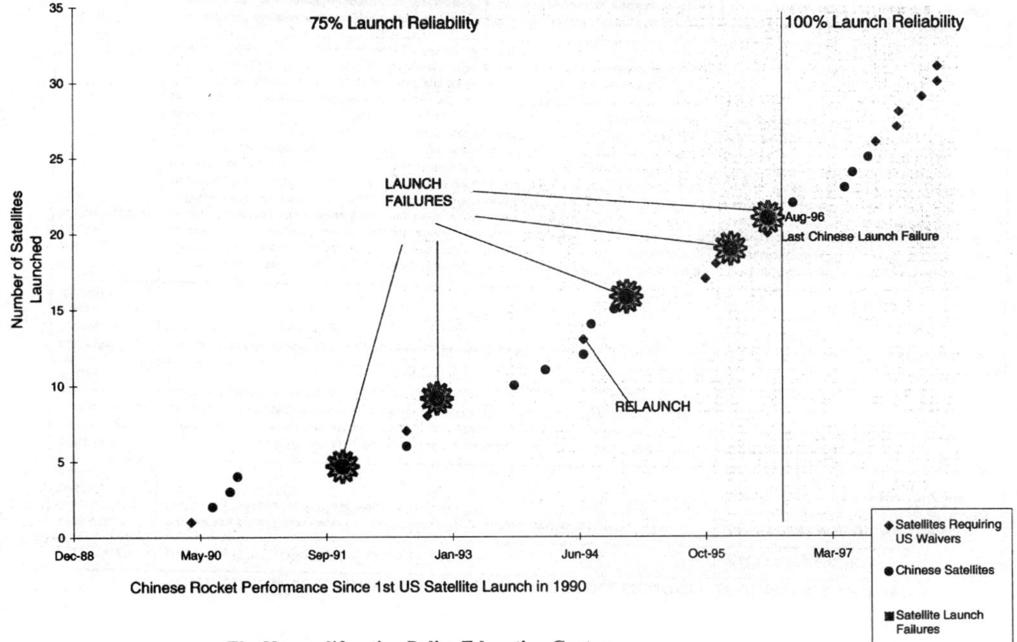

The Nonproliferation Policy Education Center
1718 M Street, N.W., Suite 244 Washington, D.C. 20036 Phone 202.466.4406/Fax: 202.659.5429 E-mail: npec@ix.netcom.com

CHART D

U.S. Satellite Transfers to China: An Increasingly Dangerous Business

LAUNCH	WHAT THE CHINESE LEARNED	MILITARY BENEFIT
APRIL 1990 (successful) AsiaSat I (Hughes)	***Why and how to build clean rooms for satellite launch integration*** and the need to protect fragile complex payloads against significant thermodyn~mic change.	***Upgraded China's ability to integrate and launch more sophisticated warhead packages*** and military communications and reconnaissance satellites.
APRIL 1992 (failure: relaunch successful 8/92) Optus B I (Hughes)	***That the launch failure was due to engine control defects associated with the launcher's first stage rather than the rocket system's attitude control system*** -- an attitude control system that the Chinese claimed was similar to that used in its M-family of theater ballistic missiles (the kind of missiles sold to Pakistan and fired against Taiwan).	***Helped confirm that the attitude control system found on their M-family theater ballistic missiles worked.***
DECEMBER 1992 (failure) Optus B 2 (Hughes)	***That Chinese-designed payload faring was adequate,*** and that launch failure was due to defective fuel tank in the satellite's upper stage.	***Upgraded China's ability to launch reconnaissance satellites reliably.***
MAY 1995 (successful) AsiaSat II (Martin Marietta)	***That their solid rocket satellite kick motor design worked.*** This motor was still in development. A test had only been attempted once before, but the results were inconclusive because the launcher on which the kick motor and the satellite were mounted failed due to a defect in the rocket's attitude control system. Given the motor's limited history, the key uncertainty was whether or not it would have sufficient quality and reliability to place the Martin Marietta satellite into its proper orbit. To help assure this, U.S. contractors completed coupled load analysis on the kick motor for the Chinese. The contractor and other foreign persons were also allowed to witness Chinese solid rocket satellite kick motor tests. The concern here was that the solid rocket propellant be configured to have exactly the right grain structure and be shaped to produce exactly the right amount of thrust for exactly the right amount of time so as 1. to avoid placing damaging amounts of force (acceleration and vibration) on the satellite and 2. to propel the satellite to the precise point it needed to go into space. Although U.S. citizens were told not to discuss what they saw with the Chinese, it is unclear whether or not foreign staff associated with AsiaSat (who were not bound by U.S. law) were briefed by the contractor or spoke with the Chinese so as to convey solid rocket propulsion know-how.	***Could help China perfect a post boost vehicle (PBV) - a system geared to evade possible missile defenses.*** In fact, the Chinese are currently working on two road-mobile solid rockets (DF-31 and DF-41) which are expected to use PBVs. PBVs work by blasting the warhead payload off its predictable ballistic trajectory down through space and the atmosphere. Such kick motors also can be used to help delivery of military communication and reconnaissance satellites.
1997-98 (4 successful launches) Motorola Iridium (Lockheed)	***That they coulo dispense multiple satellites precisely with a kick motor and new satellite dispenser of Chinese design.*** This required demonstrating that Motorola's two Iridium satellites would be released at the proper time, that they would not break from their moorings due to vibration or improper mounting, that the delivery bus's attitude control would not be destabilized by the release of the first satellite, and that the system's kick motor would generate the right amount of thrust.	***Helped China master the technology needed to develop its own multiple independently targetable reentry vehicles (MiRVs),*** technology which is interchangeable with highly precise satellite dispensers. China's new solid rocket intercontinental ballistic missiles (DF-31 and DF-41) are expected to employ such systems.

The Nonproliferation Policy Education Center
1718 M Street, N.W., Suite 244 Washington, D.C. 20036 Phone 202.466.4406/Fax: 202.659.5429 E-mail: npec@ix.netcom.com

CHART E

Date of Transfer or Report	Reported Transfer by China	Possible Violation	Administration's Response
Nov. 1992	M-11 missiles or related equipment to Pakistan (The Administration did not officially confirm reports that M-11 missiles are in Pakistan.)	MTCR Arms Export Control Act Export Administration Act	sanctions imposed on Aug. 24, 1993, for transfers of M-11 related equipment (not missiles); waived on Nov. 1, 1994
Mid-1994 to mid-1995	dozens or hundreds of missile guidance systems and computerized machine tools to Iran	MTCR Iran-Iraq Arms Nonproliferation Act Arms Export Control Act Export Administration Act	no sanctions
2nd quarter of 1995	parts for the M-11 missile to Pakistan	MTCR Arms Export Control Act Export Administration Act	no sanctions
Dec. 1994 to mid-1995	5,000 ring magnets for an unsafeguarded nuclear enrichment program in Pakistan	NPT Export-Import Bank Act Nuclear Proliferation Prevention Act Arms Export Control Act	considered sanctions under the Export-Import Bank Act; but announced on May 10, 1996, that no sanctions would be imposed
July 1995	more than 30 M-11 missiles stored in crates at Sargodha Air Force Base in Pakistan	MTCR Arms Export Control Act Export Administration Act	no sanctions
Sept. 1995	calutron (electromagnetic isotope separation system) for uranium enrichment to Iran	NPT Nuclear Proliferation Prevention Act Export-Import Bank Act Arms Export Control Act	no sanctions

1995-1997	C-802 anti-ship cruise missiles and C-801 air-launched cruise missiles to Iran	Iran-Iraq Arms Nonproliferation Act	no sanctions
before Feb. 1996	dual-use chemical precursors and equipment to Iran's chemical weapon program	Arms Export Control Act Export Administration Act	sanctions imposed on May 21, 1997
summer 1996	400 tons of chemicals to Iran	Iran-Iraq Arms Nonproliferation Act[1] Arms Export Control Act Export Administration Act	no sanctions
Aug. 1996	plant to manufacture M-11 missiles or missile components in Pakistan	MTCR Arms Export Control Act Export Administration Act	no sanctions
Aug. 1996	gyroscopes, accelerometers, and test equipment for missile guidance to Iran	MTCR Iran-Iraq Arms Nonproliferation Act Arms Export Control Act Export Administration Act	no sanctions
Sept. 1996	special industrial furnace and high-tech diagnostic equipment to unsafeguarded nuclear facilities in Pakistan	NPT Nuclear Proliferation Prevention Act Export-Import Bank Act Arms Export Control Act	no sanctions
July-Dec. 1996	Director of Central Intelligence (DCI) reported "tremendous variety" of technology and assistance for Pakistan's ballistic missile program	MTCR Arms Export Control Act Export Administration Act	no sanctions

[1] Additional provisions on chemical, biological, or nuclear weapons were not enacted until February 10, 1996.

July-Dec. 1996	DCI reported "tremendous variety" of assistance for Iran's ballistic missile program	MTCR Iran-Iraq Arms Nonproliferation Act Arms Export Control Act Export Administration Act	no sanctions
July-Dec. 1996	DCI reported principal supplies of nuclear equipment, material, and technology for Pakistan's nuclear weapon program	NPT Nuclear Proliferation Prevention Act Export-Import Bank Act Arms Export Administration Act	no sanctions
July-Dec. 1996	DCI reported key supplies of technology for large nuclear projects in Iran	NPT Iran-Iraq Arms Nonproliferation Act Nuclear Proliferation Prevention Act Export-Import Bank Act Arms Export Administration Act	no sanctions
July-Dec. 1996	DCI reported "considerable" chemical weapon-related transfers of production equipment and technology to Iran	Iran-Iraq Arms Nonproliferation Act Arms Export Control Act Export Administration Act	no sanctions
Jan. 1997	dual-use biological items to Iran	BWC Iran-Iraq Arms Nonproliferation Act Arms Export Control Act Export Administration Act	no sanctions
1997	chemical precursors, production equipment, and production technology for Iran's chemical weapon program, including a plant for making glass-lined equipment	Iran-Iraq Arms Nonproliferation Act Arms Export Control Act Export Administration Act	no sanctions

Sept. to Dec. 1997	China Great Wall Industry Corp. provided telemetry equipment used in flight-tests to Iran for its development of the Shahab-3 and Shahab-4 medium range ballistic missiles	MTCR Iran-Iraq Arms Nonproliferation Act Arms Export Control Act Export Administration Act	no sanctions
Nov. 1997/ April 1998	may have transferred technology for Pakistan's Ghauri medium-range ballistic missile that was flight-tested on April 6, 1998	MTCR Arms Export Control Act Export Administration Act	no sanctions

CHART F

How is Commercial Launch Technology Related to Ballistic Missile Technology?

The Administration's technology transfer policies have allowed China to obtain sensitive rocket and satellite technology and apply it to enhancing the reliability, accuracy, and potency of its ballistic missiles. The following are some of the ways that the Administration's technology transfer policies could aid Chinese ballistic missile development.

Provided by HNSC Chairman Floyd D. Spence

Payload Dispersal Technology

Payload dispersal technology allows single commercial rockets to deliver more than a single satellite into space per each launch. This same technology can be used to develop multiple independently targetable reentry vehicles (MIRVs) that allow ballistic missiles to carry several warheads into space and deliver each one to different targets.

Kick Motor Technology

Kick motors are used to propel satellites precisely into their desired orbits. This same technology can be applied to warhead delivery systems to enable them to evade ballistic missile defense systems.

Radiation-hardened Electronics

These specialized chips are designed to resist electromagnetic interference in space, as well as electromagnetic pulses in a nuclear combat environment.

Encryption Devices

In both commercial and military applications, encryption devices allow only authorized users to control the system.

Launcher Attitude Control

Attitude control systems regulate the angle of the rocket or missile. Reliable attitude controls are critical for accurate and reliable missile and rocket systems.

Stage Separation Systems

These systems hold missile and rocket stages together, and release on command to allow the stages to separate. Consistent stage separation is a key component of reliable missile and rocket systems.

Stage Coupling Load Analysis

Stage coupling load analysis allows missile and rocket technicians to determine forces exerted on stage couplings, as well as shocks to sensitive payloads during stage separation. These analyses allow payload designers to build warheads and satellites that will survive the rigors of being launched into space.

Business Development

By facilitating U.S. satellite companies' use of Chinese space launch facilities, the Administration ensured business for Chinese space and rocket operations that would otherwise be seriously underutilized. In addition, these launches provided hard currency (approximately $500 million to date) for development of China's own missile systems.

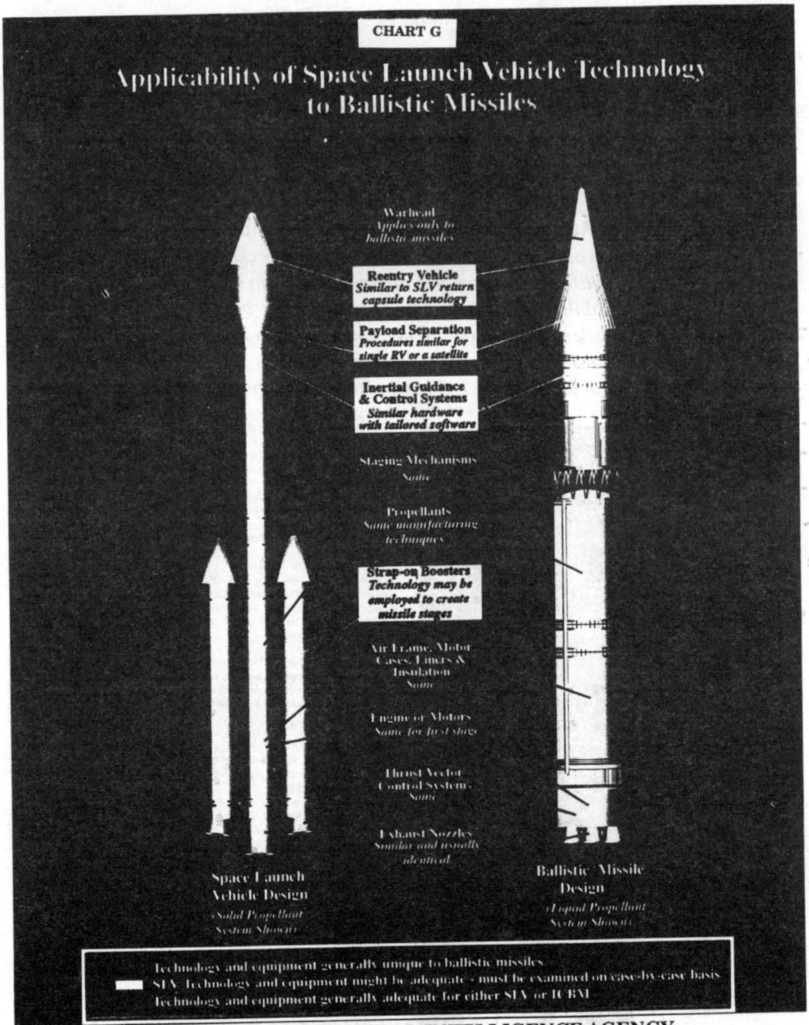

PROVIDED BY CENTRAL INTELLIGENCE AGENCY
5/14/98

CHART H

A comparison of Reagan/Bush Administration and Clinton Administration satellite export control policies

Reagan/Bush Administrations	Clinton Administration
Three waivers granted; one denial for satellite components	Ten waivers granted
DOD effective veto over U.S. satellite exports	DOD downgraded to one of four agencies, a majority of which is required to stop an export
National security principal factor in evaluating satellite export licenses	National security no longer preeminent
Missile proliferation sanctions applied to all satellite exports	All commercial communications satellites effectively exempted from missile proliferation sanctions
DOD monitors required for all launches	Inconsistent policy regarding launch monitors
The licensing requirements for technical data were clear and unambiguous	The Commerce system does not clearly control technical data

Provided by Congressman Duncan Hunter

SEP 25 1998

COMMERCIAL COMMUNICATION SATELLITES TO CHINA

SATELLITE PROGRAM	Satellite Manufacturer	License Agency	All Agencies Approve	State License	TTCP	Monitoring			Waiver Agency	Waiver Date	Launch Date	Launch Outcome
						Pre-Launch	Launch	Post-Launch Failure Analysis				
Asiasat 1	Hughes	State	Yes	Yes	Yes	Yes	Yes	n/a	State	12/89	4/90	normal
Optus B1	Hughes	State	Yes	Yes	Yes	Yes	Yes	n/a	State	4/91	8/92	normal
Optus B2	Hughes	State	Yes	Yes	Yes	Yes	Yes	Yes	State	12/89	12/92	failed
Apstar 1	Hughes	State	Yes	Yes	Yes	Yes	Yes	n/a	State	9/92	7/94	normal
Intelsat VII-A	Loral	State	Yes	Yes	Yes	Yes	Yes	No	State	9/92	2/96	failed
Optus B3	Hughes	Commerce	Yes	No	No	No	Yes	n/a	State	12/89	8/94	normal
Apstar 2	Hughes	Commerce	Yes	No	No	No	No	No	State	9/92	1/95	failed
Asiasat 2	Martin-Marietta	Commerce	Yes	Yes	Yes	Yes	Yes	n/a	State	9/92	11/95	normal
Echostar 1	Martin-Marietta	Commerce	Yes	No	No	Yes	Yes	n/a	Commerce	7/94	12/95	normal
Echostar 2	Martin-Marietta	Commerce	Yes	No	Yes	Cancelled	Cancelled	n/a	Commerce	7/94	Cancelled	Cancelled
Intelsat 805	Martin Marietta	Commerce	Yes	No	No	Cancelled	Cancelled	n/a	State	7/93	Cancelled	Cancelled
Apstar 1A	Hughes	Commerce	Yes	No	No	No	No	n/a	State	9/92	7/96	normal
Apstar 2R	Loral	Commerce	Yes	Yes	Yes	Yes	Yes	n/a	State	9/92	10/97	normal
Chinasat 7	Hughes	Commerce	Yes	No	No	No	No	n/a	Commerce	2/96	8/96	failed in orbit
Mabuhay	Loral	Commerce	Yes	Yes	Yes	Yes	Yes	n/a	Commerce	2/96	8/97	normal
Chinastar 1	Lockheed Martin	Commerce	Yes	No	No	No	Yes	n/a	Commerce	2/96	5/98	normal
Iridium	Motorola	Commerce	Yes	Yes	Yes	Yes	Yes	n/a	State	7/93	8/98	normal

Satellite Program: name of program and date of license approval. Satellites with the same name (and different numbers/letters) were bundled into a single Tiananmen waiver. (E.g., Apstar 2, Apstar 1A, and Apstar 2R have a single waiver.)
All Agencies Approve: State, Defense and Commerce all recommended approval of the license.
State License: TAA or other approval
TTCP: Technology Transfer Control Plan.
Monitoring: **Pre-launch monitoring** includes a requirement for submission of technical data to DoD for approval and monitors at all technical interchange meetings.
Launch monitors from DoD were required/present during launch. The U.S. has the right to monitor any launch from China.
Post-launch Failure Analysis - this column indicates whether companies participated in a post-launch failure analysis and whether this process was monitored by DoD. This column only applicable to launch failures.

CHART I

Date of Transfer or Report	Reported Transfer by China	Possible Violation	Administration's Response
Nov. 1992	M-11 missiles or related equipment to Pakistan (The Administration did not officially confirm reports that M-11 missiles are in Pakistan.)	MTCR Arms Export Control Act Export Administration Act	sanctions imposed on Aug. 24, 1993, for transfers of M-11 related equipment (not missiles); waived on Nov. 1, 1994
Mid-1994 to mid-1995	dozens or hundreds of missile guidance systems and computerized machine tools to Iran	MTCR Iran-Iraq Arms Nonproliferation Act Arms Export Control Act Export Administration Act	no sanctions
2nd quarter of 1995	parts for the M-11 missile to Pakistan	MTCR Arms Export Control Act Export Administration Act	no sanctions
Dec. 1994 to mid-1995	5,000 ring magnets for an unsafeguarded nuclear enrichment program in Pakistan	NPT Export-Import Bank Act Nuclear Proliferation Prevention Act Arms Export Control Act	considered sanctions under the Export-Import Bank Act; but announced on May 10, 1996, that no sanctions would be imposed
July 1995	more than 30 M-11 missiles stored in crates at Sargodha Air Force Base in Pakistan	MTCR Arms Export Control Act Export Administration Act	no sanctions
Sept. 1995	calutron (electromagnetic isotope separation system) for uranium enrichment to Iran	NPT Nuclear Proliferation Prevention Act Export-Import Bank Act Arms Export Control Act	no sanctions

1995-1997	C-802 anti-ship cruise missiles and C-801 air-launched cruise missiles to Iran	Iran-Iraq Arms Nonproliferation Act	no sanctions
before Feb. 1996	dual-use chemical precursors and equipment to Iran's chemical weapon program	Arms Export Control Act Export Administration Act	sanctions imposed on May 21, 1997
summer 1996	400 tons of chemicals to Iran	Iran-Iraq Arms Nonproliferation Act[1] Arms Export Control Act Export Administration Act	no sanctions
Aug. 1996	plant to manufacture M-11 missiles or missile components in Pakistan	MTCR Arms Export Control Act Export Administration Act	no sanctions
Aug. 1996	gyroscopes, accelerometers, and test equipment for missile guidance to Iran	MTCR Iran-Iraq Arms Nonproliferation Act Arms Export Control Act Export Administration Act	no sanctions
Sept. 1996	special industrial furnace and high-tech diagnostic equipment to unsafeguarded nuclear facilities in Pakistan	NPT Nuclear Proliferation Prevention Act Export-Import Bank Act Arms Export Control Act	no sanctions
July-Dec. 1996	Director of Central Intelligence (DCI) reported "tremendous variety" of technology and assistance for Pakistan's ballistic missile program	MTCR Arms Export Control Act Export Administration Act	no sanctions

[1] Additional provisions on chemical, biological, or nuclear weapons were not enacted until February 10, 1996.

July-Dec. 1996	DCI reported "tremendous variety" of assistance for Iran's ballistic missile program	MTCR Iran-Iraq Arms Nonproliferation Act Arms Export Control Act Export Administration Act	no sanctions
July-Dec. 1996	DCI reported principal supplies of nuclear equipment, material, and technology for Pakistan's nuclear weapon program	NPT Nuclear Proliferation Prevention Act Export-Import Bank Act Arms Export Administration Act	no sanctions
July-Dec. 1996	DCI reported key supplies of technology for large nuclear projects in Iran	NPT Iran-Iraq Arms Nonproliferation Act Nuclear Proliferation Prevention Act Export-Import Bank Act Arms Export Administration Act	no sanctions
July-Dec. 1996	DCI reported "considerable" chemical weapon-related transfers of production equipment and technology to Iran	Iran-Iraq Arms Nonproliferation Act Arms Export Control Act Export Administration Act	no sanctions
Jan. 1997	dual-use biological items to Iran	BWC Iran-Iraq Arms Nonproliferation Act Arms Export Control Act Export Administration Act	no sanctions
1997	chemical precursors, production equipment, and production technology for Iran's chemical weapon program, including a plant for making glass-lined equipment	Iran-Iraq Arms Nonproliferation Act Arms Export Control Act Export Administration Act	no sanctions

55-784 99 - 13

Sept. to Dec. 1997	China Great Wall Industry Corp. provided telemetry equipment used in flight-tests to Iran for its development of the Shahab-3 and Shahab-4 medium range ballistic missiles	MTCR Iran-Iraq Arms Nonproliferation Act Arms Export Control Act Export Administration Act	no sanctions
Nov. 1997/ April 1998	may have transferred technology for Pakistan's Ghauri medium-range ballistic missile that was flight-tested on April 6, 1998	MTCR Arms Export Control Act Export Administration Act	no sanctions

CHART J CHINESE MISSILE LAUNCHES

1	1	1	04/24/1970	70034	CZ-1	Jiuquan	DFH-1		
2	2	2	03/03/1971	71018	CZ-1	Jiuquan	SJ 1		
3	1	1	09/18/1973		FB-1	Jiuquan	JSSW ?		Failure
4	2	2	07/14/1974		FB-1	Jiuquan	JSSW ?	2nd stage engine lost thrust	Failure
5	3	1	11/05/1974		CZ-2	Jiuquan	FSW ?		Failure
6	3	3	07/26/1975	75070	FB-1	Jiuquan	JSSW-1		
7	4	2	11/26/1975	75111	CZ-2	Jiuquan	FSW-1		
8	4	4	12/16/1975	75119	FB-1	Jiuquan	JSSW-2		
9	5	5	08/30/1976	76087	FB-1	Jiuquan	JSSW-3		
10	5	3	12/07/1976	76116	CZ-2	Jiuquan	FSW-2		
11	6	6	11/10/1976		FB-1	Jiuquan	JSSW ?		Failure
12	6	4	01/26/1978	78011	CZ-2	Jiuquan	FSW-3		
13	7	7	07/27/1979		FB-1	Jiuquan	SJ ?		Failure
14	8	8	09/19/1981	81093	FB-1	Jiuquan	SJ 2 SJ 2A SJ 2B		
15	7	1	09/09/1982	82090	CZ-2C	Jiuquan	FSW-4		
16	8	2	08/19/1983	83086	CZ-2C	Jiuquan	FSW-5		
17	9	1	01/29/1984	84008	CZ-3	Xichang	DFH-2-0	3rd Stage early shutdown	Failure
18	10	2	04/08/1984	84035	CZ-3	Xichang	DFH-2-1		
19	11	3	09/12/1984	84098	CZ-2C	Jiuquan	FSW-6		
20	12	4	10/21/1985	85096	CZ-2C	Jiuquan	FSW-7		
21	13	3	02/01/1986	86010	CZ-3	Xichang	DFH-2-2		
22	14	5	10/06/1986	86076	CZ-2C	Jiuquan	FSW-8		
23	15	6	08/05/1987	87067	CZ-2C	Jiuquan	FSW-9		
24	16	7	09/09/1987	87075	CZ-2C	Jiuquan	FSW-10		
25	17	4	03/071988	88014	CZ-3	Xichang	DFH-2-3		
26	18	8	08/05/1988	88067	CZ-2C	Jiuquan	FSW-11		
27	19	1	09/06/1988	88080	CZ-4A	Taiyuan	FY-1A		
28	20	5	12/22/1988	88111	CZ-3	Xichang	DFH-2-4		
29	21	6	02/04/1990	90011	CZ-3	Xichang	DFH-2-5		
30	22	7	04/07/1990	90030	CZ-3	Xichang	Asiasat 1	First commercial launch	
31	23	1	07/16/1990	90059	CZ-2E	Xichang	Optus Model Badr A		
32	24	2	09/03/1990	90081	CZ-4A	Taiyuan	FY-1B DQ-1A DQ-1B		
33	25	9	10/05/1990	90089	CZ-2C	Jiuquan	FSW-12		
34	26	8	12/28/1991	91088	CZ-3	Xichang	DFH-2-6	3rd Stage early shutdown	Failure
35	27	1	08/09/1992	92051	CZ-2D	Jiuquan	FSW-13		

16 successful launches in a row

36	28	2	08/13/1992	92054	CZ-2E	Xichang	Optus B1		
37	29	10	10/06/1992	92064	CZ-2C	Jiuquan	FSW-14 Freja		
38	30	3	12/21/1992	92090	CZ-2E	Xichang	Optus B2	payload fairing collapsed	Failure
39	31	11	10/08/1993	93063	CZ-2C	Jiuquan	FSW-15		
40	32	1	02/08/1994	94010	CZ-3A	Xichang	DHF-3 Model SJ-4		
41	33	2	07/03/1994	94037	CZ-2D	Jiuquan	FSW-16		
42	34	9	07/21/1994	94043	CZ-3	Xichang	Apstar 1		
43	35	4	08/27/1994	94055	CZ-2E	Xichang	Optus B3		
44	36	2	11/29/1994	94080	CZ-3A	Xichang	DFH-3-1		
45	37	5	01/25/1995		CZ-2E	Xichang	Apstar 2	destroyed due to wind shear	Failure
46	38	6	11/28/1995	95064	CZ-2E	Xichang	Asiastar 2		
47	39	7	12/28/1995	95073	CZ-2E	Xichang	Echostar 1		
48	40	1	02/14/1996		CZ-3B	Xichang	Intelsat 708	attitude control lost	Failure
49	41	10	07/03/1996	96039	CZ-3	Xichang	Apstar 1A		
50	42	11	08/18/1996	96048	CZ-3	Xichang	Chinasat 7	3rd Stage early shutdown	Failure
51	43	3	10/20/1996	96059	CZ-2D	Jiuquan	FSW-17		
52	44	3	05/11/1997	97021	CZ-3A	Xichang	DFH-3-2		
53	45	12	06/10/1997	97029	CZ-3	Xichang	FY-2B		
54	46	2	08/19/1997	97042	CZ-3B	Xichang	Agila 2		
55	47	1	09/01/1997	97048	CZ-2C/SD	Taiyuan	Iridium MFS 1 Iridium MFS 2		
56	48	3	10/16/1997	97062	CZ-3B	Xichang	Apstar 2R		
57	49	2	12/08/1997	97077	CZ-2C/SD	Taiyuan	Iridium 42 Iridium 44		
58	50	3	03/25/1998	98018	CZ-2C/SD	Taiyuan	Iridium 51 Iridium 61		
59	51	4	05/02/1998	98026	CZ-2C/SD	Taiyuan	Iridium 69 Iridium 71		
60	52	4	05/30/1998	98033	CZ-3B	Xichang	Chinastar-1		

10 successful launches in a row

#1: all launches
#2: all launches of CZ or FB
#3: launches of specific model

Red : Failure or partial failure

CZ: Chang Zheng (Long March)
FB: Feng Bao (Storm)
DFH: Dong Fang Hong (Red is East)
SJ: Shi Jian (Practice)
FY: Feng Yun (Wind and Cloud)
DQ: Da Qi (Atmosphere)
FSW: Fanhui Shi Weixing (Recoverable Satellite)

QUESTIONS AND ANSWERS SUBMITTED FOR THE RECORD

JUNE 17, 18, 23, 1998

QUESTIONS SUBMITTED BY MR. SPENCE AND MR. GILMAN

Chairman SPENCE and Chairman GILMAN. Mr. Lodal claims that the multiple satellite dispensing technology China is mastering in launching Iridium satellites is unrelated to multiple independently targetable reentry vehicle (MIRV technology). Do you agree?

Mr. SOKOLSKI. No. As the Defense Department's own Air Force National Air Intelligence Center recently reported, a "minimally modified smart dispenser could be used to deploy multiple reentry vehicles" (See Bill Gertz, "Chinese Missile Enhanced by U.S.", The Washington Times, July 14, 1998). Their findings should hardly be surprising: MIRVs and precise satellite dispensers are interchangeable technologies. U.S. MIRVs, in fact, were derived from commercial multiple satellite dispensers and the Russians' current multiple satellite dispensers (including for Iridium launchings) are derived from its MIRVs.

Chairman SPENCE and Chairman GILMAN. Mr. Reinsch and others have pointed out that China had 16 straight launch vehicle successes from 1984 to 1990 and have argued that, therefore, the last 10 launch successes are of little significance. Do you agree?

Mr. SOKOLSKI. Not at all: What was required to secure China's launch successes before 1990 was nothing as compared to what its last ten successes required since 1996. China's pre-1990 successes, in fact, involved two of its crudest (and earliest) Long March launcher systems—the Long March 2 and 2C—launching the most primitive—relatively small and rugged—of Chinese payloads. In contrast, the recent string of launch failures that followed these successes (from 1990 through 1996) involved more sophisticated Long March rocket variants launching larger, more fragile U.S.-made commercial satellites. It is significant that China's record of failures involving such relatively advanced systems ended with the last Hughes failure in 1996 and that since then even the most complicated launchings involving dispensing of multiple U.S. communications satellites have been accomplished with a 100% success rate.

Chairman SPENCE and Chairman GILMAN. Do you agree with Mr. Reinsch's contention that because China used five different launch systems to produce the 10 launch successes since 1996, that these successes are far less significant that you have suggested in your testimony?

Mr. SOKOLSKI. Again, no. Where before Mr. Reinsch seems to have made too little of the significant difference between the very earliest variants of the Long March rocket system and the very latest, here he seems to make far to much of the relatively minor differences between the latest, most sophisticated ones used since 1996—the Long March 2D and 2D/SD, 3, 3A, and these most recent versions of the Long-March are all designed and manufactured by Chinese entities who freely share performance data relating to their launch of U.S.-made commercial satellites. Technology that might be transferred or mastered in the launching of one U.S.-made satellite on any of these most recent Long March variants can and is used to assist in the enhancement of the others.

Chairman SPENCE and Chairman GILMAN. Have any satellite launcher systems ever been used by nations other than China to help perfect their missile systems? If so, how?

Mr. SOKOLSKI. Yes. Russia currently uses its SL 8 space launch system to test modern warhead packages for its modern ballistic missiles. New, sophisticated warhead delivery systems are mounted on the SL 8, fired on a short-range, high altitude trajectory, and blasted downward toward earth by the launcher's final stage to achieve high reentry speeds. This allows test results similar to what would otherwise be gained from a full range shot but makes it more difficult for other nations to get at the telemetry information and reduces the risks of accidents that otherwise would be run downrange. India's space launch programs are also suspected of assisting in India's development of long-range missiles.

Chairman SPENCE and Chairman GILMAN. DoD has admitted that the U.S. team was kept in the bunker for hours after the 1996 Loral launch failure. The Department has also admitted that after they were released and they sought to recover the satellite's command and control device they found it intact, but with its encryption car missing. How important is it that the Chinese may have stolen this encryption technology?

Mr. SOKOLSKI. China's possible theft of these cards constitutes a serious long-term security problem. This is so even if possible duplication of the cards turns out to be technically less significant than what some experts fear. Certainly this much is clear: We chose not to make the incident public or to insist on getting the cards back before transferring additional satellites, and its timidity is only likely to embolden the Chinese to take additional steps to divert protected U.S. technology in the future.

Chairman SPENCE and Chairman GILMAN. It was recently claimed that GPS is six times more important to China's development of missiles than any technology that might have been gained from the transfer of U.S. satellite technology. Do You agree?

Mr. SOKOLSKI. As useful as GPS may be in increasing long-range missile accuracy, it is of little utility in the development or enhancement of a rocket system's attitude and engine controls, perfecting load coupling analysis (critical to launch fragile payloads), or the development of upper stage technology (needed to launching satellites and warheads precisely or in numbers that might challenge future missile defenses). U.S. satellite technology transfers, in contrast, can be helpful in all of these areas.

Chairman SPENCE and Chairman GILMAN. Is it not true that President Bush approved the transfer of one third of commercial satellites and related technology from the Munitions List to the Commodity Control list? Did not the Bush Administration plan to begin transferring more sensitive satellite technologies to Commerce in 1993?

Mr. SOKOLSKI. Yes.

Chairman SPENCE and Chairman GILMAN. You have indicated that many of the most sensitive kick-motor and encryption technologies were moved to Commerce control in 1996. Are these technologies not embedded in the satellite itself and delivered to the customer in outer-space 22,000 miles above the earth?

Mr. SOKOLSKI. Although most U.S. satellite are shipped with U.S.-made kick motors, this is not always the case. In the case of Asiasat II, a Chinese designed the kick motor was mated to a U.S.-made satellite. The concern in this case, was that U.S. aerospace contractors wanted to make sure that the Chinese motor would work and might be tempted to validate the Chinese design through discussions, analysis or data exchanges. U.S. contractors asked and were allowed to observe a test firing of this kick motor. Although State controls call for such tests to be overseen by U.S. government rocket engineer monitors, no such monitors were present. It is unknown what, if anything, U.S. contractors or foreign national (i.e., the customers who are not bound by U.S. law) might have conveyed to help validate the kick motor's design as a result of this test. The kick motor, however, did work without a hitch.

Chairman SPENCE and Chairman GILMAN. Is it not true that these satellites are under U.S. guard prior to launch? It seems that encryption technologies are software driven, not hardware driven, and would be destroyed by a massive explosion such as a launch failure. Is it possible to recover and reengineer this technology after a launch failure?

Mr. SOKOLSKI. All U.S. satellites should be under U.S. guard. Some, I understand may now, instead, be under private contractor guard. There is a concern that the Chinese might reengineer encryption chips it may have recovered from a launch failure but the real concern about this incident has to do with security of a more general sense (see question 5).

Chairman SPENCE and Chairman GILMAN. In your expert opinion, is it possible to launch U.S. commercial satellites on Chinese rockets without transferring technology and experience that can be applied to China's ballistic missile program?

Mr. SOKOLSKI. In practice no. This is why the House vote to embargo further transfers of U.S. satellite technology deserves more consideration. Certainly, transferring new types of U.S.-made satellites will transfer even more new sensitive technology. Ultimately, it would be best for U.S. security if U.S. firms could be given the incentives needed to do all of their launch work again from U.S. sites.

Chairman SPENCE and Chairman GILMAN. How much have these launches helped China? A little or a lot?

Mr. SOKOLSKI. A lot. Using $50 million per launch as an average price for the 14 launches of U.S.-made satellites China has hosted since 1990s, at least a half billion dollars in hard currency has flowed from the West to China's rocket program. Most industry experts believe that the real cost of conducting these launches was far less than the price charged. Much of the "profit" has gone into the development of China's various launch sites. In addition to this money, the U.S. aerospace industry has helped educate China's space engineers with hands-on experience launching payloads far more sophisticated than any it had every launched before.

Chairman SPENCE and Chairman GILMAN. What procedures are satellite manufacturers required to undertake to launch in China?

Mr. SOKOLSKI. The procedures now vary depending on the licensing decisions at State and Commerce and how the contractor himself decides to implement these licenses in consultation with Commerce and State officials.

Chairman SPENCE and Chairman GILMAN. How should launch failure reviews be regulated by the U.S. government.

Mr. SOKOLSKI. In the case of China, with complete, competent U.S. government oversight.

Chairman SPENCE and Chairman GILMAN. Mr. Lodal, you stated that "the standards and requirements, both the technical standards and the procedural standards and requirements for satellite release, are much lower than those required for MIRVs . . ."

Please describe in detail the precise technical and procedural standards and requirements for both multiple satellite and multiple warhead dispersals and describe their similarities and differences?

How relevant are the different standards and requirements for rogue regimes seeking to threaten the United States with multiple warhead dispensers where accuracy is not important?

Secretary LODAL. There has been one licensed program launched by China that involved a satellite dispenser system—the Motorola Iridium project. Each launch involved two satellites. The dispenser was designed, developed, and produced by the Chinese based on requirements provided by Motorola. The requirements as well as all technical interactions with the Chinese on this matter were monitored by DoD. In any case, the ultimate objectives of a MIRV capability and a satellite dispenser capability are significantly different. The objective of a satellite release into a parking or transition orbit does not require much accuracy and allows for a wide margin of error. The objective of a MIRV system is extremely accurate reentry with release timing and orientation providing the only mechanism to ensure that the warhead reaches its target. Although the tasks are similar, the standards and requirements (both technical and procedural) for satellite release are far lower than those required for a MIRV. If the error is large for MIRVs, then there is a substantial risk that the vehicle will burn up on reentry.

Chairman SPENCE and Chairman GILMAN. Mr. Lodal, you stated that a commercial communications satellite may have "technology embedded in it that could be useful [to others]. And if they could take it apart, or figure out how to build one just like it, or learn how to build more just like that, that would be useful." One of these embedded technologies with military application is an encryption chip. According to subsequent testimony, the Loral-built Intelsat-708 satellite that was destroyed in the February 1996 launch failure contained such a chip, which was never recovered and may have been confiscated by China.

Have we asked the Chinese if they have the chip?

What is your assessment of the utility of this technology to China?

Since encryption is one of the nine embedded technologies considered to be militarily sensitive, how do you reconcile your earlier statement with Mr. Reinsch's comment that loss of the encryption chip to China as a result of the failed February 1996 launch "wouldn't have had any effect on our national security?"

If this encryption device is in the hands of the Chinese government, doesn't that tell us the Chinese are willing to risk harming relations with the United States to steal our technology? Shouldn't this tell us about the questionable reliability of export control agreements with the Chinese?

Secretary LODAL. The cryptographic equipment and associated material for the Space Systems/Loral INTELSAT 708 satellite which was destroyed during the launch failure on February 15, 1996 was incorporated into two integrated circuit boards. The National Security Agency has determined that the cryptographic equipment that was aboard the INTELSAT 708 satellite also was destroyed and not recovered. NSA made this determination on the basis of reports and other information provided by SS/Loral.

DoD is not aware of any efforts to ask the Chinese if they have the cryptographic equipment.

Chairman SPENCE and Chairman GILMAN. Mr. Lodal, you testified that you "do not believe that there has been any transfer of these sensitive technologies [to China]. We do not have any evidence that there has been transfer of these sensitive technologies to their military programs that have made an impact on the Chinese military programs."

How can you assert that there has been no transfer of sensitive technologies to China when you acknowledge that DoD monitors were not always present to guard against unauthorized technology transfers?

Are you arguing that you do not believe China confiscated the missing encryption chip from the February 1996 failed launch?

Do you believe the lack of evidence of a transfer proves that a transfer did not occur?

How does your statement track with Secretary Slocombe's testimony that "sometimes the most important technology transfers take place . . . by comments that are made and information that is passed?"

Secretary LODAL. All that we are saying is that we, in the Department of Defense, are not aware of any transfers of sensitive technologies in connection with satellite launches in China that were licensed and monitored by DoD.

The Cryptographic equipment and associated material for the Space Systems/Loral Intelsat 708 satellite that was destroyed during the launch failure on February 15, 1996 was incorporated into two integrated circuit boards. The National Security Agency has determined that the cryptographic equipment that was aboard the INTELSAT 708 satellite also was destroyed and not recovered. NSA made this determination on the basis of reports and other information provided by SS/Loral.

DoD is not aware of any efforts to ask the Chinese if they have the cryptographic equipment.

Chairman SPENCE and Chairman GILMAN. Mr. Lodal, you stated that "there was a period of time until 1996 where in principle the monitoring was not always required. It was purely commercial and no State Department licenses were required, the monitoring was not always required, and therefore there were three [unmonitored] launches which took place . . . Our analysis and recommendation was that we should nonetheless require the monitoring in all cases, to eliminate any ambiguity."

What "ambiguity" existed that led to the Department to support a requirement for full monitoring in all instances?

Did the requirement for full monitoring that was put into place in 1996 result from the decision to transfer licensing jurisdiction for commercial communications satellites from State to Commerce or did it result from the lack of full monitoring in the 1996 launch failure and subsequent analysis?

Why was full monitoring not instituted after the 1992 launch failure?

Secretary LODAL. During the period 1993–1996, monitoring by the U.S. government was required in all licenses for launches in China of communications satellites that contained one or more of the identified military-related technologies, any launch vehicle integration technical data or any technical assistance—that is, monitoring by the U.S. government was required in all the licenses issued by the Department of State. There were three launches during this period that were not monitored. DoD license review procedures in the 1993–96 period anticipated that there would be at least one State license required for launch-related services of commercial satellites licensed by the Department of Commerce. However, these three launches apparently were deemed by the exporter not to involve activities that required any State Department license. Subsequently, at the behest of DoD, this ambiguity was resolved by a decision in connection with the 1996 change in license jurisdiction for comsats to impose conditions on Commerce-licensed comsats for launch in China similar to those that had been imposed on State Department comsat licenses, including requirements for DoD monitoring.

In DoD's view, there has been no ambiguity about the status of an investigation of a launch failure performed for the Chinese by a U.S. firm. Such an investigation constitutes a "defense service" which is subject to State Department licensing jurisdiction and DoD monitoring requirements.

Chairman SPENCE and Chairman GILMAN. Mr. Tarbell, how many copies of the May 1997 DTSA report were produced and how many exist today? Were any copies destroyed?

Mr. TARBELL. The original of the DoD report, which I signed on May 16, 1997, was hand delivered to Mr. Will Lowell, Director of State's Office of Defense Trade Controls (DTC), and a copy was delivered to Ms. Eva Tyler, the DTC action officer. Copies are maintained by me and certain members of my staff.

On June 17, 1998, DoD provided one copy to each of the following:

1. Senators Strom Thurmond and Carl Levin, Senate Armed Services Committee.
2. Senators Ted Stevens and Daniel Inouye, Senate Appropriations Committee, Subcommittee on Defense.
3. Representatives Bill Young and John Murtha, House Appropriations Committee, Subcommittee on National Security.

4. Representatives Floyd Spence and Ike Skelton, House National Security Committee.

Copies of this report have also been provided to the Department of Justice and to House Select Committee on National Security and Military/Commercial Concerns with the people's Republic of China.

DTSA maintains record copies of this report. No copies have been destroyed other than that which is incidental to the daily course of business (e.g., defective or incomplete copies, etc). DTSA has no knowledge of how many copies have been reproduced or destroyed by other parties.

Chairman SPENCE and Chairman GILMAN. Mr. Slocombe, Secretary Lodal stated last week that DoD monitoring was not always required for launches on Chinese launch vehicles, that there were three launches that did not require DoD monitoring, and that full monitoring has been required on all launches only since 1996. He explained this by stating, "Our analysis and recommendation was that we should nonetheless require the monitoring all cases, to eliminate any ambiguity."

Was the decision to require monitors a result of the failed February 1996 launch, subsequent unmonitored discussions between Loral and the Chinese, or the decision to move licensing jurisdiction from State to Commerce one month later?

What embedded military technologies did the Loral satellite that crashed in 1996 contain?

How many of the waivers approved by the Administration were for the launch of satellites containing embedded military technologies? Were all the satellites purely commercial?

Secretary SLOCOMBE. The decision to impose conditions on Commerce-licensed comsats, including requirements for DoD monitoring similar to those imposed on State Department comsat licenses, was made in connection with the 1996 transfer of all complete commercial comsats to Commerce licensing jurisdiction. The decision to require monitors was not the result of the failed February 1996 launch of the Loral Intelsat 708.

Category XV(f) of the U.S. Munitions List (USML) (22 C.F.R. Part 121.1 in the International Traffic in Arms Regulations (ITAR), under the regulatory authority of the Department of State, controls nine systems, components, or parts associated with satellites. They are: antijam systems, certain antennas, intersatellite data relay links, space regenerative baseband processing, certain radiation-hardened microelectronic circuits, certain propulsion, systems, certain attitude control and determination systems, orbit transfer engines, and cryptographic items. Twelve programs for launch in China used satellites that included some of these systems, components, or parts. Each of these programs were subject to the provisions of Section 902 of the Foreign Relations Authorization Act for FY 1990 and 1991, 22 U.S.C. sec. 2151 note), "Tiananmen Square sanctions." Each proceeded under a Tiananmen Square sanctions "waiver." The following is a list of the waived programs and the USML system, component, or part included with the satellite:

Aussat/OPTUS: Radiation-hardened microelectronic circuits, Encryption.

Asiasat: Radiation-hardened microelectronic circuits.

Asiasat–2: Radiation-hardened microelectronic circuits.

Apsat/Apstar: Radiation-hardened microelectronic circuits.

Iridium: Radiation-hardened microelectronic circuits, Intersatellite data relay links, Antenna.

Intelsat 708: Radiation-hardened microelectronic circuits, Encryption.

APMT: Radiation-hardened microelectronic circuits, Antenna.

Chinasat–7: Radiation-hardened microelectronic circuits.

Chinasat–8: Radiation-hardened microelectronic circuits.

COSAT/Chinastar–1: Radiation-hardened microelectronic circuits.

Echostar: Radiation-hardened microelectronic circuits.

Mabuhay: Radiation-hardened microelectronic circuits.

All of these programs are for commercial communication satellites.

As indicated in the list, most programs included certain chips primarily associated with critical functions for operating the satellite (for example, the mission computer). These chips are radiation-hardened for these critical functions to ensure that the satellite can operate in normal high-radiation environments in space. The critical technology that we seek to protect is the process for manufacturing radiation-hardened microcircuits. These chips are embedded in the satellite and the Chinese launch service provider would be unable to reverse-engineer such chips without removing them from the satellite and performing destructive testing of the chips. The export licenses associated with these programs permit some boards containing these chips to be shipped as spares along with the satellite. However, those licenses also require that these boards/chips remain under the strict supervision and control of the satellite manufacturer at all times, and be returned to the U.S. after the launch, if not used. Any attempt by the PRC to conduct destructive testing under these circumstances would be readily detectable. We have received no reports of such efforts.

The Iridium program was authorized under both State and Commerce licenses to include intersatellite data relay links (so-called "cross-links") and antennas with certain characteristics controlled under the ITAR. The "cross-links" are essential to the Iridium system to allow for transfer of communications signals between satellites because the system is designed to provide global communications. The links are embedded in the satellite payload and cannot be reverse-engineered without removing the equipment from the satellite and conducting extensive testing and analysis. All Iridium satellite launches from China were monitored by DoD, and DoD has no information that the Chinese have attempted to remove or tamper with the links.

The Iridium antennae are controlled by the ITAR because their spot beam coverage is less than 200 nautical miles in diameter. This is necessary because the Iridium constellation is in low earth orbit and signals must be received directly from individual handsets. The antenna cannot be reverse-engineered without access to the satellite and removal of the antenna system from the satellite. As stated earlier, all Iridium satellite launches from China were monitored by DoD, and DoD has no information that the Chinese have attempted to remove or tamper with the antenna.

The export license for the original APMT program authorized a large antenna controlled under the ITAR, but this satellite was never launched. The APMT program is being modified and separate license applications are now under review.

The Optus and Intelsat 708 programs that were licensed by State included encryption for telemetry, tracking and control that is uniquely keyed for each satellite. The encryption chips are embedded and, as with radiation-hardened chips, cannot be reverse-engineered without access to the satellite and significant testing and analysis. The Optus and Intelsat 708 launches were monitored by DoD, and DoD has no information that the Chinese attempted to remove or tamper with the encryption chips.

The only items controlled under ITAR Category XV (f) that were included with the failed Loral Intelsat 708 were certain radiation-hardened microelectronic circuits and encryption chips. The launch of the Intelsat 708 was licensed by the Department of State and was monitored by DoD. DoD has provided information separately about the encryption chips associated with the Loral Intelsat 708 launch failure.

Chairman SPENCE and Chairman GILMAN. Mr. Slocombe, did the change of encryption algorithms in 1996 result from the loss of the encryption chip from the failed launch of the Loral Intelsat-708 satellite in February 1996? Or was it the result of a regularly scheduled routine algorithm upgrade? If the latter, when was this upgrade originally scheduled? Why did the NSA originally indicate that the change was made as a result of the failed launch and the risk that the encryption chip could be compromised? How often are the algorithms changed or upgraded? Please provide the committees with a listing of all encryption algorithm changes and upgrades that have taken place since 1990.

Secretary SLOCOMBE. NSA periodically changes cryptographic algorithms as technology changes and performance requirements exceed the technical limits of existing devices. In the case of the algorithm used in the devices that were lost as a result of the 1996 launch failure of the Intelsat 708 satellite, work to replace that algorithm began in 1992. The technology and algorithm being used in satellites in development today is no longer the same as a result of that work and the normal process of technology upgrades. The change was not related to, or a result of, the loss of the Loral satellite.

We regret that NSA's original statement may have inadvertently led some to believe that there was a causal relationship between the crash and the change in the encryption algorithms. There was none.

Chairman SPENCE and Chairman GILMAN. Was the Defense Department aware of press reports in 1993 noting that a cooperative investigation had been conducted with China regarding the reasons for a launch failure in December 1992? Was the Department aware of similar press stories in 1995 citing a similar cooperative investigation regarding the launch failure in January 1995 and quoting a joint statement by Hughes and China Great Wall Industries Corporation noting that they "would work together to eliminate the two mentioned possible causes of failure?" If the Department was aware of these reports, why was no immediate action taken? If the Department was unaware of these reports, doesn't that reflect a complete breakdown in the system?

Mr. TARBELL. The investigation conducted following the failure of the Optus B–2 satellite launch in December 1992 was monitored by DoD. On the other hand, DoD did not monitor the Hughes' assistance to the Chinese in the investigation of the 1995 launch failure, nor did Hughes obtain a license from the State Department. However, among the DTSA documents identified in connection with congressional requests for production of documents concerning this and related matters, we found a May 1995 memo for the record written by a former DTSA staffer questioning whether Hughes was conducting an unauthorized launch failure investigation. This matter was not brought to my attention at the time, nor, to the best of my knowledge, to the attention of other senior managers in DTSA. Had this matter been brought to my attention, I would have raised this issue promptly with the State Department, which has licensing and compliance jurisdiction over launch failure investigations.

Chairman SPENCE and Chairman GILMAN. Did any commercial satellite launches for which waivers were granted involve satellites that contained any military sensitive hardware or software? Were these all "pure" communications satellites? Did any of them contain radiation-hardened chips or components?

Mr. Tarbell. Category XV(f) of the U.S. Munitions List (USML) (22 CFR Part 121.1) in the International Traffic in Arms Regulations (ITAR) (under the regulatory authority of the Department of State) controls nine systems, components, or parts associated with satellites. They are: antijam systems, certain antennas, intersatellite data relay links, space regenerative baseband processing, certain radiation-hardened microelectronic circuits, certain propulsion systems, certain attitude control and determination systems, orbit transfer engines, and cryptographic items. Twelve programs for launch in China used satellites that included some of these systems, components, or parts. Each of these programs were subject to the provisions of Section 902 of the Foreign Relations Authorization Act for FY 1990 and 1991 (22 U.S.C. sec. 2151 note), "Tiananmen Square sanctions." Each proceeded under a Tainanmen Square sanctions "waiver." The following is a list of the waived programs and the USML system, component, or part included with the satellite:

Aussat/OPTUS: Radiation-hardened microelectronic circuits, Encryption.

Asiasat: Radiation-hardened microelectronic circuits.

Asiasat-2: Radiation-hardened microelectronic circuits.

Apsat/Apstar: Radiation-hardened microelectronic circuits.

Iridium: Radiation-hardened microelectronic circuits, Intersatellite data relay links, Antenna.

Intelsat 708: Radiation-hardened microelectronic circuits, Encryption.

APMT: Radiation-hardened microelectronic circuits, Antenna.

Chinasat-7: Radiation-hardened microelectronic circuits.

Chinasat-8: Radiation-hardened microelectronic circuits.

COSAT/Chinastar-1: Radiation-hardened microelectronic circuits.

Echostar: Radiation-hardened microelectronic circuits.

Mabuhay: Radiation-hardened microelectronic circuits.

All of these programs are for commercial communications satellites.

As indicated in the list, most programs included certain chips primarily associated with critical functions for operating the satellite (for example, the mission computer). These chips are radiation-hardened for these critical functions to ensure that the satellite can operate in normal high-radiation environments in space. The critical technology that we seek to protect is the process for manufacturing radiation-hardened microcircuits. These chips are embedded in the satellite and the Chinese launch provider would be unable to reverse-engineer such chips without removing them from the satellite and performing destructive testing of the chips. The export licenses associated with these programs permit some boards containing these chips to be shipped as spares along with the satellite. However, those licenses also require that these boards/chips remain under the strict supervision and control of the satellite manufacturer at all times and be returned to the U.S. after the launch. Any attempt by the PRC to conduct destructive testing under these circumstances would be readily detectable. We have received no reports of such efforts.

The Iridium program was authorized under both State and Commerce licenses to include intersatellite data relay links (so-called "cross/links") and antennas with certain characteristics controlled under the ITAR. The "cross-links" are essential to the Iridium system to allow for transfer of communications signals between satellites because the system is designed to provide global communications. The links are embedded in the satellite payload and cannot be reverse-engineered without removing the equipment from the satellite and conducting extensive testing and analysis. All Iridium satellite launches from China were monitored by DoD, and DoD has no information that the Chinese attempted to remove or tamper with the links.

The Iridium antennae are controlled by the ITAR because their spot beam coverage is less than 200 nautical miles in diameter. This is necessary because the Iridium constellation is in low earth orbit and signals must be received directly from individual handsets. The antenna cannot be reverse-engineered without access to the satellite and removal of the antenna system from the satellite. As stated earlier, all Iridium satellite launches from China were monitored by DoD, and DoD has no information that the Chinese attempted to remove or tamper with the antenna.

The export license for the original APMT program authorized a large antenna controlled under the ITAR, but this satellite was never launched. The APMT program is being modified and separate license applications are now under review.

The Optus and Intelsat 708 programs that were licensed by State included encryption for telemetry, tracking and control that is uniquely keyed for each satellite. The encryption chips are embedded and, as with radiation-hardened chips, cannot be reverse-engineered without access to the satellite and significant testing and analysis. The Optus and Intelsat 708 launches were monitored by DoD, and DoD has no information that the Chinese attempted to remove or tamper with the encryption chips.

The only items controlled under ITAR Category XV (f) that were included with the failed Loral Intelsat 708 were certain radiation-hardened microelectronic circuits and encryption chips. The launch of the Intelsat 708 was licensed by the Department of State and was monitored by DoD. DoD has provided information separately about the encryption chips associated with the Loral Intelsat 708 launch failure.

Chairman SPENCE and Chairman GILMAN. Mr. Tarbell, we understand that Hughes Space & Communications has applied for a new export license for a satellite export to China that previously had been approved. We understand that this new license is necessary because Hughes desires to equip this satellite with an antenna of larger dimensions and increased capabilities. It has been written that this antenna will provide the Chinese with the ability to "eavesdrop" on countries in that region.

What military and intelligence capabilities will this provide the Chinese?

Will this allow the Chinese to eavesdrop on Japan, Korea or Taiwan?

Will this satellite be equipped with a kick motor? Will this allow the Chinese to reposition the satellite over North America?

Mr. TARBELL. Export licenses for the APMT program are currently under review within DoD and in the interagency process. All of the questions you ask are being investigated in the course of determining the DoD position with respect to these licenses.

Chairman SPENCE and Chairman GILMAN. Were the DTSA report's conclusions known to you at the time DoD reviewed the Chinasat-8 application?

Mr. TARBELL. Yes.

Chairman SPENCE and Chairman GILMAN. Mr. Holum, last Thursday you stated that "we are satisfied that no authorized transfers contributed to their [China's] military missile or even to their space launch vehicle program."

Are you also satisfied that no unauthorized transfers contributed to China's military or space launch vehicle program?

How can you say that when you admit that three launches were unmonitored?

Secretary HOLUM. Any unauthorized transfer would be a violation of U.S. law and could be subject to prosecution because it is possible the transfer could be a threat to national security. However, based on current analyses, it appears that China has not gained military advantage from the authorized launching of U.S. satellites, and approved information sharing. In general, China's military rockets have followed a normal evolutionary process and are not much more capable today than they were 15 years ago.

Chairman SPENCE and Chairman GILMAN. Mr. Holum, you stated that "there is a dramatic difference . . . between a smart dispenser that can dispatch multiple warheads to different orbits, as distinct from a reentry vehicle that would be able to dispense multiple warheads to different targets on land. The technologies are dramatically different."

According to a November 20, 1991 memo by Henry Sokolski, who was DOD's Deputy for Non-Proliferation Policy under President Bush, "The technology to place satellites in orbit is directly related and can be applied to development of ballistic missiles." Do you agree or disagree with this assessment?

Secretary HOLUM. There has been one licensed program launched by China that involved a satellite dispenser system—the Motorola Iridium project. Each launch involved two satellites. The dispenser was designed, developed and produced by the Chinese based on requirements provided by Motorola. The requirements as well as all technical interactions with the Chinese on this matter were monitored by DoD. In any case, the ultimate objectives of a MIRV capability and a satellite dispenser capability are significantly different. The objective of a satellite release into a parking or transition does not require much accuracy and allows for a wide margin for error. The objective of a MIRV system is extremely accurate reentry with release timing and orientation providing the only mechanism to ensure that the warhead reaches its target. Although the tasks are similar, the standards and requirements (both technical and procedural) for satellite release are far lower than those required for a MIRV. If the error is large for MIRVs, then there is a substantial risk that the vehicle will burn up on reentry.

In a different context, however, the statement attributed to Mr. Sokolski is clearly correct. Space launch vehicles themselves, for example, can be used directly to deliver weapons payloads. This is why, in authorizing launches of U.S. satellites on Chinese boosters beginning in the Reagan Administration, the U.S. does not authorize the transfer of technology useful in Chinese launch vehicles.

Chairman SPENCE and Chairman GILMAN. Mr. Sokolski goes on to state in his memo: "In order to place a satellite into orbit and maximize [its] orbital life time, the accuracy and efficiency of the launch vehicle is critical. Any time a satellite is integrated and launched, technical information has to be exchanged that can be used to upgrade launch vehicles and launch processes. Analysis and feedback required to satisfy insurance underwriters for launch of Western satellites results in the transfer of technical knowledge in the areas of increased payload, accuracy, and reliability to foreign launch vehicle designers." Do you agree or disagree with this?

Secretary HOLUM. I refer you to the analysis provided in the previous question and note that continuing U.S. policy and practice since the beginning of U.S. satellite launches on Chinese boosters in 1989 has been not to authorize the transfer of technology useful for Chinese launch vehicles.

Chairman SPENCE and Chairman GILMAN. Dr. William Graham, the former Science Advisor to Presidents Reagan and Bush and former Director of the White House Office of Science and Technology Policy has commented that "the deployment of multiple satellites from a SLV [space launch vehicle] and multiple warheads from an ICBM are also similar operations. Information and experience gained from one of these activities can be applied to the other. While MIRVed ICBMs having sufficent accuracy to attack hardened targets, such as missile silos, require higher accuracy than do SLVs that deploy muitiple satellites, such SLV satellite deployment is sufficiently accurate to be used for nuclear or biological attack of unhardened targets, such as cities." Do you agree or disagree with this statement?

Secretary HOLUM. I would disagree with the statement attributed to Dr. Graham to the extent this statement is meant to imply that the Iridium satellite dispenser, which China designed and built with its own technology, is capable of being used to dispense multiple, independently targested reentry vehicles (MIRVs). It is not.

It should be pointed out that China has not yet deployed MIRVs, much less tested them, and there is currently no evidence of any relationship between the Chinese-designed and -built Iridium satellite dispenser and any future Chinese MIRV capability that may come into existence.

Chairman SPENCE and Chairman GILMAN. Mr. Holum, in response to a question about the exchange of information between Hughes and Loral and the Chinese after the failed 1996 launch, you stated that "when information of such an exchange came to the attention of the Department of State, the matter was referred promptly to the Department of Justice for investigation."

Precisely when did the State Department learn that an information exchange may have occurred? When did State refer the matter to the Department of Justice? Was Loral notified that the matter had been referred? Why or why not?

Secretary HOLUM. The Department learned of this matter on May 15, 1996, when Loral informed the Department orally that it was involved in a review of the Long March failure. At that time, the Department informed Loral that such an activity could not be undertaken without a State Department license and that Loral should disclose any violations in which it had participated. On May 22, 1996, the Department received a letter from Loral describing its involvement, as well as that of Hughes. However, the Department found Loral's disclosure inadequate and directed both Loral and Hughes by letter dated May 29, 1996, to provide detailed and comprehensive disclosures of all of their activities associated with the Chinese launch failure analysis.

U.S. law enforcement, who, like state, awaited DOD's review of the documentation provided by the U.S. firms in late June 1996, opened a criminal investigation in May 1997 in light of DOD's detailed analysis of possible violations of the Arms Export Control Act and implications for national security.

While the Department did not refer this matter to the DOJ and U.S. Customs Service, as such, the Department met with them in May 1997 to encourage a criminal investigation, in part so that all the facts and circumstances having a bearing on U.S. national security interests could be established.

The Department does not have a policy of informing persons, including U.S. firms, when they become the subject of law enforcement investigations in order to protect the confidentiality and effectiveness of law enforcement efforts.

Chairman SPENCE and Chairman GILMAN. Mr. Holum, you and Mr. Reinsch both pointed out that the ability of U.S. companies to launch satellites from China is critical to the economic well-being of the U.S. space launch industry. Mr. Holum, you specifically state that if launches were prohibited from China, "our leadership in the satellite business would falter" and you claimed that "our satellite launches, our satellite industry is at risk." In addition, you stated that such an action would "cause enormous economic consequences here at home."

An article in the May 26, 1998 USA Today cites an aerospace industry analyst as saying that even if satellite companies were to be prohibited today from launching in China, it would only cost them an average of five percent of their total business. If true, how can the loss of such a small percentage place our satellite industry at risk?

How many launches of U.S. satellites take place in China each year? What percentage of satellite launches currently take place from China and is this percentage expected to grow, shrink, or stay the same as the number of overall satellite launches increases? How will the development by the United States of a more competitive space launch booster affect the competitiveness of U.S. launch facilities and services?

Secretary HOLUM. The commercial issue of whether to permit international customers to continue launching U.S.-made satellites from China involves maintaining the strength of the U.S. satellite industry and assuring that both the industry and U.S. workers share fully in the phenomenal growth in worldwide demand for telecommunications satellites.

The United States is presently the leading builder of satellites in the world. The Satellites Industry Association estimates the U.S. industry produced $9.1 billion in satellites in 1997, or two-thirds of total world satellite production. Counting sales of satellite-related equipment and services, U.S. satellites production is the core of an industry which generated $23.1 billion in revenues last year and provided high-paying, high-technology jobs for 101,700 U.S. workers.

Launch customers weigh a variety of factors in choosing a launch vehicle including launch availability, vehicle lift capacity, launch price, the reliability of the launch vehicle, suitability of the launch site in terms of intended satellite orbit, and market considerations. Because Asian customers appear to prefer their commercial satellites to be launched on Chinese rockets, U.S. industry could lose a substantial number of Asian satellite contracts if Chinese rockets were precluded from launching them into orbit. These losses would go beyond the value of the satellites, because the U.S. companies would also lose opportunities for sales of related equipment and services. The Satellite Industry Association has estimated that blocking the launch of U.S.-made satellites from China could cause the industry to forfeit $8.3 billion in revenue and 16,000 jobs for U.S. workers over a five-year period.

Contracts lost by U.S. industry due to a ban on Chinese launches of U.S.-made satellites could be a windfall for foreign satellite companies, mainly in Europe. These companies have long struggled to gain a financial and manufacturing base from which to challenge the dominance of the U.S. satellite industry. Shifting billions of dollars in satellite contracts from the United States to Europe would contribute to building such a financial and manufacturing base. This risks strengthening strategic competitors for the U.S. satellite industry and permanently eroding the U.S. share of world satellite market.

In both 1996 and 1997, there were three Chinese launches of U.S.-built satellites. Six launches are expected in 1998, three of which have already occurred. China accounts for roughly 10% of internationally-competed satellite launches. Chinese participation in the international launch services market is expected to continue at roughly this level, but could be influenced—up or down—by any perceived improvement or deterioration in the reliability of Chinese rockets.

Chairman SPENCE and Chairman GILMAN. Mr. Holum, you testified that "the technology safeguards agreement with China provide (sic) that no U.S. person can assist China in any way in the design, development, operation, maintenance, modification, or repair of launch vehicles." Is this not precisely what occurred after the February 1996 launch failure? How does this correlate with your assertion that "the procedures (for protecting U.S. technology) work?"

Secretary HOLUM. Most U.S. defense companies support and adhere closely to the prohibitions and constraints established in our bilateral agreement with China, as well as with the terms and conditions of U.S. Government licenses that implement U.S. policy in this area.

Unfortunately, this area, like others, is not immune from the possibility that persons may violate our laws and regulations. When this occurs, such actions need to be investigated and prosecuted in accordance with the criminal and civil remedies available in our laws.

Chairman SPENCE and Chairman GILMAN. Mr. Holum, you stated that "the communications satellites that we do authorize for sale, commercial communications satellites, we are aware, can be used by the Chinese military for communications purposes in the same way that the U.S. military frequently uses commercial telephone lines for communications."

Was the President made aware of the possibility that the People's Liberation Army could use these satellites to improve the reliability of their military communications?

Did this information factor into the President's decision to issue waivers allowing the export of satellites that could be used in this manner? Was this possible use of the satellites reflected in any of the waiver justification memos prepared for the President? If not, why not?

Secretary HOLUM. The satellites exported to China are commercial communications satellites. These kinds of satellites are used by government, military and civilian entities throughout the world.

Since the Kennedy Administration, the U.S. has supported the non-discriminatory use of international telecommunications services. For example, Intelsat, which currently has 142 member countries including China, does not dictate the use of the satellite provided that the time is paid for and the service does not result in technical harm to the satellite.

We have not recommended approval of a military communications satellite to the President.

In the cases where we recommend that the President waive sanctions for the sale of commercial communications satellites to Chinese entities, we did not advise the President that the Chinese military could use commercial telecommunications services, because such use confers no special military advantage.

Chairman SPENCE and Chairman GILMAN. Mr. Holum, you stated that it is not justifiable to deny license under the "innocent until proven guilty rubric," unless there is "sufficient probable cause to justify an indictment." Do you believe there was no such "probable cause" to deny Loral a license or waiver for its Chinasat-8 satellite in light of the May 1997 DTSA report that reportedly concluded "United States national security has been harmed?"

Secretary HOLUM. The policy we have followed in the Department of State of not denying export privileges prior to indictment is based on the discretionary authority provided to us in section 38(g)(3)(a), where the President may disapprove license applications if the applicant is the subject of an indictment under specified statues. The Department's policy is to apply that policy broadly in all such cases of indictment, but not to debar or withhold export privileges prior to indictment.

The Department's policy of denying export privileges upon an indictment was considered excessive several years ago by Congress in the wake of complaints from U.S. industry that the denial of export privileges prior to conviction might be unfair. At that time, Congress was sympathetic to industry concerns and it passed legislation (vetoed by President Bush) that would have eliminated the discretionary authority provided in section 38(g)(3)(a). Nevertheless, the Department decided, following consultation with the Department of Justice and a review of this question in 1993, to continue to follow a strict policy of denial for indicted persons. Accordingly, if Loral is indicted, it will be subject to this policy of denial. Whether there is probable cause to justify an indictment, is a matter for the judicial process to decide.

Chairman SPENCE and Chairman GILMAN. Is there any precedent for approving a license or waiver for the export of an item—whether it be commercial communications satellites or other items—by a company that is under investigation but has not yet been indicted? If so, please be specific. Is there any precedent for revoking a license for a company when that company has been indicted, or if there is sufficient probable cause to justify an indictment?

Secretary REINSCH. There is precedent for approving licenses for companies which are under investigation, in part because many investigations ultimately conclude that no wrongdoing occurred. There is also precedent for revoking or suspending a license, both before and after an indictment, where the Department has reason to believe that there is a risk of diversion. The Department may also suspend the ability of a company to ship under a license exception where it has reason to believe the company might use it improperly. All license applications are reviewed on a case-by-case basis, and information regarding an investigation is one of the factors considered during the license review process to determine whether the parties to the transaction are trustworthy and can be relied upon to carry out the export transaction in accordance with regulatory requirements, including any license conditions and provisos. The details of specific cases are subject to the protections of Section 12(c) of the Export Administration Act, which prohibits their public disclosure.

Chairman SPENCE and Chairman GILMAN. The waiver justification memos prepared by and for the NSC cited numerous reasons for approving waivers that would allow U.S.-made satellites to be exported for launch in China. Most of the reasons cited were economic, commercial, political, and diplomatic in nature. Were any national security considerations reflected in those memos? What were they? Did you ever convey an assessment of the national security risks of satellite launches to the White House for consideration in the waiver decision? Did the memos ever address the similarities between space launch vehicles and ballistic missiles? Did they ever discuss the kinds of technical data that could be transferred to China and whether that data could have an impact on China's ballistic missile program? Did they ever discuss what military useful capabilities China might learn from launching American satellites?

Secretary REINSCH. Satellites are controlled for reasons for national security and foreign policy. The reviews by Defense, State, Commerce and other agencies of satellite license applications are focused on any national security and foreign policy concerns which such launches might raise. No agency would recommend a waiver to the White House until such time as all national security concerns had been addressed, through the licensing process, and satisfactorily resolved. Agencies engage in an intensive review of national security concerns as part of the license approval process and the waiver recommendation is based on that extensive review. The questions you pose regarding discussion of the similarities between space launch vehicles and ballistic missiles usefulness have been discussed extensively during the formulation of the general policy in the Reagan and Bush administrations, and in the Clinton Administration review of these policies.

Chairman SPENCE and Chairman GILMAN. The waiver justification memos note that some satellites would be owned and operated by the Chinese for the purpose of providing voice, data, and television transmissions to China. Did they ever discuss the national security implications of having the People's Liberation Army use those satellites to transmit encrypted military communications more reliably? Did the waiver justification memos every discuss the national security implications of a failed launch and the prospect that military sensitive items might fall into Chinese hands?

Secretary REINSCH. The Executive Branch reviewed the question of Chinese access to normal civil telecommunications equipment in the early 1990s and concluded that this was in the U.S. national interest. Since the 1960's, the U.S. has supported the non-discriminatory use of International Telecommunications Services. For example, Intelsat, which currently has 142 member countries, including China, does not dictate how the satellite may be used or by whom. The U.S. has the same policy for satellites sold for national systems to foreign countries and assumes in making these sales that all portions of the purchasing government may have access to the system. It would be unrealistic to seek to block access to Commercial Telecommunications Services by the Chinese military given the widespread availability of other satellite option around the world.

Chairman SPENCE and Chairman GILMAN. In 1994, the waiver justification memo prepared for the NSC regarding the Echostar satellite export noted that "the Eschostar is a commercial communications satellite under the export control jurisdiction of the Commerce Department. . . . " Identical language is also included in memoranda prepared regarding the Cosat project in 1995 and the China sat project in 1996. All of these waivers occurred prior to the 1996 transfer of export licensing jurisdiction from State to Commerce. Why were these satellite projects under the jurisdiction of the Commerce Department? Were they "pure" commercial communications satellites, or did they contain any of the nine military-sensitive technologies that would place them under the licensing jurisdiction of State?

Secretary REINSCH. The ECHOSTAR, COSAT and CHINASAT were commercial communications satellites that had none of the nine technical parameters that would have made them subject to State Department licensing jurisdiction. These satellites were covered by the Bush Administration decision to transfer jurisdiction in 1992.

Chairman SPENCE and Chairman GILMAN. Did any of these satellites for which waivers were approved prior to November 1996 contain radiation-hardened chips or components?

Secretary REINSCH. Most commercial communications satellites do use such radiation-hardened chips, and I expect that all of these did. Number six of the nine technical parameters referred to reads "employing radiation-hardened devices controlled elsewhere in section 121.1 that are not 'embedded' in the satellite in such a way as to deny physical access (here 'embedded' means that the device cannot feasibly either be removed from the satellite or be used for other purposes)."

Chairman SPENCE and Chairman GILMAN. Of all the waivers issued by this Administration, how many were for the export of satellites containing any of the nine militarily-sensitive technologies? When were these waivers issued and which agency was the licensing agency?

Secretary REINSCH. The waivers issued by this administration for satellites containing any of the nine military sensitive technologies were initiated by the State Department. In several cases (IRIDIUM, APMT and CHINASAT–8) the programs were moved under Commerce jurisdiction in November 1996 after initial licensing actions had been completed by the State Department. Commerce licensed the hardware after interagency review using the authority of the State-secured Presidential Waiver for the program.

Chairman SPENCE and Chairman GILMAN. Please provide for the record a listing of all launches from China since 1990 involving U.S.-made satellites, and in each case please note whether the launch was successful or failed, whether U.S. monitors were present, whether those monitors were present at the launch or also during technical discussions with China or during any post-launch failure activities, which agency licensed each launch, whether there were Technical Assistance Agreements for each launch, which launches were bundled in a single waiver, which satellites have had licenses and waivers granted but have not yet been launched, and which companies are under investigation for activities related to which satellite exports?

Secretary REINSCH. [See Chart on page 374.]

Chairman SPENCE and Chairman GILMAN. Mr. Tarbell testified that DOD should have had a monitor participating in the investigation of the failed 1995 launch of the Apstar-2 satellite. You have stated that you did not give DOD copy of the failure report for that launch because the Department of Commerce was sufficiently "competent" to judge the content of the report and whether it adhered to the terms of the Commerce license. Do you believe that, regardless of the content of the report, DOD monitoring should have taken place to ensure against the unlicensed transfer of export-controlled technical information? Doesn't this reflect a breakdown or failure in the technology safeguards process?

Secretary REINSCH. A 1998 Commerce Department review of the report, conducted at my request, confirmed the original conclusion that the analysis contained no information specific to the launch vehicle or the satellite and that its release to the insurance companies involved in the launch and the Chinese was appropriate. It also determined that it would have been more appropriate to refer the failure report to the State Department prior to providing Hughes with any approval. This is another area where the President's 1996 decision strengthened controls on satellite exports, and since the 1996 transfer of jurisdiction, all agencies including Commerce agree that such failure analyses are best conducted under the aegis of a State license and with appropriate monitoring by the Department of Defense.

Chairman SPENCE and Chairman GILMAN. Mr. Reinsh, please provide the committees with a copy of the launch failure analysis for the January 1995 Apstar-2 satellite launch and any Commerce Department assessment of that analysis showing that it did not violate the terms of the Commerce-granted license.

Secretary REINSCH. This has been provided separately as part of the Committees' request for various documents. The report was submitted as a Commodity Classification Request. A written response was provided by the Department of Commerce that stated that "these two reports and executive summary have been reviewed and determined to contain no design or production data specific to the spacecraft, the launch vehicle or the interface of those two systems."

Chairman SPENCE and Chairman GILMAN. Mr. Reinsch, in October, 1997 Hughes Space & Communications sent a letter to the Commerce Department stating their belief that technical information associated with the export and launch of a U.S.-made satellite by a foreign launch provider does not require an export license. Hughes also asks how technical information for satellite operation (telemetry, tracking, control, and maintenance data) is export controlled. They ask if this information can be exported without an export license, or if it requires an export license, what control number should be used when applying for a license. It is our understanding that the Commerce Department has not responded to this letter. Has the Commerce Department responded to the Hughes letter of October 8, 1997?

Secretary REINSCH. Hughes asked Commerce to reconfirm a State Department commodity jurisdiction decision and a Commerce Department commodity classification from 1992 and 1993 which informed Hughes as to which technology was controlled by Commerce and what the licensing requirements for this technology were. The issues of whether, in light of the President's 1996 decision to transfer jurisdiction and strengthen controls on communications satellites, to change practices inherited from the Bush Administration remains under review and we have not yet responded to Hughes.

Chairman SPENCE and Chairman GILMAN. Is it true that technical information associated with the export and launch of a U.S.-made satellite by a foreign launch provider does not require an export license? Please provide us with the Commerce regulation, along with the date of its implementation, that identifies how this information is controlled.

Secretary REINSCH. The text of the regulation has been provided separately. At this time, most data associated with a satellite export requires a license and all data requires U.S. Government review. Marketing data, and data necessary and reasonable for a purchaser to have assurance that a U.S.-built item intended to operate in space has been designed, manufactured and tested in conformance with specified contract requirements as well as data necessary for normal in-orbit satellite operations, to evaluate in-orbit anomalies, and to operate and maintain associated ground station equipment has been classified by the Department as EAR99. Data required by the launch service provider limited to form, fit, function, mass, electrical, mechanical, dynamic/environmental, telemetry, safety, facility, launch pad access, and launch parameters was also classified as EAR99. Such data is restricted to the embargoed countries, denied parties and certain named missile, chemical and biological and nuclear end users (EAR, Part 736). In practice, the above noted data is released only to endusers who are already covered by a hardware license. Since all development and production data for the satellite as well as all data related to the launch vehicle are under State Department jurisdiction, the use data for the satellite which is under Commerce jurisdiction has little utility beyond providing information as to how to fit the satellite to the launcher and how to use it once in orbit. Since 1996, we have also required exporters to provide to DOD copies of all technical data released to the Chinese launch service provider. We are continuing to review these earlier decisions to see how they should be adjusted, barring any change in jurisdiction which would render action unnecessary.

Chairman SPENCE and Chairman GILMAN. If Hughes is now asking how telemetry, tracking, control, and maintenance data is export controlled, and since they have been exporting satellites to China, does that mean that Hughes has exported this information without a license in the past? Please provide us with the Commerce regulation, along with the date of its implementation, that identifies how this information is controlled.

Secretary REINSCH. The regulation is attached. Hughes had previously received written confirmation from the Departments of Commerce and State (Commerce Commodity Classification 33173 dated January 25, 1993 and State Department Commodity Jurisdiction decision CJ 116–93 dated Sep 17, 1993) that the specific data that they presented was classified as 9E96G (EAR99 under our current regulation). Subsequent letters from Commerce dated Dec 23, 1993, and May 31, 1994, confirmed that classification. In March 1996, Commerce published a new Export Administration Regulation that revised the classification of most products in order to bring them into closer alignment with the practice in the multilateral regimes. By their letter of Oct 8, 1997, Hughes asked for verification that the data previously classified as 9E96G was now properly classified as EAR99. We are continuing to review these earlier decisions to see how they should be adjusted, barring any change in jurisdiction which would render action unnecessary. [See regulation on page 347.]

QUESTIONS SUBMITTED BY MR. SKELTON

Mr. SKELTON. Could you please explain what you meant between providing the Chinese the “know how” and providing them the “know why” in terms of sensitive technology as a result of our satellite export policy?

Mr. SOKOLSKI. “Know how” conveys a given technical procedure, such as satellite integration for a particular satellite and rocket launcher. “Know why” goes further to explain in engineering and scientific terms of why a given procedure is arranged the way it is (i.e., why certain steps must be followed in a given order by others and what fundamental problems or risks these procedures are designed to mitigate or resolve). Such “know why” would enable the Chinese on their own to engineer around such problems for other rocket or satellite systems. In short, know how is relevant only to a particular system, know why empowers the student to engineer around similar problems for a variety of systems.

QUESTIONS SUBMITTED BY MR. HAMILTON

Mr. HAMILTON. It might be helpful if our witnesses would just respond item by item to these various technologies, and let us know what the facts are, how you would respond to that chart. Could we have them in writing, please?

Mr. TARBELL. Export of a satellite to China for launch involves transfer of the satellite, fuel for the satellite, some ground support equipment belonging to the satellite manufacturer (exported temporarily), and, sometimes, an additional motor needed to "kick" the satellite into its final orbit (whenever the launch vehicle cannot do so itself). When such launches are monitored by the Department of Defense, the fuel, satellite, kick motor, and ground support equipment are monitored by DoD personnel to prevent unauthorized access or tampering by the Chinese. As a result of these precautions and monitoring of specific satellite launch programs, it is highly unlikely that the Chinese have legally gained any significant technology in any of the technologies outlined in the Chart titled "How is Commercial Launch Technology Related to Ballistic Missile Technology." Below is a point-by-point analysis of each technology area.

Payload Dispersal System: There has been one licensed program launched by China that involved a satellite dispenser system—the Motorola Iridium project. Each launch involved two satellites. The dispenser was designed, developed and produced by the Chinese based on requirements provided by Motorola. The requirements as well as all technical interactions with the Chinese on this matter were monitored by DoD. In any case, the ultimate objectives of a MIRV capability and a satellite dispenser capability are significantly different. The objective of a satellite release into a parking or transition orbit does not require much accuracy and allows for a wide margin for error. The objective of a MIRV system is extremely accurate reentry with release timing and orientation providing the only mechanism to ensure that the warhead reaches its target. Although the tasks are similar, the standards and requirements (both technical and procedural) for satellite release are far lower than those required for a MIRV. If the error is large for MIRVs, then there is a substantial risk that the vehicle will burn up on reentry.

Kick Motor Technology: For some satellites, it has been necessary for the satellite manufacturer to provide a kick motor that would push the satellite into its final orbit. The kick motors exported are subject to strict export licensing conditions. The Chinese are prohibited access to this equipment during the transportation of the kick motor to China and prohibited unescorted access to it at the launch site. Technical data authorized for export to the Chinese is limited to that general "form, fit and function" data required to mate the kick motor to the launch system. The kick motors themselves are integrated to the satellite by the satellite manufacturer and the complete system (satellite *and* kickmotor) is encased in a payload fairing prior to turnover to the Chinese for mating to the launch vehicle. The Chinese could only acquire useful information on the kick motor through a detailed internal examination of the motor which would likely render it unserviceable.

Radiation-Hardened Electronics: All satellites contain certain embedded radiation hardened electronic circuits. Such embedded circuits cannot be tampered with or removed without destroying the satellite. For that reason, the Chinese cannot obtain any useful information about such circuits in the process of launching a U.S. satellite without the knowledge of the U.S. satellite manufacturer. Useful technology transfer to the Chinese would require destructive testing and extensive laboratory analysis which renders the chips themselves unserviceable. It is germane to note that the Europeans, who have had virtually unlimited access to advanced US-manufactured radiation hardened chips, have been unable to reliably duplicate the processes required to manufacture them indigenously.

Encryption Devices: Since 1990, there have been four failures of Chinese launches of US-built satellites. Two of the satellites did not have any encryption devices on board. In one of the two other cases (the Optus B–2 in December 1992), the devices were recovered and returned to the US. In the other case (the Intelsat 708 failure in February 1996), the cryptographic equipment and associated material for the Space Systems/Loral Intelsat 708 which was destroyed during the launch failure on February 15, 1996, was incorporated into two integrated circuit boards. The National Security Agency has determined that the cryptographic equipment that was aboard the INTELSAT 708 satellite also was destroyed and not received. NASA made this determination of the basis of reports and other information provided by SS/Loral.

Launcher Attitude Control: Because it is their own launch vehicle, the Chinese provide all equipment related to the launcher and its components, including the attitude control system on board the launch vehicle which guides the vehicle to placing the satellite into the correct orbit. No information or assistance may be provided lawfully by US personnel to the Chinese on that system without obtaining an export license that the State Department and no State license has ever been granted that would authorize export to the Chinese of such assistance.

Stage Separation Systems: Because it is their own launch vehicle, the Chinese provide all equipment related to the launcher and its components, including the system that causes each individual stage of the launcher to separate once its fuel is expended. US exporters may provide explosive bolts to be used by the Chinese for separation systems between the launch vehicle and satellite under an approved export license. However, the license authorizes bolts to be provided as end items with only that technical data which is required to interface them with the existing payload separation system. Useful technology transfer could only take place through destructive testing of the bolts, which was not authorized under the export license.

Stage Coupling Load Analysis: To ensure safety of flight and mission success, the satellite manufacturer is required to provide a "finite element model" of the satellite which the launch vehicle provider uses to perform the coupled bloods analysis. The launch service provider then conducts the analysis and provides the results to the US manufacturer for analysis by the US participants. Export licenses prohibit the provision to the Chinese of the methodology used to perform coupled loads analysis; only the results of analysis are authorized for export. Satellite exporters are similarly prohibited from correcting or recommending changes to Chinese methodology. Complete coupled-loads-analysis results on a satellite could only be useful to a Chinese ballistic missile program if the warhead had the same mass, size and environmental requirements of a particular satellite.

Mr. SOKOLSKI. *General:* Export of a satellite to China for launch involves transfer of the satellite, fuel for the satellite, some ground support equipment belonging to the satellite manufacturer (exported temporarily), and, sometimes, an additional motor needed to "kick" the satellite into its final orbit (whenever the launch vehicle cannot do so itself). When such launches are monitored by the Department of Defense, the fuel, satellite, kick motor, and ground support equipment are under constant surveillance by DoD personnel to prevent unauthorized access or tampering by the Chinese. As a result of these precautions and monitoring of specific satellite launch programs, it is highly unlikely that the Chinese have legally gained any significant technology in any of the technologies outlined in the Chart titled "How is Commercial Launch Technology Related to Ballistic Missile Technology." Below is a point-by-point analysis of each technology area.

Payload Dispersal System: There has been one licensed program launched by China that involved a satellite dispenser system—the Motorola Iridium project. Each launch involved two satellites. The dispenser was designed, developed and produced by the Chinese based on requirements provided by Motorola. The requirements as well as all technical interactions with the Chinese on this matter were monitored by DoD. In any case, the ultimate objectives of a MIRV capability and a satellite dispenser capability are significantly different. The objective of a satellite release into a parking or transition does not require much accuracy and allows for a wide margin for error. The objective of a MIRV system is extremely accurate reentry with release timing and orientation providing the only mechanism to ensure that the warhead reaches its target. Although the tasks are similar, the standards and requirements (both technical and procedural) for satellite release are far lower than those required for a MIRV. If the error is large for MIRVs, then there is a substantial risk that the vehicle will burn up on reentry.

Kick Motor Technology: For some satellites, it has been necessary for the satellite manufacturer to provide a kick motor that would push the satellite into its final orbit. The kick motors exported are subject to the U.S.-China Technology Safeguards Agreement and the Chinese are prohibited access during the transportation of the kick motor to China and prohibited unescorted access at the launch site. Technical data authorized for export to the Chinese is limited to general "form, fit and function" required to mate the kick motor to the launch system. The kick motors themselves are integrated to the satellite by the satellite manufacturer and the complete system (satellite and kick motor) encased in a payload fairing prior to turnover to Chinese for mating to the launch vehicle. The Chinese could only acquire useful information on the kick motor through a detailed internal examination of the motor which would likely render it unserviceable.

Radiation-Hardened Electronics: All satellites contain certain embedded radiation hardened electronic circuits. Such embedded circuits cannot be tampered with or removed without destroying the satellite. For that reason, the Chinese cannot obtain any useful information about such circuits in the process of launching a U.S. satellite without the knowledge of the U.S. satellite manufacturer. Useful technology transfer to the Chinese would require destructive testing and extensive laboratory analysis which renders the chips themselves unserviceable. It is germane to note that the Europeans, who have had virtually unlimited access to advanced US-manufactured radiation hardened chips, have been unable to reliably duplicate the processes required to manufacture them indigenously.

Encryption Devices: Since 1990, there have been failures of Chinese launches of US-built satellites. Two of the satellites did not have any encryption devices on board. In one of the two cases (the Optus B–2 in December 1992), the devices were recovered and returned to the US. In the other case (the Intelsat 708 failure in February 1996), the command processor boxes as a whole were recovered, but not the circuit cars which contained the encryption. In both cases the encryption involved embedded, single chip devices. However, any loss of the chips and associated encryption algorithms would have only minimal impact on national security because the older satellite in the Intelsat case used older algorithms for which the keys are unique. There is some chance that a third-party could reverse-engineer such devices to gain some knowledge about the engineering principles for making such chips. If this were to occur, the impact on national security would not be significant.

Launcher Attitude Control: Because it is their own launch vehicle, the Chinese provide all equipment related to the launcher and its components, including the attitude control system onboard the launch vehicle which guides the vehicle to placing the satellite into the correct orbit. No information or assistance may be provided by US personnel to the Chinese on that system without obtaining a munitions license from the State Department, and no State license has ever been granted that would allow export to the Chinese of such assistance.

Stage Separation Systems: Because it is their own launch vehicle, the Chinese provide all equipment related to the launcher and its components, including the system that causes each individual stage of the launcher to separate once its fuel is expended. The US has provided explosive bolts to be used by the Chinese for separation systems between the launch vehicle and the satellite. However, the bolts were provided as end items with only that technical data required to interface them with the existing payload separation system. Useful technology transfer could only take place through destructive testing, which was not authorized.

Stage Coupling Load Analysis: To ensure safety of flight and mission success, the satellite manufacturer is required to provide a "finite element model" of the satellite which the launch vehicle provider uses to perform the coupled loads analysis. The launch service provider then conducts the analysis and provides the results to the US manufacturer for analysis by the US participants. Export licenses prohibit the provision of the methodology used to perform coupled loads analysis, only the results of analysis are authorized for export. Satellite providers are similarly prohibited from correcting or recommending changes to Chinese methodology. Complete coupled loads analysis results on a satellite could only be useful to a Chinese ballistic missile program if the warhead had the same mass, size and environmental requirements of a particular satellite.

QUESTIONS SUBMITTED BY MR. HUNTER

Mr. HUNTER. Did you ever inquire of them (Hughes) when they made the public statement that they would work together with the Chinese to eliminate the two mentioned possible causes of launch failure? Did you inquire of Hughes as to whether or not they were, in fact, meeting with the Chinese?

Secretary REINSCH. Neither I nor my staff recall making such an inquiry. The Hughes correspondence provided in conjunction with the investigation of the 1995 APSTAR 2 failure documents that certain meetings were held with the Chinese launch service provider. Those meetings were represented as data gathering meetings. Subsequent to the release of the Hughes identified potential failure modes, we are not aware of any additional meeting where Hughes worked with the Chinese to redesign and thus overcome the identified deficiencies.

QUESTIONS SUBMITTED BY MR. BEREUTER

Mr. BEREUTER. How were we involved in monitoring and can I have a copy of that monitoring report?

Secretary LODAL. The failure review in 1992 was conducted under the umbrella of the Hughes Technical Assistance Agreement authorized by the Department of State. The 1992 failure analysis meetings were monitored by an Air Force officer assigned to DTSA. We have searched our files and do not have copies of any monitor reports from that failure review.

QUESTIONS SUBMITTED BY MR. BATEMAN

Mr. BATEMAN. The witnesses stated that the panel was briefed by the National Security Council in preparation for their testimony. Identify any member of the NSC or its staff who discussed your testimony with you. Why did you or they believe you needed such discussion? Did their advice influence in any way your testimony?

Mr. TARBELL. We met with the Deputy National Security Adviser and his staff to review testimony and to ensure full compliance with Committee requests. It is customary for agencies to coordinate written remarks and to prepare for testimony prior to Congressional hearings in order to ensure that information provided is complete and accurate.

Mr. BATEMAN. What did the Clinton Administration do to promote, encourage, and support the commercial launch facilities within the United States before 1998?

Mr. TARBELL. I have consulted with the officials who are responsible for such matters in DoD. They informed me that the Clinton Administration issued a Presidential Decision Directive on National Space Transportation Policy (NSTC–4) in August 1994. In part, the directive encouraged the cost-effective use of commercially provided U.S. products and services, to the fullest extent feasible, that meet mission requirements. It also fostered the international competitiveness of the U.S. commercial space transportation industry, actively, considering commercial needs and factoring them into decisions on improvements in launch facilities and launch vehicles.

Moreover, in a August 1997 a Memorandum of Agreement (MOA) on Federal Launch Site Operations was approved by the Department of Defense, the Federal Aviation Administration, and the National Aeronautics and Space Administration. The primary objective of the MOA is to assist private sector, and state local government use of federal launch property, facilities and equipment. The term "spaceport" was used in the MOA to describe commercial, and state and local entities either currently, or potentially, involved in launch-site-related activities.

So far, several steps have been taken to implement this MOA. Two spaceports have been created—one in Florida and the other in California—on excess federal property at the Cape Canaveral Air Station and Vandenberg Air Force Base. Two additional spaceports are under construction—one in Virginia and the other in Alaska. Another spaceport in New Mexico is being actively considered.

Mr. BATEMAN. What role did the Department of Defense play in granting of a waiver in February of this year for the export of the Chinasat-8 satellite? Were the national security implications of this action adequately presented before a decision was made? Did the DoD recommend the waiver be granted, oppose it, or make no recommendation?

Mr. TARBELL. DoD conducted a thorough review of all Chinasat-8 associated license applications referred by the State and Commerce Departments. DoD recommended approval of the licenses, but only if safeguards and other conditions were applied, including a requirement for DoD monitoring of the satellite launch and associated technical meetings, and DoD review of technical data prior to its transfer of China.

The State Department is responsible for preparing and recommending waivers (of Tiananmen Square sanctions) to the President. DoD does not comment on the policy aspects of the national-interest waiver recommendation prepared by State. DoD's role is only to provide a technical review of the waiver memorandum to ensure that it accurately characterizes the safeguards that are to be applied to the licenses for the satellite export. In the case of Chinasat-8, the Department asked DoD to review the waiver to ensure that it accurately reflected the conditions and safeguards associated with the licenses. DoD conducted a technical review of the waiver memorandum and determined that it accurately reflected those safeguards.

Mr. BATEMAN. Several times witnesses declined to answer pertinent questions because the subject was under investigation by the Department of Justice. Cite the legal authority for your declining to answer.

Secretary HOLUM. The Department of Justice's authority in this regard derives from the Constitution and principles of separation of powers. the constitutional doctrine of executive privilege protects the Executive's ability to discharge its constitutional responsibilities. Pursuant to the Executive's constitutional responsibility to "take care that the Laws be faithfully executed," the Department of Justice has the obligation under the criminal justice system to protect the confidentiality, integrity and independence of ongoing criminal investigations and prosecutions. This constitutional position is evidenced by the Department's long-standing policy and practice of not providing details of ongoing criminal matters to Congress. Congress has been respectful of the Department's policy, which has been applied consistently during Administrations of both parties and which occasionally requires that the Department ask other executive branch agencies not to make a disclosure to Congress that might compromise an ongoing criminal matter.

Mr. BATEMAN. The witnesses stated that the panel was briefed by the National Security Council in preparation for this testimony. Identify the member of the NSC or its staff who discussed this testimony with you. Why did you or they believe that you needed such discussion? Did this advice influence in any way your testimony?

Secretary HOLUM. We met with the Deputy National Security Advisor and NSC staff to review testimony to ensure full compliance with Committee requests and to exchange relevant information concerning the subject matter of the hearing. It is customary for agencies to coordinate written remarks and to prepare testimony prior to Congressional hearings in order to ensure that information provided is complete and accurate. My testimony reflected my best judgment of the questions presented, based on the facts available to me.

Mr. BATEMAN. What did the Clinton Administration do to promote, encourage, and support the commercial launch facilities within the United States before 1998?

Secretary HOLUM. The USG has consistently encouraged growth in the domestic launch industry.

The Administration, particularly through the Departments of Defense and Transportation and NASA, has been working with industry on the Evolved Expendable Launch Vehicle (EELV) and the Reusable Launch Vehicle (RLV) programs to develop the next generation of launch vehicles.

We at the Department of State have met with several U.S. companies working to develop new space launch vehicles, both domestically and in joint ventures with foreign companies.

The Commercial Space Launch agreements negotiated with China, Russia, and Ukraine address the conditions of competition faced by U.S. launch services providers in the international marketplace as these economies continue their transition to a market-directed basis.

For additional information, we refer you to the Department of Defense and NASA.

Mr. BATEMAN. The witnesses stated that the panel was briefed by the National Security Council in preparation for their testimony. Identify the member of the NSC or its staff who discussed your testimony with you. Why did you or they believe you needed such discussion? Did their advice influence in anyway your testimony?

Secretary REINSCH. We met with the Deputy National Security Advisor and his staff to review testimony and to ensure full compliance with Committee requests. It is customary for agencies to coordinate written remarks and to prepare for testimony prior to Congressional hearings in order to ensure that information provided is complete and accurate.

Mr. BATEMAN. Several times witnesses declined to answer pertinent questions because the subject was under investigation by the Department of Justice. Cite the legal authority for your declining to answer.

Secretary REINSCH. The Department of Commerce recognizes and defers to the Department of Justice's (DOJ) responsibilities for enforcing criminal laws. In that regard, the Department of Justice states that its authority in this regard derives from the Constitution and principles of separation of powers. Pursuant to the Executive's constitutional responsibility to "take care that the Laws be faithfully executed, "DOJ has the obligation under the criminal justice system to protect the confidentiality, integrity and independence of ongoing criminal investigations and prosecutions. This constitutional position is evidenced by DOJ's longstanding policy and practice of not providing details of ongoing criminal matters to Congress. Congress has honored DOJ's policy, which has been applied consistently during Administrations of both parties and which occasionally requires that DOJ ask other executive branch agencies not to make a disclosure to Congress that might compromise an ongoing criminal matter.

Mr. BATEMAN. What did the Clinton Administration do to promote, encourage and support commercial launch facilities within the United States before 1998?

Secretary REINSCH. The Clinton Administration has promoted, encouraged and supported commercial launch facilities in the United States in several ways. As directed by the President's 1994 National Space Transportation Policy and the Commercial Space Launch Act, DOD and NASA have continued providing use of active launch and range facilities to support over 50 U.S. commercial space launches between January 1993 and September 1998—all on a direct cost basis. To manage the scheduling and planning required to accommodate these high commercial launch rates, commercial space launch operators have participated very substantially with DOD and NASA in each quarterly Current Launch Schedule Review Board meeting and in regularly scheduled Chief Executive Officer-level meetings with senior Air Force officials. Executive Branch agencies also took several meaningful steps in the past two years to clarify procedures for commercial users of government launch facilities. For instance, in 1996, the Air Force developed a policy to clarify the process for state and private entities to use in arranging to use its real property for commercial space activities. In 1997, the Department of Defense, Federal Aviation Administration and NASA completed a memorandum of agreement to clarify federal government interactions with state government and private launch site operators.

Since 1993, the Air Force has invested over $1 billion to continue the operation and maintenance of launch facilities and ranges at both cape Canaveral Air Station, Florida and Vandenberg Air Force Base, California to sustain the ability to support both government and commercial space launch activities. The Air Force has also worked closely with both Boeing and Lockheed Martin to schedule and assist in privately-financed modifications and upgrades to DOD Delta and Atlas launch facilities, to accommodate new commercial launch systems. In addition, DOD spent almost $400 million to construct a new launch pad on the site of an older pad at Vandenberg Air Force Base to accommodate larger Atlas launch vehicles for DOD missions, and about $50 million for a new Delta operations building at Cape Canaveral, but the facilities can also be used for NASA and commercial launches. Similarly, NASA funded a $25 million upgrade to the Vandenberg Delta launch complex for their missions, which enabled Boeing also jointly funded a remote blockhouse after the January 1997 Delta launch failure at Cape Canaveral, which benefits DOD, NASA and commercial users.

DOD also worked closely with Spaceport Florida Authority (a state government entity) to arrange for them to modify a navy Trident ballistic missile test pad for small commercial space launch vehicles and the Air Force entered into long-term leases of land and facilities to allow private entities—the Western Commercial Space Center (WCSC) and Astrotech—to construct and operate launch vehicle and spacecraft processing contracts with Astrotech and WCSC have enabled them to recoup part of their facility investments with income from paying customers. NASA also recently arranged for the Virginia Spaceport to construct and operate a state-sponsored, privately operated launch facility at its Wallops Island launch center, and the FAA licensed the operation of all three.

For the future, DOD has committed roughly $1.6 billion to the Evolved Expendable Launch Vehicle (EELV) program to field two competing facilities of modernized, lower-cost U.S. launch systems by 2002. EELV will require construction of several new processing and launch facilities on both coasts and the companies involved in the program are expected to contribute up to $2 billion in private investments. Not only that, but the Air Force has already invested $300 million toward the 15-year, $650 million Range Standardization and Automation (RSA) program which is aimed at modernizing the range infrastructure to improve the capacity, reliability, and operation and maintenance costs for both government and commercial range users, including the U.S. commercial space launch industry.

Mr. BATEMAN. Could the Administration provide the specifics of the bilateral agreements, the regulations, the standards, everything procured side-by-side to this committee, in order that we can determine whether those witnesses who appeared yesterday are correct or whether what we are hearing today is correct, that there is really no difference between the one administration's policy and the other.

Secretary REINSCH. The text of the bilateral agreement, license conditions and the regulations are being provided separately.

QUESTIONS SUBMITTED BY MRS. FOWLER

Mrs. FOWLER. Last Saturday's New York Times contained a piece indicating that the PLA has obtained dedicated access to transponders on the U.S.-built satellites that have been sold to China. In response, according to the Los Angeles Times of June 14, 1998, a White House spokesman, rather than denying this usage, instead sought to trivialize it, stating that "militaries around the world use commercial satellites" and "we do not believe it provides any particular military benefit." Do you agree with the statement that providing the PLA with the ability to conduct military communications does not provide them with "any particular military benefit?"

Mr. TARBELL. Commercial communications satellites are used by government, military, and civilian entities throughout the world to handle both sensitive and non-sensitive communication needs. These satellites are used by the U.S. military for many of its communications. For particularly sensitive military communications needs, the U.S. uses satellites with many special military characteristics which are not contained in commercial communications satellites that have been exported to Chinese customers.

Examples of these special military features include: anti-jamming antennae features to protect against intentional jamming by an adversary; frequency hopping to protect against interception of communications; intersatellite relay links to enable strategic communications; advanced baseband processing to protect against interception and jamming of communications and improve the quality of the signals; and precise attitude control systems that are necessary features for high resolution imaging.

License applications for the export of commercial communications satellites to China are reviewed by DoD to ensure that sensitive military capabilities are not transferred to the Chinese. In addition, the licenses for the export of satellites to China contain conditions and safeguards to prevent China form gaining physical access to the satellite prior to launch. This includes DoD monitoring at the launch site.

Mrs. FOWLER. Was there any Defense, State, or Commerce Department analysis of the possibility that the PLA would obtain dedicated access to transponders on these satellites for its military communication prior to licensing of these satellite deals?

Mr. TARBELL. Yes. Various DOD and Intelligence Community organizations review export licenses for satellites. That review takes into account that the PLA could be a user for these satellites.

Mrs. FOWLER. Has the U.S. government licensed the sale of any ground station technology to any Chinese entity that would support the satellite user's ability to encrypt its transmissions prior to sending them to the satellite or decrypt them once they have been transmitted? Under what control regime would such licenses be issued, and who would oversee the issuance of such licenses? If such sales were licensed, who received permission to proceed with these sales and to what entities were these technologies sold? What specific technologies, if any, were approved for sale? Do we have any information revealing links between any recipients of such ground station technologies and the PLA?

Mr. TARBELL. The response to this question will be provided directly to the Committee by the National Security Agency.

Mrs. FOWLER. Last Saturday's New York Times contained a piece indicating that the PLA has obtained dedicated transponders on the U.S. built satellites that have been sold to China. In response, according to the Los Angeles Times of June 14, 1998, a White House spokesman, rather than denying this usage, instead sought to trivialize it, stating that "militaries around the world use commercial satellites" and "we do not believe it provides any particular military benefit." Do you agree with the statement that providing the PLA with the ability to conduct military communications does not provide them with "any particular military benefit?"

Secretary HOLUM. The satellites exported to China are commercial communications satellites. These kinds of satellites are used by government, military and civilian entities throughout the world.

The Chinese military's use of a commercial communications satellite in this manner is analogous to the way that national telephone systems are used by both the military and civilian sectors. The systems confer no particular significant military advantage to the user.

For sensitive military communications, the U.S. uses satellites with many special characteristics. We do not provide such satellites to China.

Mrs. FOWLER. Was there any State Department analysis of the possibility that the PLA would obtain dedicated access to transponders that the PLA would obtain for its military communications prior to the licensing of these satellite deals?

Secretary. HOLUM. The satellites licensed for export to China are commercial communications satellites, not military communications satellites, and convey no significant military benefit. There was no need for such analysis.

Mrs. FOWLER. Has the U.S. government licensed the sale of any ground station technology to any Chinese entity that would support the satellite user's ability to encrypt its transmissions prior to sending them to the satellite or decrypt them once they have been transmitted? Under what control regime would such licenses be issued, and who would oversee the issuance of such licenses? If such sales were licensed, who received permission to proceed with these sales and to what entities were the technologies sold? Do we have any information revealing the links between any recipients of such ground station technologies and the PLA?

Secretary HOLUM. In 1994, following a Presidential Determination that the exports were in the national interest, the State Department licensed ground stations for the Intelsat and Inmarsat programs.

In 1995, following a determination by the President that the export of cryptographic items on the U.S. munitions list was in the national interest, State licensed ground stations that support encryption.

In December, 1996, licensing jurisdiction for encryption was transferred from the State Department to the Commerce Department. The Commerce Department then picked up the control of these items pursuant to new procedures established in 1996 in conjunction with transfer including the provisions of the Executive Order and applying the special controls for "Significant Items".

Mrs. FOWLER. Last Saturday's New York Times contained a piece indicating that the PLA has obtained dedicated access to transponders on the U.S.-built satellites that have been sold to China. In response, according to the Los Angeles Times of June 14, 1998, a White House spokesman, rather than denying this usage, instead sought to trivialize it, stating that "militaries around the world use commercial satellites" and "we do not believe it provides any particular military benefit." Do you agree with the statement that providing the PLA with the ability to conduct military communications does not provide them with "any particular military benefit?"

Secretary REINSCH. Dedicated access to telecommunications satellite channels is a service readily available on the world market from U.S. and non-U.S. suppliers. Commercial communications satellites are used by the government, military, and civilian entities throughout the world, and many militaries use such dedicated channels for routine traffic. Access to normal civil telecommunications equipment, including transponders on communications satellites, which has long been available from U.S. allies, does not provide the Peoples Liberation Army with any particular military benefit. Even if the U.S. chose to deny the sale of American-made equipment, foreign producers face no such prohibition from their governments, and in several telecommunications sectors in China the lead is held by foreign firms, not the U.S. For particularly sensitive military communications, the U.S. uses satellites with many special characteristics. We do not provide these satellites to China. Use of commercial communications satellites confers no significant military advantage for the user.

Mrs. FOWLER. Was there any Defense, State, or Commerce Department analysis of the possibility that the PLA would obtain dedicated access to transponders on these satellites for its military communications prior to the licensing of these satellite deals?

Secretary REINSCH. All three agencies received the question of Chinese access to normal civil telecommunications equipment in the early 1990s and concluded that this was in the U.S. national interest. The U.S. has supported for more than 30 years the non-discriminatory use of International Telecommunications Service. Intelsat, for example, currently has 142 member countries, including China, and does not dictate how the satellite may be used or by whom. The U.S. has the same policy for satellites sold to foreign countries and assumes in making these sales that all portions of the purchasing government may have access to the system. It would be unrealistic to block access to Commercial Telecommunications Services by the Chinese military given the widespread availability of other satellite options around the world.

Mrs. FOWLER. Has the U.S. government licensed the sale of any ground station technology to any Chinese entity that would support the satellite user's ability to encrypt its transmissions prior to sending them to the satellite or decrypt them once they have been transmitted? Under what control regime would such licenses be issued, and who would oversee the issuance of such licenses? If such sales were licensed who received permission to proceed with these sales and to what entities were these technologies sold? What specific technologies, if any, were approved for sale? Do we have any information revealing links between any recipients of such ground station technologies and the PLA?

Secretary REINSCH. The decision to release civil telecommunications equipment, such as ground station technology, to China was made prior to the termination of the COCOM regime by the Bush Administration, in concert with our NATO allies and Japan. Even if the U.S. chose to deny the sale of American-made equipment, foreign producers face no such prohibition from their governments, Access to normal civil telecommunication equipment has long been available to China and does not provide the Peoples Liberation Army with any particular military benefit. Use by the PLA of low-level, unclassified technology which is freely available from a number of foreign sources was not determined to be contrary to U.S. national interests. Additional information in response to this question will be provided under separate cover.

QUESTIONS SUBMITTED BY MR. WELDON

Mr. WELDON. Was there an encrypted device in that (Loral) satellite?

Secretary REINSCH. The satellite in question did not contain equipment to encrypt telecommunications data. It did contain two embedded encryption devices used to protect the command uplink. These devices used older algorithms and their keys were unique to the Loral satellite. NSA has determined that the loss of such chips poses no risk to national security.

Mr. WELDON. Will you tell us for the record whether or not there was an encrypted device in that satellite?

Secretary REINSCH. The cryptographic equipment and associated material for the Space Systems/Loral Intelsat 708 which was destroyed during the launch failure on February 15, 1996, was incorporated into two integrated circuit boards. The National Security Agency has determined that the cryptographic equipment that was aboard the INTELSAT 708 satellite also was destroyed and not recovered. NSA made this determination on the basis of reports and other information provided by SS/Loral.

QUESTIONS SUBMITTED BY MR. SHERMAN

Mr. SHERMAN. A former CIA Official, Dr. Gordon Oehler, testified before the Senate Foreign Relations Committee that since the shift from State to Commerce, intelligence analysts now have better opportunity to review satellite export license applications. In your view, has the transfer from State to Commerce in any way impaired our national security interests? Has it enhanced interdepartmental oversight?

Mr. TARBELL. In 1996, when President Clinton decided to transfer additional jurisdiction for exports of commercial communications satellites to Commerce, DoD supported this transfer because the transfer did not involve certain sensitive technologies associated with satellites and launch vehicles. It was accompanied by several changes in procedures that protect DoD's ability to ensure that transfers are consistent with U.S. national security. First, licensing determinations are now made by majority vote of the reviewing agencies, unlike the previous system in which the initial determination was always made by commerce. Second, an export license for a commercial communications satellite can now be denied for broad national security reasons even though comsats are not on the State Department munitions list. Third, communications satellites are not subject to formal foreign availability determinations unlike many other dual use items. Fourth, since the transfer of jurisdiction, all agencies involved in export licensing agreed to incorporate into the Commerce licenses the strong safeguards that had been imposed on State Department commercial communications satellite licenses, including DoD monitoring.

Mr. SHERMAN. What procedures are satellite manufacturers required to undertake to launch in China? In your view, do these procedures provide sufficient safeguards? What is DoD's current policy on requiring monitors at launches in China and Russia?

Mr. TARBELL. Since 1996, all comsat licenses have contained safeguards that include:

A requirement that every U.S. satellite exporter produce a technology control plan that must be approved by the Department of Defense.

The technology control plan must include: (1) a detailed transportation plan for shipping the satellite to ensure that only U.S. personnel have access to the satellite at all times; and (2) a detailed physical and operational security plan including procedures for the supervised mating of the satellite to the launch vehicle.

A requirement that a Department of Defense monitor be present at every technical meeting between the U.S. company and Chinese launch officials to ensure that no information or technical data is exchanged that would improve Chinese missile capabilities.

A requirement that all technical data and information that the U.S. company wants to transfer to the Chinese launch service provider must be approved in advance by the Defense Department.

A requirement that DoD monitors be present at the launch site in China to oversee physical site security, and to ensure that launch operations are conducted without any transfer of technology that would improve China's missile capabilities or give Chinese officials access to the satellite.

Mr. SHERMAN. How should launch failure reviews be regulated by the U.S. Government?

Mr. TARBELL. Launch failure investigations are defense services and are controlled under the International Traffic in Arms Regulations, 22 C.F.R. Parts 120–128.

Mr. SHERMAN. What procedures are satellite manufacturers required to undertake to launch in China? In your view, do these procedures provide sufficient safeguards?

Secretary HOLUM. We have a very strict policy regarding what types of cooperation are permitted and very clear laws governing our export licensing process. Our laws and policy are consistent with and meet our obligations under the multilateral export control regime, the MTCR.

The Department of Defense provides the technical expertise required to monitor satellite launch programs. DOD's Defense Technology Security Administration (DTSA) ensures the protection of U.S. technology through a variety of means.

After a U.S. exporter receives a license the exporter if required to provide a technology transfer control plan to DTSA for approval. This is an extensive plan which specifically identifies technical data to be released, covers that data's administrative control and release, and details transportation and security arrangements at the launch site.

DTSA provides oversight and monitoring of launch activities, including pre-launch meetings in the United States and China, the launch itself, and any post-launch activities. DTSA oversees every technical exchange between the U.S. satellite manufacturer and any Chinese entity involved in the launch.

DTSA enforces the clear technical data limits set out in each export license. DTSA monitors are present to ensure, to the greatest extent possible, that exporters do not, even inadvertently, provide the Chinese with any information other than that which is approved. This is strong protection against U.S. companies involved in the Chinese launch of U.S. satellites assisting China with the design, development, operation, maintenance, modification or repair of launch vehicles.

Overall, the Administration is fully satisfied with our safeguard regime. We are of course consistently reviewing it to see if there are areas in which we can make improvements.

Mr. SHERMAN. A former CIA official, Dr. Gordon Oehler, testified before the Senate Foreign Relations Committee that since the shift from State to Commerce, intelligence analysts now have a better opportunity to review satellite export license applications. In your view, has the transfer from State to Commerce in any way impaired our national security interests? Has it enhanced interdepartmental oversight?

Secretary HOLUM. We do not believe the transfer has impaired national security interests; certainly there was care taken in the 1996 transfer to ensure Commerce's regulations were appropriately modified in order to ensure national security interests were not impaired and it was on this basis that the Department was able to support the transfer of jurisdiction.

There is a well-established process by which CIA reviews information concerning all parties (i.e., not just the end user, but also the foreign and intermediate consignees, *inter alia* when munitions license applications are received in the Department. In the case of satellite licenses, CIA, in particular its Non-Proliferation Center, has always been involved in an additional review of satellite license applications submitted to State through its membership in the Missile Technology Export Committee, to which all such applications are referred.

Mr. SHERMAN. How should launch failure reviews be regulated by the U.S. Government?

Secretary HOLUM. A launch failure review by its very nature implies a defense service because space launch vehicles are defense articles included within the coverage of the U.S. Munitions List at Category IV. Therefore, such reviews involving foreign launch vehicles require the prior written approval of the Department of State in accordance with the International Traffic in Arms Regulations. The Commerce rules do not cover launch failures and any technology transfer involving the actual launch vehicle would require a separate State license.

Mr. SHERMAN. U.S. satellites manufacturers currently have $1.7 billion in contracts with Chinese and Asian customers for launch in China, and potentially $8 billion in contracts over the next five years. An estimated 8,000 U.S. jobs are supported by the current contracts. It is my understanding that Europeans satellite manufacturers such as Daimler-Benz are competing with U.S. companies for Chinese contracts. When did the Europeans begin cooperating with China on satellite technology and to what extent do our European Allies supply satellites and technology to China?

Secretary REINSCH. U.S. satellite manufacturers face vigorous competition from European aerospace consortia. Aerospatiale, Daimler-Benz, and Matra Marconi are among the world's top ten revenue-producing companies in the commercial space sector. Sino-European space cooperation dates from 1987 when China Great Wall Industry Corporation signed an agreement with Messerschmitt Bolkow Blohm (MBB) of West Germany to jointly develop the Dong Fang Hong 3 satellite. MBB supplied design expertise, in addition to solar array and antenna technology. During the late 1980s, several French and German scientific packages were launched on Long March rockets from China. A follow-on to the Dong Fang Hong 3 was built by the Chinese Academy of Space Technology with assistance from Daimler-Benz

(MBB's competitor) in 1997. This work was, in part, an outgrowth of the 1994 Daimler-Benz joint venture with the Chinese Aerospace Corporation (CASC) called "Euraspace," under which contracts were agreed for two SINOSAT satellites. The first SINOSAT satellite was recently delivered to China. Some European countries do not share our concerns over technology transfer to China and have been more willing to supply advanced technology to that country.

Mr. SHERMAN. If the U.S. bans commercial satellite exports to China, will this give European manufacturers a competitive edge in the Asian market?

Secretary REINSCH. Several Asian customers have required U.S. satellite companies to launch their satellites on Chinese rockets. If a ban on U.S. exports were imposed, these customers will have no choice but to purchase commercial satellites from foreign suppliers who would allow launches from China. The domestic industry estimates that blocking satellite exports to China would put an estimated 16,000 U.S. jobs at risk and forfeit $8.3 billion in revenue to European satellite manufacturers over the next five years.

Mr. SHERMAN. It is my understanding that the Bush Administration transferred one third of commercial satellites and related technology from the U.S. Munitions List to the Commodity Control list. In the last ten years, how many items have been moved from the U.S. Munitions List to the Commodity dual-use list? Can you enumerate these items?

Secretary REINSCH. A number of items have been moved from the U.S. Munitions List to the Commerce Control List, as a part of the Bush and Clinton Administration's commitment to eliminate overlap between the two control lists unless national security or intelligence interests warranted continued USML control. These include focal plane arrays, low level image intensification tubes and commercial products incorporating advanced image intensifier tubes; commercial communications satellites (first moved to CCL from USML in October 1992); hot section technology for development, production or overhaul of commercial aircraft engines, satellite fuels to accompany a specific Commerce licensed launch, licenses associated with the International space station, and commercial encryption items. Examples of items that have moved from Commerce licensing jurisdiction to the Munitions List are L–100 components and third generation night vision equipment.

Mr. SHERMAN. What procedures are satellite manufacturers required to undertake to launch in China? In your view, do these procedures provide sufficient safeguards?

Secretary REINSCH. Commerce licenses for communications satellites contain numerous conditions and provisos developed in conjunction with the Departments of Defense and State. Under Commerce licenses since 1996, exporters are obligated to comply with the terms of the satellite technology safeguards agreement between the U.S. and China, which requires them to develop a technology transfer control plan which identifies the level and extent of technical data to be released, and which also includes plans for securing the satellite during its transportation to the launch site; have all technical data under the license reviewed by the Defense Department prior to its release to the launch service provider and have a Defense Department monitor present at technical meetings and launch activities with the Chinese launch service provider, a requirement for a secure transportation plan; have a separate cryptographic equipment safeguard plan for communications security equipment; limit technology which can be released under the Commerce license to only form, fit and function data used to mate the satellite to the rocket and require the exporter, in the event of a launch failure, to obtain a license from State. There have been no allegations regarding export control violations of Commerce satellite licenses since the transfer of jurisdiction in 1996.

Mr. SHERMAN. How should launch failure reviews be regulated by the U.S. government?

Secretary REINSCH. Since the President's 1996 decision to transfer remaining jurisdiction of commercial communications satellites to the Commerce Department for export licensing, we have strengthened controls on satellite exports in several ways, including the regulation of launch failure reviews. In regards to the participation of U.S. firms in launch failure analyses, all agencies agree that such failure analyses are best conducted under the aegis of a State license and with appropriate monitoring by the Department of Defense. Our licenses issued now to satellite exporters require them to obtain State Department approval prior to any participation in launch failure analyses.

QUESTIONS SUBMITTED BY MR. GIBBONS

Mr. GIBBONS. Could Under Secretary Reinsch provide three memos which came out of his office: one dated October 24, 1995, consisting of 10 pages, memo to Betty

Baca from you regarding the national interest waiver for—and it is redacted, I presume that is Loral Communications, satellite project with attachment; another memo dated December 8, 1995, consisting of 17 pages, again, to Betty Baca regarding a national interest waiver satellite project with attachments; a third document dated December 4, 1995, a memo to Betty Baca regarding a national interest waiver for—and it is redacted again, and I presume that is Loral Communications satellite projects with attachments, just so that we can see what your involvement was with the national waiver for that satellite to confirm that you had nothing to do with it.

Secretary REINSCH. The answer to this question and the requested documents were provided separately, in a July 6, 1998, letter from Under Secretary Reinsch to Chairman Spence.

[See memo on page 344.]

Mr. GIBBONS. Please provide us with a copy of each CEO organization or individual that has been a member or is a member of CSPP or participant in one of those meetings and what the subject of their discussions may have been at those meetings.

Secretary REINSCH. [See Computer Systems Policy Project on page 346.]

○

ISBN 0-16-058445-0

9 780160 584459 90000